Trends and Prospects in Applied Electromagnetics

Trends and Prospects in Applied Electromagnetics

Guest Editors

Yating Yu
Baolin Nie

MDPI

Basel • Beijing • Wuhan • Barcelona • Belgrade • Novi Sad • Cluj • Manchester

Guest Editors

Yating Yu
School of Mechanical and
Electrical Engineering
University of Electronic
Science and Technology of
China
Chengdu
China

Baolin Nie
School of Mechanical and
Electrical Engineering
University of Electronic
Science and Technology of
China
Chengdu
China

Editorial Office
MDPI AG
Grosspeteranlage 5
4052 Basel, Switzerland

This is a reprint of the Special Issue, published open access by the journal *Applied Sciences* (ISSN 2076-3417), freely accessible at: https://www.mdpi.com/journal/applsci/special_issues/L77J2F25K1.

For citation purposes, cite each article independently as indicated on the article page online and as indicated below:

Lastname, A.A.; Lastname, B.B. Article Title. *Journal Name* **Year**, *Volume Number*, Page Range.

ISBN 978-3-7258-5435-6 (Hbk)
ISBN 978-3-7258-5436-3 (PDF)
https://doi.org/10.3390/books978-3-7258-5436-3

Contents

About the Editors

Yating Yu

Yating Yu is a full professor in the School of Mechanical and Electrical Engineering, the University of Electronic Science and Technology of China, in Chengdu, China. Her main research interests include electromagnetic nondestructive testing and evaluation as well as structure health monitoring. She was selected as a "Outstanding Reviewer" for the journal *NDT &E international* in 2024.

Baolin Nie

Baolin Nie is an associate professor in the School of Mechanical and Electrical Engineering, the University of Electronic Science and Technology of China, in Chengdu, China. His main research interests include electromagnetic compatibility, signal and power integrity, and multi-physics modeling and simulation in electronic and electrical systems.

applied sciences

MDPI

Editorial

Special Issue: Trends and Prospects in Applied Electromagnetics

Yating Yu * and Baolin Nie

School of Mechanical and Electrical Engineering, University of Electronic Science and Technology of China, Chengdu 611731, China; blnie@uestc.edu.cn
* Correspondence: wzwyyt@uestc.edu.cn

1. Introduction

With the rapid advancement of information technology, humans are living in an increasingly intricate electromagnetic environment [1–3]. Applied electromagnetics, a crucial branch of electromagnetics, serves as a link between theoretical electromagnetics and engineering requirements. It focuses on applying the fundamental theories and principles of electromagnetic fields and waves (characterized by Maxwell's equations) to address practical engineering issues, design diverse devices and systems, and develop novel technologies. These applications span a wide range of fields, including electrical engineering [4], communication engineering [5,6], railway transportation engineering [7,8], nondestructive testing and evaluation [9,10], and biomedical engineering [11].

In recent years, applied electromagnetics has witnessed significant progress in various domains, such as computational electromagnetics [12], Electromagnetic Compatibility and Interference (EMC&EMI) [13], electromagnetic nondestructive testing and evaluation [14], electromagnetic wave propagation [15], and bio-electromagnetics [16]. This Special Issue, entitled "Trends and Prospects in Applied Electromagnetics," aims to present the latest research findings in applied electromagnetics.

2. Overview of Contributions

2.1. Electromagnetic Compatibility and Interference

Yang et al. [17] carried out an analysis of the mechanism and coupling paths of radiation interference in a high-speed PWM axial flow fan. An integrated and non-inductive filtering circuit was put forward. This circuit was designed to absorb the peak voltage that enters the windings and to determine the filter parameters. The findings indicate that the filtering method has been proven to decrease overall electromagnetic interference. The maximum peak reduction reaches 41.9 dB, and this method does not affect the desired signals. Tripathi et al. [18] developed a numerical model to explore the ability of bucky paper (BP), a porous membrane composed of a highly cross-linked network of carbon nanotubes, to enhance the electromagnetic interference (EMI) shielding properties of carbon-fiber-reinforced polymer (CFRP) composites. The numerical simulation results showed that the incorporation of BP significantly improves the EMI shielding effectiveness (SE) of CFRP composites above 2 GHz, attributed to its high conductivity within that frequency range.

To address the EMI generated by switching operations in high-voltage DC (HVDC) converter stations, Schneider et al. [19] proposed a finite-difference time-domain method to study the characteristics of single switching events. They utilized a realistic model of an HVDC converter station, with a particular focus on determining the influence of the valve hall on EMI shielding. The investigation revealed that the valve hall does not restrict high-frequency electromagnetic fields within the valve hall. Instead, it delays their exit through the bushings in the wall and spreads them out over time. Fortunately, incorporating electromagnetic absorbing materials into the design of the valve hall can significantly reduce the EMI outside the converter station.

2.2. Computational Electromagnetics

Tripathi et al. [18] explored a quantum tunneling-based equivalent electrical circuits and Monte Carlo method to predict the frequency-dependent electrical conductivity and EMI shielding effectiveness (SE) of the hybrid BP/CFRP composites. Damian et al. [20] described a simulation approach for the S (scattering)-parameters and the absorbed energy of polymeric nanocomposites with metallic inclusions (iron and aluminum, with two particle sizes) by considering the Nicolson–Ross–Weir procedure in infinite media. The results showed that due to higher conductivity, nanocomposites with Al particles could absorb a greater amount of energy compared to those with Fe particles in the composite materials, at both inclusion sizes. The significant reduction in transmission confirmed that for composites with added metallic powder, microwave energy was extensively absorbed by the materials. Yelkenci et al. [21] proposed a method to determine the matching media parameters that maximize the electromagnetic energy penetrating into a human arm modeled as a radially stratified cylinder. The accuracy of this approach was demonstrated by calculating the electric field amplitudes inside and outside the layers for the determined parameters. The numerical results indicated that when a matching medium was considered, the penetrating field increased by a factor of 1.3 to 13.96 compared to the case without a matching medium. Due to the nonlinear nature in energy harvesting, Abramovitz et al. [22] developed a clamped-type overhead line magnetic energy harvester equipped with a controlled active rectifier, which can generate a significant DC output power. Moreover, a piecewise nonlinear analytical model of the magnetic harvester was derived. This research presented a closed-form solution that takes into account both the nonlinearities of the core and rectifiers, enabling an accurate quantitative prediction of the harvester's key parameters.

2.3. Design of Key Electrical Devices

To mitigate electromagnetic interference (EMI) in radio frequency (RF) and Wi-Fi (wireless fidelity standard) 5 and 6E wireless communication applications, Lin et al. [23] devised an ultra-wideband common-mode (CM) filter for a gigahertz (GHz) data rate signal. This filter employed an asymmetrical magnified coupled DGS to generate a second-order transmission zero, thereby extending the suppression bandwidth. The results showed that the CM noise was suppressed by over 10 dB within the frequency range of 2.9 GHz to 16.2 GHz. In the time domain measurement, the proposed filter was able to block 62.3% of the CM noise magnitude. Fan et al. [24] put forward a new filtering switch with outstanding working performance, which was fabricated using an optimized coupled microstrip line. The switching device was connected to the microstrip circuit to achieve the filtering and shutdown functions.

To reduce the insertion loss, Liang et al. [25] proposed an ultra-wideband, low-insertion-loss, and high-precision 6-bit digital step attenuator (DSA). In the proposed DSA, a series inductive compensation structure (SICS) was designed to address the high-frequency attenuation values, and a small bit compensation structure (SBCS) was intended for large attenuation bits. Using the 0.25 μm GaAs p-HEMT process, it exhibited the highest attenuation accuracy, the lowest insertion loss, the best IP1dB, and good matching performance within the 2–22 GHz range.

In addition, Karpuz et al. [26] developed a novel single/double-layer N-way Wilkinson power dividers (WPDs) by employing slow-wave structures to downsize them. Based on the proposed methods, two-, four-, and eight-way power dividers were designed and measured at the center frequencies of 2.03, 1.77, and 1.73 GHz. Leveraging meandered transmission lines, three- and five-way WPDs with new input/output port configurations were designed, enabling the input power to be halved at the next output port.

2.4. Electromagnetics in Interdisciplinary

Aiming at the threat posed by the highly leaked magnetic field from electric vehicles (EVs) charging to humans, Okada et al. [27] investigated the suitability of spatial averaging methods for non-uniform exposure and made contributions to the smooth assessment in Wireless Power Transfer (WPT) systems. To enhance the identification accuracy of metal information, Xie et al. [28] proposed a metal material analysis model based on the discrete wavelet transform, using a Constant Phase Angle Pulse Eddy Current (CPA-PEC) sensor. The experimental analysis shows that the stability of the quantitative evaluation index of eddy current features reaches 97.1%, and the comprehensive accuracy error is less than 0.32%.

3. Conclusions

The articles compiled in our Special Issue spotlight the significant progress achieved in electromagnetic compatibility and interference, computational electromagnetics, the design of key electrical devices, and electromagnetics in interdisciplinary domains. These articles offer a valuable frontier perspective and technical guidance for applied electromagnetics.

Currently, applied electromagnetics still confronts numerous challenges. As the number of applied electrical devices continues to grow, the active and passive electromagnetic fields render the electromagnetic environment intricate. In the future, several valuable topics include how to control electromagnetic interference from passive sources to achieve electromagnetic compatibility, and how to utilize the active electromagnetic fields for the benefit of humanity. For instance, advanced electromagnetic materials such as metamaterials will be employed; higher accuracy and resolution electromagnetic nondestructive testing and evaluation are needed; new computational electromagnetics models will be established due to the application of new materials or novel electronic devices; EMC and EMI models, EMC standards, and EMC measurement methods can be developed to address the more complex magnetic environment; and an increasing number of artificial intelligence techniques will be introduced into electromagnetic applications.

Author Contributions: Conceptualization, Y.Y. and B.N.; writing—original draft preparation, Y.Y.; writing—review and editing, Y.Y. and B.N. All authors have read and agreed to the published version of the manuscript.

Funding: This research received no external funding.

Acknowledgments: Thanks to all of the authors and peer reviewers for their valuable contributions to this Special Issue 'Trends and Prospects in Applied Electromagnetics'. We would also like to express our gratitude to all the staff and people involved in this Special Issue.

Conflicts of Interest: The author declares no conflicts of interest.

References

1. Liu, Q.-F.; Zheng, S.-Q.; Zuo, Y.; Zhang, H.-Q.; Liu, J.-W. Electromagnetic Environment Effects and Protection of Complex Electronic Information Systems. In Proceedings of the 2020 IEEE MTT-S International Conference on Numerical Electromagnetic and Multiphysics Modeling and Optimization (NEMO), Hangzhou, China, 7–9 December 2020; pp. 1–4.
2. Li, M.; Wei, G. A Review of Quantitative Evaluation of Electromagnetic Environmental Effects: Research Progress and Trend Analysis. *Sensors* **2023**, *23*, 4257. [CrossRef] [PubMed]
3. Wang, H.; Li, S.; Liu, M.; Li, J.; Zhou, X. Review on shielding mechanism and structural design of electromagnetic interference shielding composites. *Macromol. Mater. Eng.* **2021**, *306*, 2100032. [CrossRef]
4. Sarker, N.; Podder, P.; Mondal, M.R.H.; Shafin, S.S.; Kamruzzaman, J. Applications of Machine Learning and Deep Learning in Antenna Design, Optimization, and Selection: A Review. *IEEE Access* **2023**, *11*, 103890–103915. [CrossRef]
5. Khalifa, M.O.; Yacoub, A.M.; Aloi, D.N. A Multiwideband Compact Antenna Design for Vehicular Sub-6 GHz 5G Wireless Systems. *IEEE Trans. Antennas Propag.* **2021**, *69*, 8136–8142. [CrossRef]
6. Arani, M.S.; Shahidi, R.; Zhang, L. Dielectric Resonator Antenna Design for 5G mm-Wave Band. In Proceedings of the 2025 IEEE 20th International Symposium on Antenna Technology and Applied Electromagnetics (ANTEM), St. John's, NL, Canada, 20–23 July 2025; pp. 1–4.
7. Wang, F.; Yu, Y.; Luo, Z.; Wei, L.; Zeng, J.; Tian, G. An integrated wireless system for dynamic strain monitoring of Einel-rad-Einelfahrwerk bogies for high-speed rail transport. *Commun. Eng.* **2025**, *4*, 87. [CrossRef]
8. Yuan, F.; Yu, Y.; Wu, J.; Peng, J.; Jiang, C.; Fan, X.; Li, J. RCF crack direction assessment in moving ferromagnetic material by DC electromagnetic NDT technique. *NDT E Int.* **2023**, *138*, 102882. [CrossRef]
9. Yu, Y.; Sun, C.; Wang, L.; Fan, Y. A dual-parameters microwave measurement approach for spruce moisture content. *Nondestruct. Test. Eval.* **2024**, *40*, 867–885. [CrossRef]
10. Wang, Z.; Yu, Y. Thickness and Conductivity Measurement of Multilayered Electricity-Conducting Coating by Pulsed Eddy Current Technique: Experimental Investigation. *IEEE Trans. Instrum. Meas.* **2019**, *68*, 3166–3172. [CrossRef]
11. Huang, P.; Xu, L.; Xie, Y. Biomedical Applications of Electromagnetic Detection: A Brief Review. *Biosensors* **2021**, *11*, 225. [CrossRef]
12. Hoefer, W.J.R. A Brief History of Computational Electromagnetics in Microwave Engineering—A Personal Retrospective. *IEEE J. Microw.* **2023**, *3*, 338–348. [CrossRef]
13. Katiyar, A.K.; Prakash, A.; Dubey, S.K. Magnetic field generator for the calibration of magnetometers and electromagnetic compatibility (EMC) and electromagnetic interference (EMI) compliance. *Instrum. Sci. Technol.* **2024**, *53*, 141–156. [CrossRef]
14. Yu, Y.; Gao, K.; Liu, B.; Li, L. emi-Analytical Method for Characterization Slit Defects in Conducting Metal by Eddy Current Nondestructive Technique. *Sens. Actuators A Phys.* **2020**, *301*, 111739. [CrossRef]
15. Li, C.; Xu, X.; Lv, C.; Wu, X.; Jiao, Y.; Ren, Y.; Yan, Z.; Ren, Q. Electromagnetic Wave Propagation in Axisymmetric Time-Varying Plasma Using Newton-ADE BOR-FDTD. *IEEE Antennas Wirel. Propag. Lett.* **2025**, 1–5. [CrossRef]
16. Zhu, K.; Kiourti, A. Detection of Extremely Weak and Wideband Bio-Magnetic Signals in Non-Shielded Environments Using Passive Coil Sensors. *IEEE J. Electromagn. Rf Microw. Med. Biol.* **2022**, *6*, 501–508. [CrossRef]
17. Yang, J.; Wei, Y.; Zhao, X.; Wang, C.; Du, P. A Peak Absorption Filtering Method for Radiated EMI from a High-Speed PWM Fan. *Appl. Sci.* **2025**, *15*, 4568. [CrossRef]
18. Tripathi, K.; Hamza, M.H.; Morales, M.A.; Henry, T.C.; Hall, A.; Chattopadhyay, A. Electromagnetic Interference Shielding Analysis of Hybrid Buckypaper-Reinforced Polymer Matrix Composites: A Quantum Tunneling-Informed Equivalent Circuit Approach. *Appl. Sci.* **2024**, *14*, 8960. [CrossRef]
19. Schneider, J.B.; Olsen, R.G. Observations about Propagation, Attenuation, and Radiation of EMI from HVDC Converter Stations. *Appl. Sci.* **2024**, *14*, 6740. [CrossRef]
20. Damian, R.F.; Ciobanu, R.C. Modeling Absorbed Energy in Microwave Range for Nanocomposite Hot Melts Containing Metallic Additives. *Appl. Sci.* **2025**, *15*, 541. [CrossRef]
21. Yelkenci, T. Analysis of the Matching Media Effects by Microwave Field Distribution Simulations for the Cylindrically Layered Human Arm Model. *Appl. Sci.* **2025**, *15*, 268. [CrossRef]
22. Abramovitz, A.; Shvartsas, M.; Kuperman, A. Nonlinear Analysis and Solution for an Overhead Line Magnetic Energy Harvester with an Active Rectifier. *Appl. Sci.* **2024**, *14*, 11178. [CrossRef]
23. Lin, D.-B.; Wang, M.-H.; Pramudita, A.A.; Adiprabowo, T. Design of a Novel Ultra-Wideband Common-Mode Filter Using a Magnified Coupled Defected Ground Structure. *Appl. Sci.* **2023**, *13*, 7404. [CrossRef]
24. Fan, X.; Chen, X.; Xu, W.; Feng, L.; Yu, L.; Yuan, H. A Filtering Switch Made by an Improved Coupled Microstrip Line. *Appl. Sci.* **2023**, *13*, 7886. [CrossRef]
25. Liang, Q.; Wang, K.; Wang, X.; Yan, Y.; Liang, X. An S–K Band 6-Bit Digital Step Attenuator with Ultra Low Insertion Loss and RMS Amplitude Error in 0.25 μm GaAs p-HEMT Technology. *Appl. Sci.* **2024**, *14*, 3887. [CrossRef]

26. Karpuz, C.; Cakir, M.; Gorur, A.K.; Gorur, A. Design of N-Way Wilkinson Power Dividers with New Input/Output Arrangements for Power-Halving Operations. *Appl. Sci.* **2023**, *13*, 6852. [CrossRef]
27. Okada, M.; Miwa, K.; Kodera, S.; Hirata, A. Compliance Assessment of the Spatial Averaging Method for Magnetic Field Leakage from a Wireless Power Transfer System in Electric Vehicles. *Appl. Sci.* **2024**, *14*, 2672. [CrossRef]
28. Xie, Y.; Yu, Y.; Li, L. Discrete Wavelet Transform—Based Metal Material Analysis Model by Constant Phase Angle Pulse Eddy Current Method. *Appl. Sci.* **2023**, *13*, 3207. [CrossRef]

applied sciences

MDPI

Article

A Peak Absorption Filtering Method for Radiated EMI from a High-Speed PWM Fan

Jinsheng Yang [1], Yanhong Wei [1], Xuan Zhao [1,2], Chulin Wang [1] and Pingan Du [1,*]

1 School of Mechanical and Electrical Engineering, University of Electronic Science and Technology of China, Chengdu 611731, China; jsyanglj@std.uestc.edu.cn (J.Y.); 202022040407@std.uestc.edu.cn (Y.W.); 202112040906@std.uestc.edu.cn (X.Z.); 202311040615@std.uestc.edu.cn (C.W.)

2 AECC Aero Engine Control System Institute, Wuxi 214000, China

* Correspondence: dupingan@uestc.edu.cn

Abstract: Axial flow fans are widely used for heat dissipation in electronic devices. Due to its frequent speed-regulation to adapt to the change in heat load, a fan can cause significant electromagnetic radiation interference. In this study, a peak absorption filtering method is proposed to address the radiation interference issue in a high-speed PWM axial flow fan. The mechanism and coupling paths of radiation interference were analyzed, and a radiation interference calculation using finite integration technique by a hybrid field-circuit model and experimental measurement were conducted to identify the winding as the main source of radiation in PWM fan. Considering the limited space inside the fan, an integrated, non-inductive filtering circuit was designed to absorb the peak voltage entering the windings and the filter parameters are determined via circuit simulation. The measurement results indicate that the filtering method can reduce overall electromagnetic interference with a maximum peak reduction of 41.9 dB, without affecting the useful signals.

Keywords: filter design; finite integration technique; interference suppression; peak radiated interference; radiated modeling; radiated source

1. Introduction

Axial flow fans are extensively employed in electronic devices for thermal management due to their compact structure, high efficiency, and reliable airflow generation. A typical axial flow fan comprises three key components: a brushless DC (BLDC) motor, an impeller, and a fan frame, as illustrated in Figure 1. The impeller is directly coupled to the motor shaft, and when energized, the motor drives the impeller to rotate, producing an axial airflow that enhances convective heat dissipation within electronic enclosures.

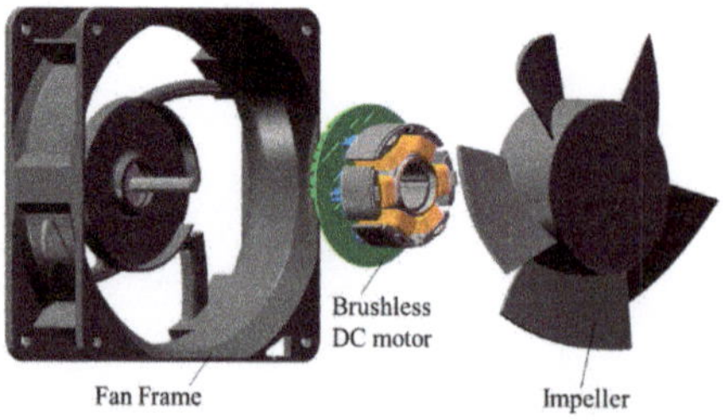

Figure 1. Structure of the axis flow fan.

To accommodate varying thermal loads under different operating conditions, axial flow fans often employ dynamic speed regulation. The most prevalent control method is pulse width modulation (PWM), which adjusts the motor's rotational speed by modulating the duty cycle of the input voltage. While PWM ensures efficient thermal management, it introduces rapid switching transients in the motor windings, leading to abrupt changes in current and voltage. These high-frequency switching actions generate conducted and radiated electromagnetic interference (EMI), manifesting as distinct spectral peaks at specific harmonic frequencies [1–4], as evidenced by the measurement results in Figure 2.

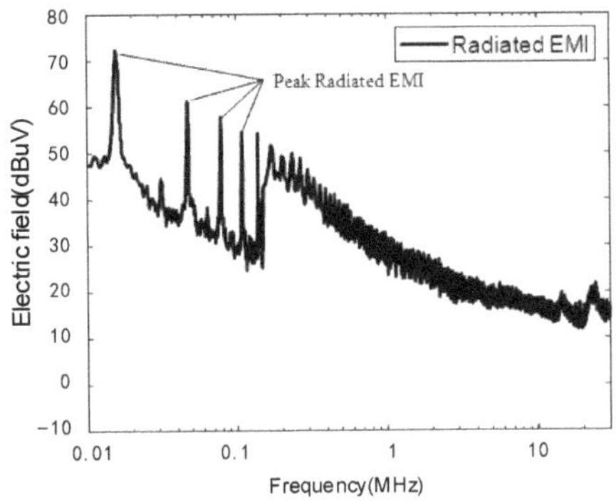

Figure 2. Radiated interference generated in the axial flow fan.

The EMI emissions from BLDC motors pose a critical challenge in modern electronic systems, particularly in applications where electromagnetic compatibility (EMC) is strictly regulated. In industries such as aviation, aerospace, and medical electronics, even minor EMI disturbances can disrupt sensitive communication systems, navigation equipment, or diagnostic instruments [5,6]. Consequently, regulatory bodies have established stringent EMI limits for cooling fans and other electromechanical components. Additionally, standardized testing methodologies, including MIL-STD-461G, CISPR 25, and DO-160, have been developed to assess and mitigate EMI risks in critical environments. Given the increasing demand for high-performance and low-EMI cooling solutions, a thorough investigation into the generation mechanisms, propagation paths, and suppression techniques of motor-induced EMI is imperative. It is of great significance to study the mechanism of motor radiated interference and develop effective suppression methods.

At present, research mainly focuses on the mechanism of conducted interference, focusing on the modeling and suppression of conducted interference in electrical machines. Conducted interference models for switch devices were established in [7], and a numerical model of a power transfer system was developed and used in simulation [8]. The conducted interference of a motor drive system was analyzed and modeled in [9–11]. The EMI suppression method of the DC motor was studied in [12,13] to attenuate both common-mode (CM) and differential-mode (DM) interference.

Research on radiated interference has mostly focused on modeling and predicting the radiated interference from components in power systems [14,15]. Ref. [16] presents an equivalent modeling method for the electromagnetic radiation of PWM fan. Many publications analyzing EMI sources and propagation provide a better understanding of the EMI generation mechanism [17–19]. For suppression of radiated interference, the main methods include changing the PWM control strategy and optimizing the system structure [20–23]. The design process of the methods above is complex due to the redesign of the system structure.

Designing EMI filters is an effective approach to mitigate radiated interference. Chuang C [24] designed the filter to attenuate both conducted and radiated interference. And the installation location and topology of the filter are also a problem worth investigating. The prevalent filter topologies often incorporate inductive components, leading to bulky filter designs that are not suitable for small-sized axial flow fans.

Therefore, this paper focuses on the radiated interference exceeding the standard of a small axial flow fan. The generation mechanism of the radiated interference and the coupling path were analyzed based on the principle of PWM motor speed control. The main radiation source of the fan was clarified through the calculation of the radiated interference model by the finite integration technique (FIT) [25]. Taking into consideration the spatial structure of the fan and the frequency characteristics of the interference, a non-inductive filter circuit was designed. The measurement results showed that the filter circuit can effectively suppress the radiated interference without increasing the effective signal insertion loss and meet the standard limit requirements.

2. Circuits Structure and Radiated EMI Measurement

2.1. Drive Circuit Structure and Speed Control Strategy

The axial flow fan is driven by a PWM permanent magnet brushless DC motor. The drive circuit mainly consists of a control chip, a MOSFET drive circuit, an inverter circuit, a protection circuit, and a winding coil, as shown in Figure 3.

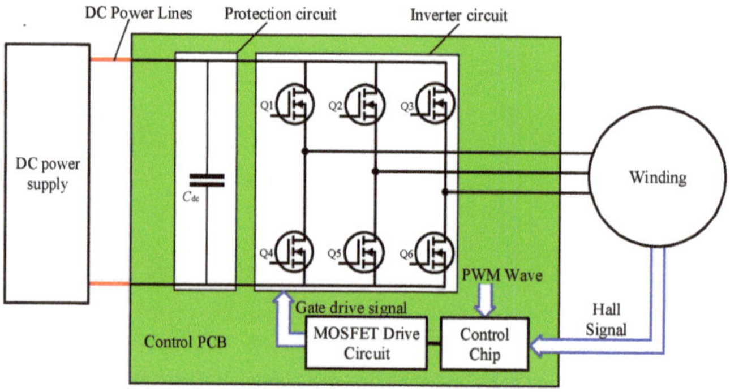

Figure 3. Schematic diagram of the fan drive system.

The inverter circuit adopts a three-phase half-bridge inverter topology consisting of six power MOSFETs. The control chip outputs a gate drive signal by recognizing the duty cycle of the modulated square wave to control the switching of the MOSFETs, thus changing the winding terminal voltage to adjust the motor speed. In this paper, the fan adopted the HPWM-LON modulation mode, i.e., the power MOSFETs in the upper bridge arm were controlled by the PWM signal, while the MOSFETs in the lower bridge arm were continuously conducting within one cycle of phase alternation. The modulation frequency was 15.5 kHz. Figure 4 shows the waveforms of each bridge arm drive signal in the time domain during the modulation process. Q1, Q2, and Q3 are the drive signals for MOSFET in the upper bridge arm, and Q4, Q5, and Q6 are the drive signals for the ones in the lower bridge arm.

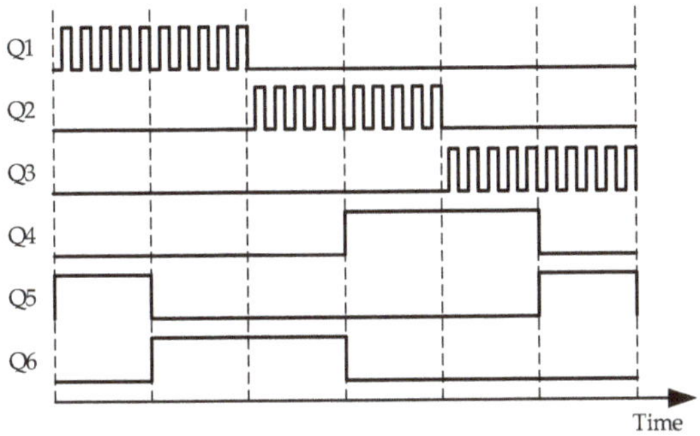

Figure 4. Driving signal of each bridge arm.

2.2. Radiated EMI Measurement and Analysis of Results

According to MIL-STD-461G, the radiated emission measurement configuration of the fan was as is shown in Figure 5. The measurement layout consisted of a DC power supply, two DC line impedance stabilization networks (LISN), two power lines, a receiver antenna, an EMI receiver, and the fan under test. The measurement environment was a shielded dark room, and the receiving antenna was placed 1 m away from the device under test, as shown in Figure 6. Within the rated operating range of the fan, the EMI receiver antenna measured the amplitude of the electric field in the frequency range of 10 kHz–30 MHz.

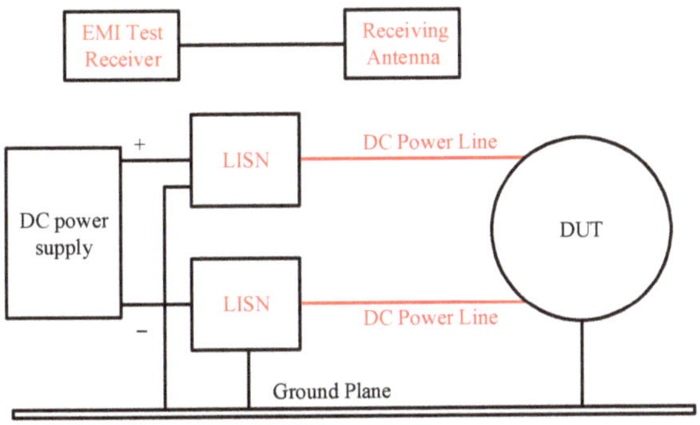

Figure 5. Radiation emission measurement configuration.

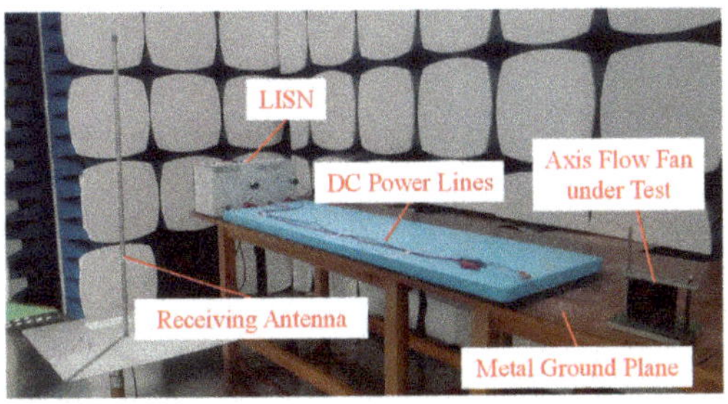

Figure 6. Radiation emission measurement site.

The measurement procedures were as follows:

a. Ambient Condition Check: Confirm that the environmental conditions comply with the specified standards to prevent any impact on test accuracy.

b. Equipment Initialization: Power on the measurement instruments and allow adequate stabilization time.

c. Device Preparation: Switch on the test equipment and ensure it stabilizes properly before proceeding.

d. Emission Assessment: Following the configuration in Figure 5, measure the radiated emissions from the test device and its connected cables.

The fan under test had a rated voltage of 24 V, a rated power of 103.8 W, a peak speed of 10,500 rpm, and a pole pair number of 2. Under the PWM condition, the results of the electric field radiation measurement on the fan were as shown in Figure 7. At the frequency of 15.5 kHz, the intensity of the radiated electric field measured was 72.41 dBμV/m, which was higher than the limit value of 57.02 dBμV/m required by the standard. According to the measurement results, it can be seen that there were a number of peaks of interference exceeding the standard limits in the measurement band at a series of peaks with a fundamental frequency of 15.5 kHz, which is consistent with the modulation frequency of the PWM, indicating that these frequency peaks were indeed caused by the switching of the MOSFETs in the inverter circuit with their frequency harmonics.

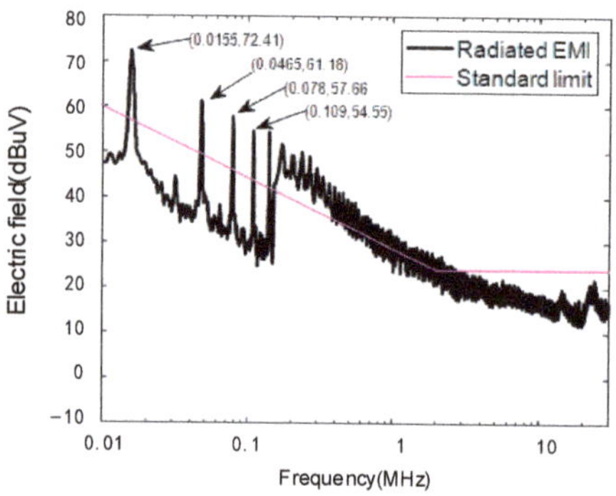

Figure 7. Measurement results of radiated interference.

3. Mechanism of Fan Radiation and Analysis of Radiated Source

3.1. Generation Mechanism and Path Analysis of Conducted Interference

In EMC design, the three elements of interference sources, propagation paths, and sensitive equipment are included. Radiated interference is generated externally by conducted interference in circuits through components with an antenna effect. The generation of conducted interference in the fan is closely related to the high-speed on/off characteristics of power MOSFETs. Assuming that the amplitude of the PWM wave is U, the period is T, the rising and falling edge times are t, and the turn-on time of the switching device in one cycle is ton, the duty cycle can be expressed as

$$d = t_{\text{on}}/T \tag{1}$$

According to the Fourier transform, the amplitude-frequency characteristics of the interference source can be obtained:

$$|V(f)| = 2dU\left|\frac{\sin(n\pi d)}{n\pi d}\right|\left|\frac{\sin(\pi tf)}{\pi tf}\right| \tag{2}$$

where n is the harmonic number. According to Equation (2), the PWM signal contains many high harmonics in addition to the fundamental waveform for normal operation, and these harmonics propagate through the conduction path to other components.

The conducted paths of EMI can be divided into common mode conducted paths and differential mode conducted paths. Each time a switching action occurs, the voltage at the midpoint of the bridge arm relative to the negative line of the power supply changes rapidly, thus generating common mode interference; at the same time, the switching process of the switching device also causes a sharp change in the current of the bridge arm, resulting in differential mode interference.

The common mode interference conduction path is shown as the red dashed line in Figure 8. Common mode interference is generated by transient voltage trips during switching dynamics, and the resulting current to ground forms a closed loop through the windings, ground, and DC line.

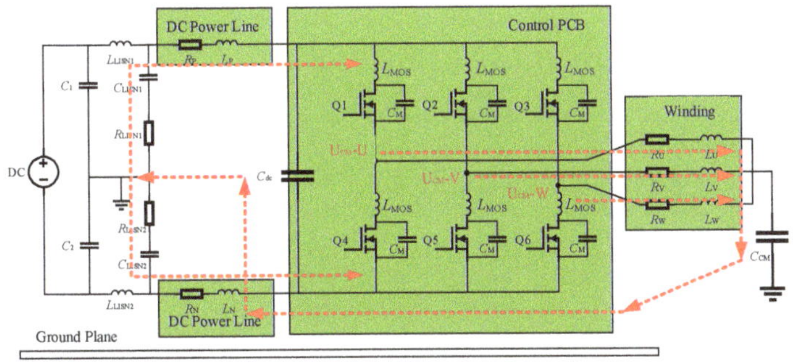

Figure 8. Paths of CM interference.

The differential mode interference conduction path is shown by the dotted line in Figure 9.

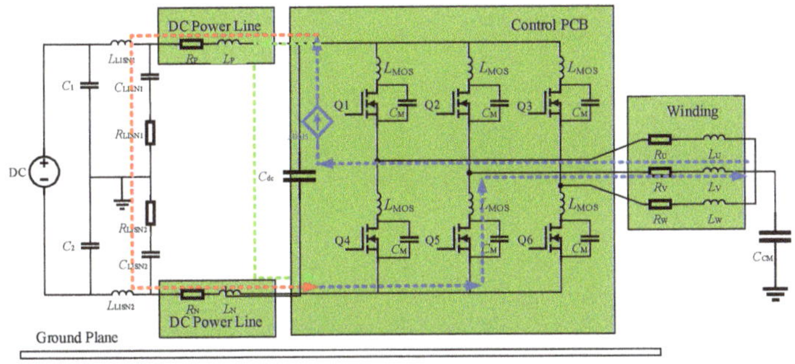

Figure 9. Paths of DM interference.

Path 1: Assuming that the switching state is that Q1 and Q5 are open, the differential mode current flows into the DC side from the U phase; flows through the DC cable, LISN, and flows out from the V phase; and finally passes through the winding, forming a closed loop.

Path 2: Differential mode current flows from the U phase into C_{dc} and out from the V phase, and it finally passes through the winding, forming a closed loop.

In the conducted path of electromagnetic interference, the DC power lines, windings, and PCB have antenna effects and can be highly efficient sources of electromagnetic radiation, and the interference current will radiate energy outward through these pathways, as shown in Figure 10. Therefore, before investigating the suppression methods of radiated interference, the intensity of radiated interference from the components in the loop needs to be further analyzed to target the proposed interference suppression methods.

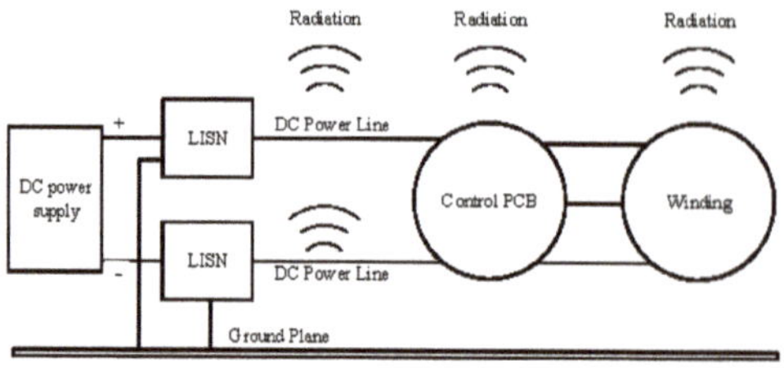

Figure 10. Schematic diagram of radiated emissions from a fan.

3.2. Analysis of Radiated Interference from DC Power Lines

To analyze the electric field radiation intensity of the DC line, the UL1332 20# power lines without shielding layer were replaced by a double shielded line of RG316D, and the shielding layer was well grounded. The results of the radiated emission measurement are shown in Figure 11. There was no significant decrease in the amplitude of the peak field in the tested frequency band, which indicates that shielding the DC cables did not reduce the system radiated interference, i.e., for lower frequency bands, the DC power lines were not the main source of radiation for the axial flow fan.

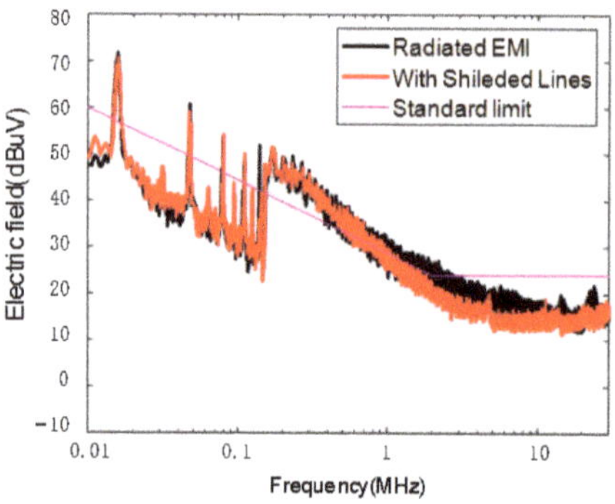

Figure 11. Comparison of radiated interference before and after the use of shielded lines.

3.3. Analysis of Radiated Interference from Winding and Control PCB

The windings and PCB were integrated inside the axial flow fan, and it was impractical to experimentally measure their radiation intensity separately. Hence, the radiated interference from the winding and control PCB was modeled, calculated, and analyzed by finite integration technique in the time domain, respectively.

According to the finite integral theory, the Maxwell mesh equation was obtained as

$$\mathbf{Ce} = -\frac{d}{dt}\mathbf{b}, \widetilde{\mathbf{C}}\mathbf{h} = \mathbf{i}_C + \mathbf{i}_S + \frac{d}{dt}\mathbf{d} \tag{3}$$

where C and $\widetilde{C}$ are the discrete curl matrices of the main grid and the dual grid, **e** is the voltage along the edge, and **b** is the magnetic flux through the face elements in the main grid. **h** is the magnetic field along the edge of the dual grid. **d**, $\mathbf{i}_C$, and $\mathbf{i}_S$ are the electric flux, loss current, and external current, respectively.

Using the central difference discretization as well as the frog-leaping scheme for the time derivative terms of the two spinodal equations in the above equation, the explicit time-domain recursive equation of the finite integration method is given by

$$\begin{cases} \mathbf{e}^{n+1} = C_{AE}\mathbf{e}^{n} + C_{AH}\left(\widetilde{\mathbf{C}}\mathbf{h}^{n+\frac{1}{2}} - \mathbf{i}_S^{n+\frac{1}{2}}\right) \\ \mathbf{h}^{n+\frac{1}{2}} = \mathbf{h}^{n-\frac{1}{2}} - \Delta t D_{\mu}^{-1}\mathbf{C}\mathbf{e}^{n} \end{cases} \quad (4)$$

where

$$\begin{cases} C_{AE} = \left(\frac{D_\varepsilon}{\Delta t} + \frac{D_\sigma}{2}\right)^{-1}\left(\frac{D_\varepsilon}{\Delta t} - \frac{D_\sigma}{2}\right) \\ C_{AH} = \left(\frac{D_\varepsilon}{\Delta t} + \frac{D_\sigma}{2}\right)^{-1} \end{cases} \quad (5)$$

where D_ε, D_σ and D_μ are the permittivity matrix, electrical conductivity matrix, and magnetic reluctivity matrix, respectively. The transient electromagnetic process of the winding and PCB can be calculated according to Equations (4) and (5) by CST Studio Suite. To achieve field-circuit coupling, the excitation current is represented by adding the current $\mathbf{i}_L$ to the right side of Ampere's law, i.e.,

$$\widetilde{\mathbf{C}}\mathbf{h}^{n+\frac{1}{2}} = D_\sigma\frac{\mathbf{e}^{n+1} + \mathbf{e}^{n}}{2} + D_\varepsilon\frac{\mathbf{e}^{n+1} - \mathbf{e}^{n}}{\Delta t} + {\mathbf{i}_L}^{n+\frac{1}{2}} \quad (6)$$

The field-circuit coupled transient process can be calculated according to (6).

The axial flow fan has a winding structure with a three-phase star structure and six slots. To assess the radiated interference of the winding, this paper employed a hybrid modeling approach that integrated field and circuit elements. Specifically, during the winding modeling process, 3D modeling was conducted based on the actual winding layout. The R and L parameters of the winding were then compensated using a circuit model, as illustrated in Figure 12. As CST's computational algorithms have difficulty in precisely modeling the influence of magnetic poles on winding parameters, we intentionally omitted the material properties of this component during the 3D modeling phase. Consequently, parameter compensation was applied in the circuit domain with $R = 0.287\ \Omega$ and $L = 250\ \mu H$ according to reference [16]. This approach not only aligns with the actual 3D layout of the winding but also ensures that the circuit parameters of the winding meet the requirements.

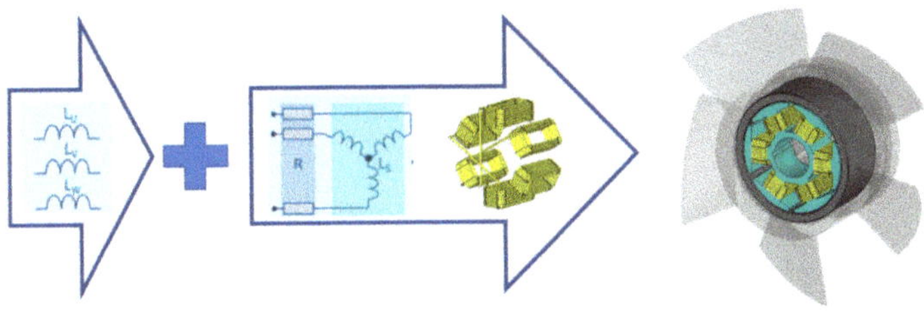

Figure 12. Hybrid modeling approach for windings.

The stator winding was fabricated using 0.71 mm diameter enamel-coated copper wire, with each slot containing 25 precisely wound turns. The 3D modeling process was carried out in SolidWorks 2021, which included detailed representations of the stator slots, coil windings, and permanent magnet components. For the electrical connections, a star-type arrangement was implemented for the three-phase winding system, with the completed 3D winding assembly illustrated in Figure 12. The magnetic circuit incorporated high-performance NdFeB permanent magnets, while the core structure was assembled from stacked silicon steel laminations to optimize magnetic effects. The components in the circuit were linked to the field model through an external circuit, where the FIT method in the time domain with field-circuit coupling was employed for a numerical solution. The full-wave field-circuit simulation model of the winding and inverter circuit was established in CST, as shown in Figure 13.

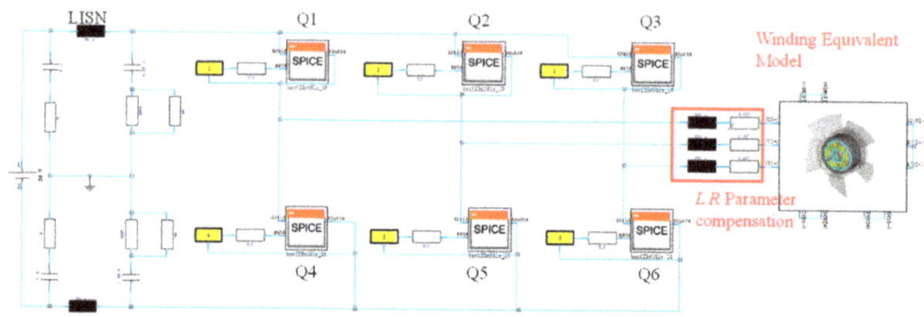

Figure 13. Field-circuit radiation simulation model for winding-inverter circuits.

The control PCB was then imported into CST Microwave Studio in ODB++ format. The winding port Touchstone model obtained through parameter extraction was imported into CST Design Studio as a circuit block to establish the field-circuit radiation simulation model of the PCB and the inverter circuit, as shown in Figure 14.

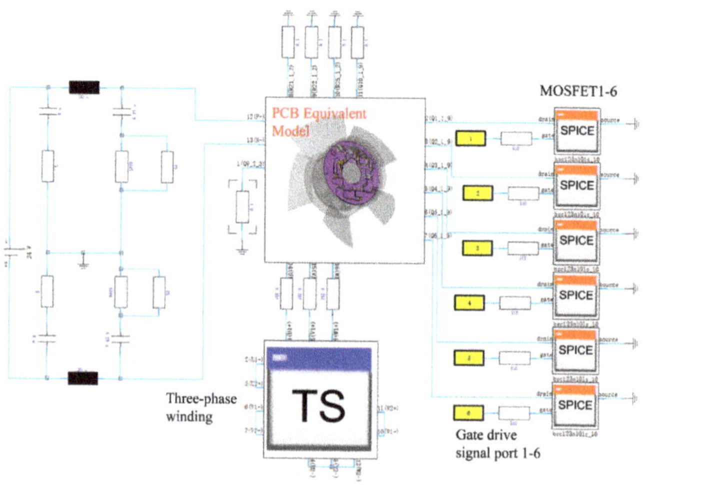

Figure 14. Field-circuit radiation simulation model for control PCB-inverter circuits.

The influence of the fan frame and fan impeller on the radiation of the internal winding and PCB was also not negligible. Therefore, the 3D model of the fan was imported into CST for combined calculations. The HPWM-LON control strategy for the fan drive system was implemented using MATLAB/Simulink2021b to generate the gate signals for the MOSFETs.

The simulation conditions were configured according to the experimental measurement setup. The measurements were conducted in an anechoic chamber, with the boundary conditions set as a "Conducting Wall" for the bottom surface and "Open" for the remaining boundaries. An adaptive meshing method was employed, with the convergence of electromagnetic energy serving as the stopping criterion for the simulation iterations. A time-domain solver was used for the simulation, and the accuracy threshold was set to −35 dB, meaning the iterations stopped once the post-calculation energy decay fell below −35 dB. The electric field probe was set at the same position as the receiving antenna in the measurement. Considering the influence of the impeller and the fan frame on the internal radiation, the full-wave simulation models of the winding and the control PCB as the radiation sources are shown in Figure 15, respectively. A comparison of the winding and PCB radiation simulation results was obtained as shown in Figure 16.

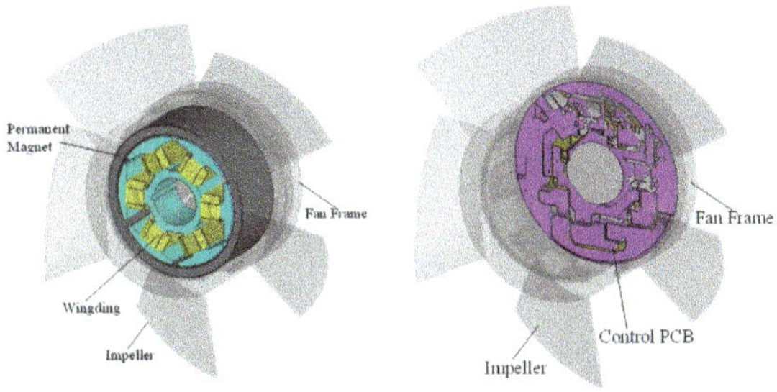

Figure 15. Full-wave simulation models of the winding and the control PCB with the impeller and the fan frame.

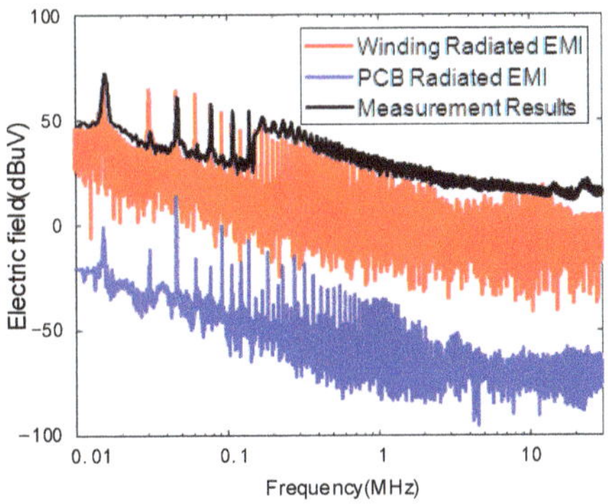

Figure 16. Comparison of the simulation results of the winding-radiated EMI and the PCB-radiated EMI with the measurement results.

According to the simulation results, the peak frequency point of the radiated electric field was consistent with the experimental measurements. The radiated electric field intensity of the windings was much higher than that of the control PCB, which was close to the radiated electric field measurement results of the overall fan. The calculated surface current of the PCB at f = 15.5 kHz is shown in Figure 16. The position of maximum surface current is where the PCB was connected to the three-phase winding, i.e., the output side of the inverter circuit, marked by the red box in Figure 17. Therefore, it can be concluded that the windings were the main component that caused excessive radiated interference from the fan.

The input voltage to the ground of the three-phase winding was measured using an oscilloscope, and the time-domain measurement results are shown in Figure 18. Each time the MOSFET turned on and off, there was a significant voltage peak, as marked by the red circle in the figure. Therefore, a filter circuit was able to be designed at the winding input port to absorb voltage peaks and prevent them from entering the winding and causing radiated interference.

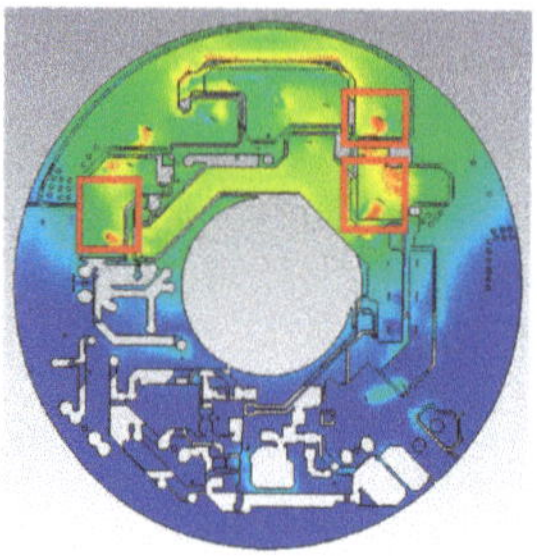

Figure 17. Simulation results of PCB surface current.

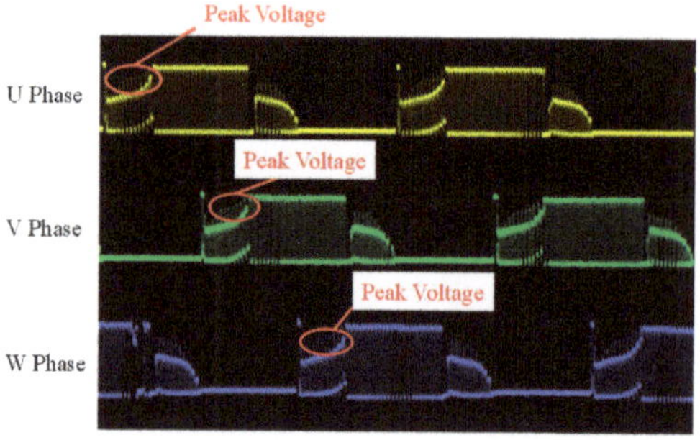

Figure 18. Voltage at three-phase winding ports in the time-domain.

4. Filter Suppression of Radiated Interference

According to the analysis of the interference path and the main source of radiation, the interference voltage was generated by the switching device in the inverter circuit, which entered the windings of the fan and caused radiated interference.

The voltage for the port of the windings was calculated via the field-circuit simulation model shown in Figure 13. Taking the U phase as an example, the simulation results are presented in Figure 19, compared with the spectrum of the voltage at the winding port obtained by FFT. The calculation results demonstrate a consistency between the simulation and measurement. Therefore, the circuit simulation model can effectively guide the design of EMI filters.

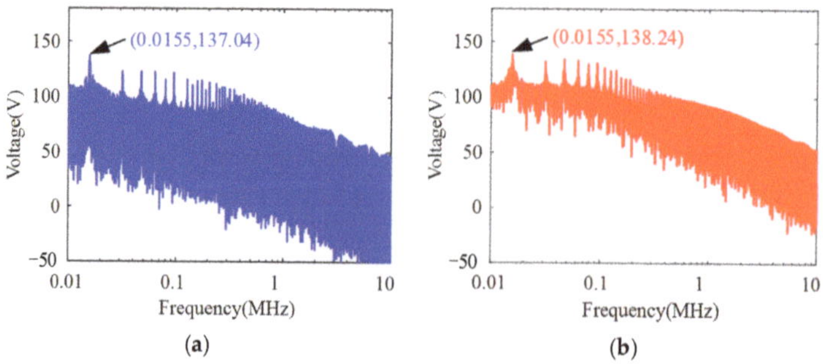

Figure 19. Comparison of the simulation results of the winding voltages with the measurement results. (**a**) Measurement results; (**b**) simulation results.

For the filter circuit topology design, if the low-pass filter includes an inductive component, the inductance value of the selected device will be at the mH level due to the lowest frequency of the interference peak of 15.5 kHz, resulting in a large filter size. As a miniaturized motor, axial flow fans have limited internal space, which makes it challenging to accommodate larger-sized components. Hence, this paper proposes a non-inductive RC snubber circuit connected in parallel between the drain and source of each MOSFET, absorbing the voltage peaks due to high-frequency switching. The circuit topology is shown, and the RC snubber circuit in parallel with the MOSFET is marked in red in Figure 20.

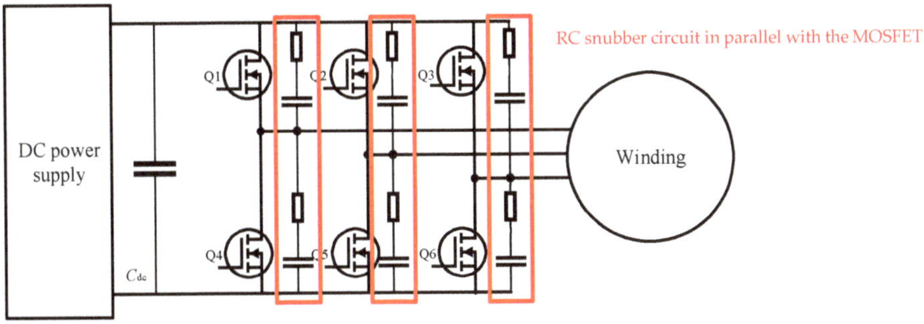

Figure 20. RC filter circuit topology.

The determination of RC parameters was achieved through circuit simulation. By adding parallel resistor and capacitor elements in the circuit simulation of Figure 13, the parameter combinations of resistors and capacitors in the filter were varied separately, and the absorption effect of the filter was measured by calculating the frequency-domain voltages of the winding. Using this simulation model, we calculated the winding voltage under two scenarios:

1. With a fixed capacitance of $C = 10\ \mu F$, the resistance values were varied as $R = 1\ \Omega$, $10\ \Omega$, and $100\ \Omega$.

2. With a fixed resistance of $R = 10\ \Omega$, the capacitance values were varied as $C = 1\ \mu F$, $10\ \mu F$, and $100\ \mu F$.

As depicted in Figures 21 and 22, the comparison of peak voltages with the voltage without filter across the winding terminals revealed that a smaller resistor and a larger capacitor yielded a more effective suppression of peak voltages. Additionally, to preclude overcurrent in the circuit, the resistance should not be chosen to be excessively small. Furthermore, considering the spatial constraints of the PCB, the capacitor value should also not be excessively large. The parameters of the filter were determined as $R = 10\ \Omega$, $C = 10\ \mu F$.

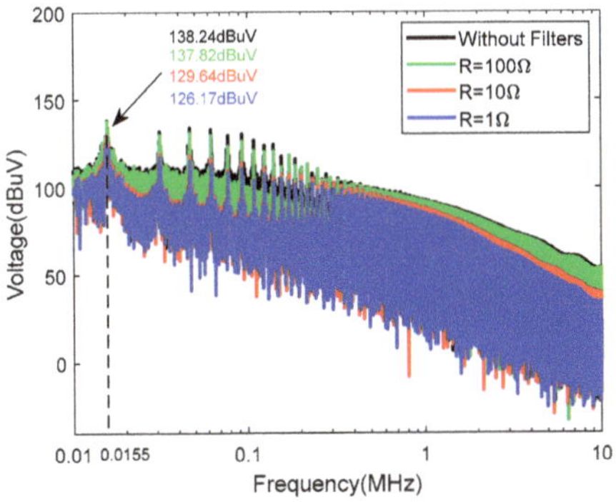

Figure 21. Simulation results of winding voltages with different R when $C = 10\ \mu F$.

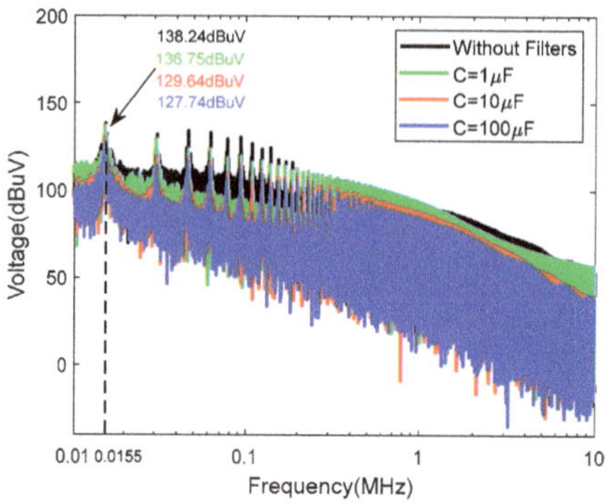

Figure 22. Simulation results of winding voltages with different C when $R = 10\ \Omega$.

The device met the space size requirements of the fan and can be integrated into the PCB as shown in Figure 23. The red squares mark where the RC devices were installed, in parallel with each MOSFET. The control PCB was designed and directly processed by the authors for the integration of the complete fan, which is a four-layer board including key components such as MCUs, inverter circuits, and RC filter circuits.

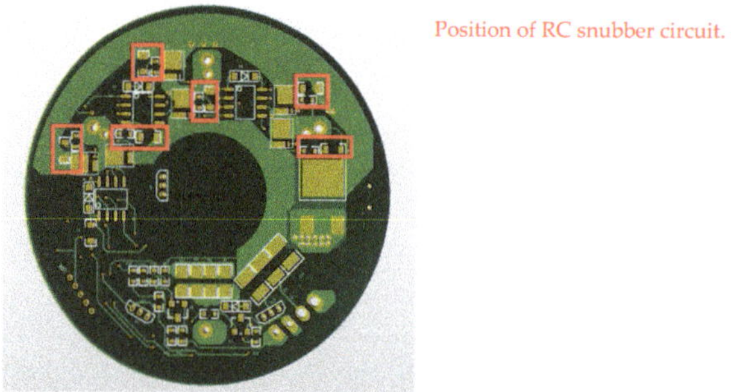

Figure 23. Control PCB with integrated RC filter.

The comparative results before and after filtering for the fan radiated interference measurement are shown in Figure 24. The measurement results show that the designed RC filter had a significant absorption effect on the radiated interference in the frequency band. Radiated interference attenuation reached a maximum of 41.9 dB. Peak radiated interference in the frequency band was reduced below the standard limits. Meanwhile, during the experimental measurements, the speed of the fan was not affected by the access to the filter, which indicates that the designed filter can suppress radiated interference without affecting the effective signal insertion loss.

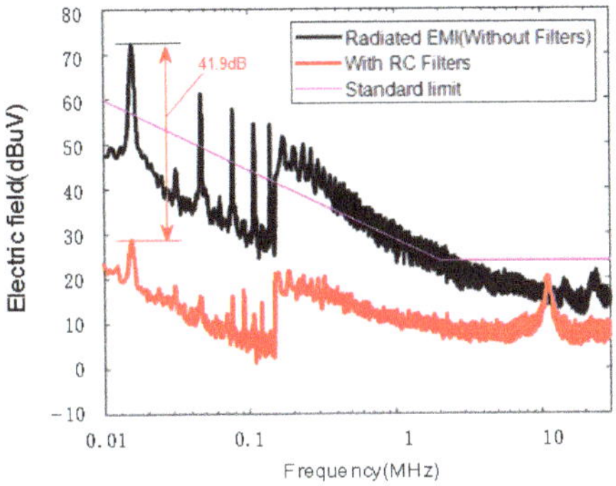

Figure 24. Comparison results of radiated interference before and after filtering.

5. Conclusions

EMC standards require stricter radiated emission limits for a PWM axial fan and its drive system. Meanwhile, due to the limited internal space of the small fan, there are also certain size requirements for the interference suppression measures. Therefore, this paper aimed to identify the main sources of radiated interference in PWM-regulated fans through measurements and simulations of radiated emissions. Additionally, a radiated interference suppression method that can be integrated into a PCB was proposed.

The propagation paths of the interference were determined through the study of the fan speed control mechanism. By modeling the radiated interference of individual components in the fan, the primary sources of radiated interference were identified. Subsequently, a small-sized peak-absorbing circuit was designed to address the issue. Experimental verification was conducted to assess the effectiveness of the radiated interference suppression measures. The experimental results demonstrate that the integrated non-inductive RC filter effectively suppressed peak radiated interference from the fan without impacting the fan's speed. The insertion loss achieved by the designed filter for radiated interference reached 41.9 dB.

Author Contributions: Conceptualization, J.Y., Formal analysis, Y.W.; Investigation, C.W.; Methodology, J.Y. and X.Z.; Resources, X.Z.; Supervision, P.D., Validation, J.Y. and Y.W., Writing—original draft, J.Y., Writing—review & editing, P.D. All authors have read and agreed to the published version of the manuscript.

Funding: This work was supported by the National Natural Science Foundation of China under Grant 52175218.

Institutional Review Board Statement: Not applicable.

Informed Consent Statement: Not applicable.

Data Availability Statement: The data supporting the findings of this study are available from the corresponding author upon reasonable request.

Conflicts of Interest: The authors declare no conflict of interest.

References

1. Zhai, L.; Hu, G.; Song, C.; Lv, M.; Zhang, X. Comparison of Two Filter Design Methods for Conducted EMI Suppression of PMSM Drive System for Electric Vehicle. *IEEE Trans. Veh. Technol.* **2021**, *70*, 6472–6484. [CrossRef]
2. Nan, C.; Ayyanar, R.; Xi, Y. A 2.2-MHz Active-Clamp Buck Converter for Automotive Applications. *IEEE Trans. Power Electron.* **2018**, *33*, 460–472. [CrossRef]
3. Carobbi, C.F.M.; Izzo, D. Evaluation and Improvement of the Reproducibility of CISPR 25 ALSE Test Method. *IEEE Trans. Electromagn. Compat.* **2018**, *60*, 1069–1077. [CrossRef]

4. Zhang, B.; Wang, S. A Survey of EMI Research in Power Electronics Systems With Wide-Bandgap Semiconductor Devices. *IEEE J. Emerg. Sel. Top. Power Electron.* **2020**, *8*, 626–643. [CrossRef]
5. Leuchter, J.; Stekly, V.; Blasch, E. Investigation of avionics power switch loading versus aircraft electromagnetic compatibility. *IEEE Aerosp. Electron. Syst. Mag.* **2015**, *30*, 24–34. [CrossRef]
6. Touré, B.; Schanen, J.L.; Gerbaud, L.; Meynard, T.; Roudet, J.; Ruelland, R. EMC Modeling of Drives for Aircraft Applications: Modeling Process, EMI Filter Optimization, and Technological Choice. *IEEE Trans. Power Electron.* **2013**, *28*, 1145–1156. [CrossRef]
7. Jing, L.; Du, M.; Wei, K.; Hurley, W.G. An Improved Behavior Model for IGBT Modules Driven by Datasheet and Measurement. *IEEE Trans. Electron Devices* **2020**, *67*, 230–236. [CrossRef]
8. Hikage, T.; Yamagishi, M.; Shindo, K.; Nojima, T. Active implantable medical device EMI estimation for EV-charging WPT system based on 3D full-wave analysis. In Proceedings of the 2017 Asia-Pacific International Symposium on Electromagnetic Compatibility (APEMC), Seoul, Republic of Korea, 20–23 June 2017; pp. 87–89.
9. Bellan, D. Circuit models for EMC analysis of CM current in three-phase motor drive systems. In Proceedings of the 2014 6th IEEE Power India International Conference (PIICON), Delhi, India, 5–7 December 2014; pp. 1–5.
10. Zhai, L.; Song, C. Conducted EMI from motor drive system of electric vehicle under load operation. In Proceedings of the 2017 Asia-Pacific International Symposium on Electromagnetic Compatibility (APEMC), Seoul, Republic of Korea, 20–23 June 2017; pp. 181–183.
11. Moreau, M.; Idir, N.; Moigne, P.L. Modeling of Conducted EMI in Adjustable Speed Drives. *IEEE Trans. Electromagn. Compat.* **2009**, *51*, 665–672. [CrossRef]
12. Ji, J.; Wu, W.; He, Y.; Lin, Z.; Blaabjerg, F.; Chung, H.S.H. A Simple Differential Mode EMI Suppressor for the LLCL-Filter-Based Single-Phase Grid-Tied Transformerless Inverter. *IEEE Trans. Ind. Electron.* **2015**, *62*, 4141–4147. [CrossRef]
13. Kumar, M.; Jayaraman, K. Design of a Modified Single-Stage and Multistage EMI Filter to Attenuate Common-Mode and Differential-Mode Noises in SiC Inverter. *IEEE J. Emerg. Sel. Top. Power Electron.* **2022**, *10*, 4290–4302. [CrossRef]
14. Costa, F.; Gautier, C.; Revol, B.; Genoulaz, J.; Démoulin, B. Modeling of the near-field electromagnetic radiation of power cables in automotives or aeronautics. *IEEE Trans. Power Electron.* **2013**, *28*, 4580–4593. [CrossRef]
15. Wei, S.; Pan, Z.; Yang, J.; Du, P. A Fast Prediction Approach of Radiated Emissions From Closely-Spaced Bent Cables in Motor Driving System. *IEEE Trans. Veh. Technol.* **2022**, *71*, 6100–6109. [CrossRef]
16. Yang, J.; Zhao, X.; Xia, J.; Zhang, W.; Du, P.; Nie, B. An Equivalent Modeling Method for Electromagnetic Radiation of PWM Fans with Multiple Radiation Sources. *Appl. Sci.* **2025**, *15*, 2887. [CrossRef]
17. Huibin, Z.; Jih-Sheng, L.; Hefner, A.R.; Yuqing, T.; Chingchi, C. Modeling-based examination of conducted EMI emissions from hard- and soft-switching PWM inverters. *IEEE Trans. Ind. Appl.* **2001**, *37*, 1383–1393. [CrossRef]
18. Chang, C.-H.; Cheng, H.-L.; Chang, E.-C. Using the buck-interleaved buck–boost converter to implement a step-up/down inverter. *Eng. Comput.* **2017**, *34*, 272–284. [CrossRef]
19. Guo, X.; Yang, Y.; Zhu, T. ESI: A Novel Three-Phase Inverter With Leakage Current Attenuation for Transformerless PV Systems. *IEEE Trans. Ind. Electron.* **2018**, *65*, 2967–2974. [CrossRef]
20. Huang, J.; Shi, H. Reducing the Common-Mode Voltage through Carrier Peak Position Modulation in an SPWM Three-Phase Inverter. *IEEE Trans. Power Electron.* **2014**, *29*, 4490–4495. [CrossRef]
21. Ramaiah, V.J.; Keerthipati, S. Hybrid PWM Scheme for Pole-Phase Modulation Induction Motor Drive Using Carrier-Based Hexagonal and Octadecagonal SVPWM. *IEEE Trans. Ind. Electron.* **2020**, *67*, 7312–7320. [CrossRef]
22. Jiang, Y.; Wu, W.; He, Y.; Chung, H.S.H.; Blaabjerg, F. New Passive Filter Design Method for Overvoltage Suppression and Bearing Currents Mitigation in a Long Cable Based PWM Inverter-Fed Motor Drive System. *IEEE Trans. Power Electron.* **2017**, *32*, 7882–7893. [CrossRef]
23. Han, D.; Morris, C.T.; Sarlioglu, B. Common-Mode Voltage Cancellation in PWM Motor Drives with Balanced Inverter Topology. *IEEE Trans. Ind. Electron.* **2017**, *64*, 2683–2688. [CrossRef]
24. Chuang, C.-C.; Hua, C.-C.; Hung, C.-W.; Yao, C.-J. Study on Evaluating the EMI Filter for Electronic Communication Fans. *J. Robot. Netw. Artif. Life* **2022**, *8*, 269–272. [CrossRef]
25. Qiu, H.; Wang, S.; Sun, F.; Wang, Z.; Zhang, N. Transient Electromagnetic Field Analysis for the Single-Stage Fast Linear Transformer Driver with Two Different Configurations Using the Finite-Element Method and Finite Integration Technique. *IEEE Trans. Magn.* **2020**, *56*, 7515805. [CrossRef]

applied sciences

MDPI

Article

Electromagnetic Interference Shielding Analysis of Hybrid Buckypaper-Reinforced Polymer Matrix Composites: A Quantum Tunneling-Informed Equivalent Circuit Approach

Kartik Tripathi [1], Mohamed H. Hamza [1], Madeline A. Morales [2], Todd C. Henry [2], Asha Hall [2] and Aditi Chattopadhyay [1,*]

[1] School for Engineering of Matter, Transport and Energy, Arizona State University, Tempe, AZ 85287, USA; ktripat3@asu.edu (K.T.); mhamza1@asu.edu (M.H.H.)

[2] DEVCOM Army Research Laboratory, Aberdeen Proving Ground, Aberdeen, MD 21005, USA; madeline.a.morales.civ@army.mil (M.A.M.); todd.c.henry2.civ@army.mil (T.C.H.); asha.j.hall.civ@army.mil (A.H.)

* Correspondence: aditi@asu.edu

Abstract: A novel modeling approach is developed for investigating the effectiveness of buckypaper (BP), a porous membrane made of a highly cross-linked network of carbon nanotubes, in improving the electromagnetic interference (EMI) shielding properties of carbon fiber-reinforced polymer (CFRP) composites. The methodology uses quantum tunneling-based equivalent electrical circuits and Monte Carlo simulations to predict the frequency-dependent electrical conductivity and EMI shielding effectiveness (SE) of the hybrid BP/CFRP composites. The study examines a signal frequency range of 50 MHz to 12 GHz that includes the very high and X-band. The results show that at a frequency of 12 GHz, the transverse conductivity increases to approximately 12.67 S/m, while the longitudinal conductivity decreases to about 3300 S/m from an initial value of 40,000 S/m. These results are then integrated into the ANSYS High-Frequency Structure Simulator to predict SE by simulating the propagation of electromagnetic waves through a semi-infinite composite shield element. The numerical simulations illustrate that incorporating BP significantly improves the SE of CFRP composites beyond 2 GHz owing to its high conductivity in that frequency range. For instance, at 12 GHz signal frequency, adding a single BP interleaf enhances the SE of a [90, 0] laminate by up to ~64%.

Keywords: carbon nanotubes; buckypaper; EMI shielding; electrical conductivity; quantum tunneling; CFRPs; polymer composites

1. Introduction

The increased use of electronics in aerospace structures and other applications underscores the need to develop effective electromagnetic interference (EMI) shielding strategies for protecting sensitive electronic systems from external electromagnetic signals. Carbon nanotubes (CNTs) have emerged as a promising material due to their remarkable properties, such as high tensile strength (100 GPa) and stiffness (1 TPa), excellent thermal conductivity (~3500 W/m/K) and electrical conductivity (~105 S/m) and have many potential applications in aerospace such as de-icing and lightning strike protection [1,2].

For EMI shielding applications at the structural scale, harnessing the full potential of CNTs poses a substantial challenge [3]. Researchers have explored the utilization of CNT-doped resins as a matrix in composite material systems to attain desired electrical properties [4–9]. Du et al. [5] utilized a coagulation technique to fabricate a nanocomposite consisting of single-walled carbon nanotubes (SWCNT) and polymethyl methacrylate (PMMA), achieving an electrical conductivity of 10^{-3} S/m with a 2% SWCNT weight fraction. Sawi et al. [6] extensively investigated the electrical conductivity of double-walled carbon nanotubes within an epoxy-based RTM6 matrix, analyzing its frequency and temperature dependence. Although the doping approach enhances overall properties, it is limited in harnessing the full potential of CNTs due to challenges in incorporating high CNT weight fractions. Primarily, CNTs tend to agglomerate due to van der Waals forces, resulting in property variations at larger scales [10–14].

Additionally, high CNT weight fractions elevate resin viscosity, impeding resin flow during fabrication. Porous membranes composed of interconnected CNTs, known as buckypaper (BP), offer a viable solution by enabling the incorporation of high weight fractions of CNTs and ensuring their uniform dispersion throughout the composite structure [15]. These films facilitate strong interlaminar bonds, making them compatible for integration into laminate structures, hence this approach overcomes the issues associated with CNT-doped resin in composite applications [16]. Due to its high electrical conductivity, homogenous dispersion through the resin microstructure, and carbon fiber reinforced polymer (CFRP) stiffness and strength enhancement [17], BP is selected as a promising candidate to investigate its EMI shielding performance within various CFRP laminate configurations.

The most common EMI shielding strategy involves utilizing conductive barriers to block unwanted external electromagnetic signals that might disrupt the internal circuitry. CFRP composites, which are increasingly being used in aerospace applications, are capable of EMI shielding due to the presence of electrically conductive fibers [18]. However, unidirectional CFRP composites exhibit significant anisotropy in electrical conductivity, attributed to the distinct mechanisms manifesting in the longitudinal and transverse directions. Along the fiber direction, the electrical conductivity of the lamina closely aligns with that of the fibers, whereas in the transverse direction, obtaining consistent conductivity values is challenging.

In contrast to longitudinal conductivity, transverse conductivity exhibits significant variability and unpredictability, which is primarily attributed to the stochastic arrangement of fibers and the variability in fiber volume fraction. The existing literature demonstrates a notable lack of consensus, as diverse studies offer dissimilar predictions for transverse conductivity values ranging from 10^{-5} to 30 S/m [8,9,19–22]. Banerjee et al. [19] used a tunneling model based on thermal activation energy and reported a transverse conductivity of ~0.04 S/m. Wasselynck et al. [20,21] used a theoretical model of impedance analysis with Monte-Carlo simulations to account for microstructure variability and predicted transverse conductivity to be approximately ~8 S/m. Accounting for the variability in microstructure and quantum tunneling phenomena is crucial for the accurate prediction of transverse conductivity [23]. When estimating the EMI shielding effectiveness (SE) at high frequencies (>1 GHz), relying on DC conductivity leads to inaccuracies due to the significant rise of the skin effect in an AC field as a result of induced eddy currents [24]. This phenomenon limits the flow of current to the outer surface of the conductor rather than a uniform distribution across the cross-section. Hence, an accurate prediction of the EMI shielding effectiveness of composites across a broad frequency spectrum is achievable solely through AC conductivity, and very limited work is reported in this area. Mohan et al. [8] employed a micromechanics model to predict the electrical conductivity in CNT-doped epoxy composites, showing a direct correlation between transverse electrical conductivity and frequency. The lack of a comprehensive model for predicting AC conductivity necessitates the need for a more accurate modeling approach that accounts for both microstructure variability and frequency dependence.

Various techniques have been investigated to enhance the EMI shielding performance of composite structures. Some researchers have used metallic layers or conductive coatings to enhance the reflective component of EMI shielding [24–26]. As observed in a previous study by the co-authors [24], the incorporation of copper mesh improves the reflection component of EMI shielding effectiveness. Yazdi et al. [25] have characterized the EMI shielding effectiveness of glass fiber reinforced polymer composites coated with aluminum, aluminum bronze and zinc. Their investigations show an enhanced SE of up to 60 dB from the baseline value of 7 dB. The major drawback associated with including metallic layers is that the structural strength is compromised due to its poor bonding with the composite ply.

The use of a modified epoxy with magnetic or conductive filler materials such as CNTs has been investigated [27,28]. Liu et al. [27] studied the EMI SE of nanocomposites by varying the weight fractions of different filler materials, including multi-walled carbon nanotubes (MWCNTs), Fe_3O_4, and Fe within the epoxy resin. Their work reported an upward trend of shielding effectiveness with increasing frequency by using CNTs, while the Fe and Fe_3O_4-doped composites showed a less frequency dependence. Zhang et al. [29] explored strategies for assembling two-dimensional metal carbide sheets (MXenes) to improve electromagnetic interference (EMI) shielding. They found that rGO/CoNi nanosheets exhibited a high absorption shielding effectiveness, achieving a value of 54.1 dB. Although these methods can be beneficial in enhancing both the electrical and mechanical properties of composites, they do not harness the full potential of carbon nanotubes at the laminate level [10–14]. This paper presents a novel approach for modeling the AC conductivity of BP-reinforced CFRP laminates accounting for quantum tunneling and skin effects.

The paper is structured into two main sections: the first section introduces an equivalent circuit method for characterizing the frequency-dependent electrical conductivity of carbon fiber lamina. The CFRP microstructure is transformed into an equivalent circuit, and the effective impedance is analyzed to estimate the longitudinal conductivity. The transverse conductivity is determined using Monte Carlo simulations informed by quantum tunneling principles. The second section focuses on coupling the estimated conductivity values with electromagnetic wave propagation simulations to investigate the EMI SE of hybrid BP/CFRP samples. Finally, the EMI SE results are compared with an alternative shielding technique using copper mesh layers.

2. Materials and Methods

To assess the AC conductivity, each CFRP lamina is modeled using polyacrylonitrile-based unidirectional IM7 carbon fiber and EPON-862 epoxy system comprising diglycidyl ether of bisphenol F (DGEBF) resin and diethylenetriamine (DETA) hardener. In the case of hybrid samples, Miralon S-T01AVB-12 sheets provided by Nanocomp Technologies (Merrimack, NH, USA) were employed as BP interleaves between the CFRP plies. A schematic of the hybrid composite and the BP morphology is shown in Figure 1 [17]. Relevant properties of fiber, resin, and BP are summarized in Table 1 [30–33]. For constructing the transverse representative volume element and establishing the equivalent circuit model, various parameters outlined in Table 2 were used. Table 3 contains some physical constants used in this study.

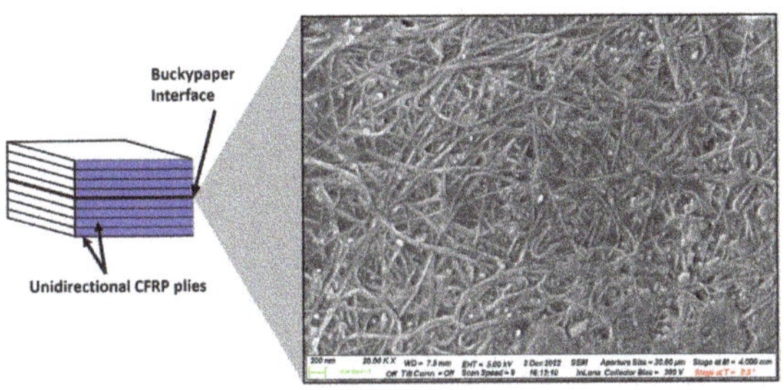

Figure 1. Hybrid laminate schematic showing BP microstructure obtained through SEM [17].

Table 1. Constituent electrical properties [30–33].

	DC Resistivity ρ_{DC} (Ω-m)	Relative Permittivity ε_r	Refractive Index n	Relative Permeability μ_r
Fiber	Longitudinal: 1.5×10^{-5} Transverse: 9.3×10^{-5}	1	1.6	1
Epoxy	10^8	3	1.55	1
BP	1.3×10^{-5}	1	-	1

Table 2. Relevant composite sample parameters used in the study.

Parameter (Unit)	Value
Fiber mean diameter (μm)	5.24
Fiber diameter standard deviation (μm)	0.295
Fiber volume fraction	0.6
CFRP lamina thickness (mm)	0.25
BP thickness (mm)	0.1

Table 3. Physical constants used in this study.

Parameter (Unit)	Value
Electron charge (C)	1.6×10^{-19}
Electron mass (kg)	9.11×10^{-31}
Permittivity of free space (F/m)	8.854×10^{-12}
Permeability of free space (H/m)	$4\pi \times 10^{-7}$
Reduced Planck's Constant (Js)	1.05×10^{-34}
Speed of light (m/s)	3×10^{8}

2.1. Electrical Conductivity Characterization of CFRPs

To characterize the frequency-dependent electrical conductivity of individual CFRP plies, longitudinal and transverse CFRP representative volume elements (RVE) are transformed into an equivalent electrical circuit consisting of inductors, capacitors, and resistors. Next, the impedance is analyzed, which is subsequently converted into complex conductivity based on the RVE dimensions.

2.1.1. Longitudinal Conductivity

For the longitudinal case, the analysis is based on the premise that in addition to their internal electrical resistance, the fibers also possess inductive effects, while the matrix acts as an insulating material. Figure 2 presents a graphical depiction of a unidirectional (UD) CFRP lamina, showcasing its microstructure and the associated equivalent circuit for the longitudinal orientation. As depicted, individual fibers are represented as a combination of a resistor (R) and an inductor (L) arranged in series. For the N parallel fibers, the overall circuit consists of N parallel resistor-inductor elements.

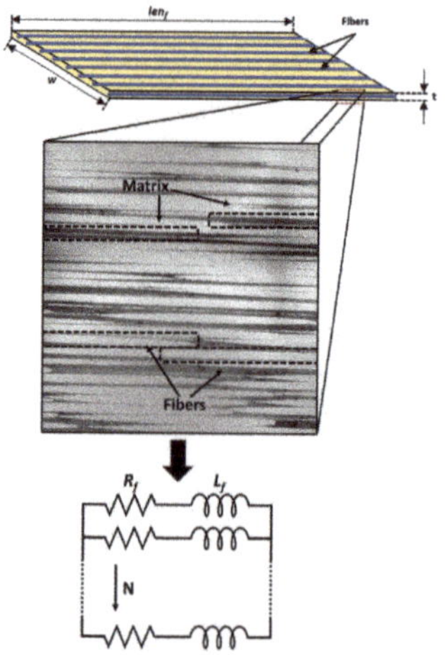

Figure 2. Microstructure of a unidirectional CFRP lamina and its equivalent electrical circuit in the longitudinal direction consisting of the N fibers, each having their own resistance (R_f) and inductance (L_f).

The resistance of each wire R_f can be expressed in terms of fiber resistivity ρ_f, length len_f, and radius r_f as follows:

$$R_f = \frac{\rho_f len_f}{\pi r_f^2} \quad (1)$$

The inductance of each fiber, denoted L_f, is expressed as a sum of its self-inductance and the mutual inductance of the neighboring fibers. The self inductance represented as $\left(L_f\right)_S$ can be expressed in terms of the fiber dimensions and vacuum permeability μ_o [34]:

$$\left(L_f\right)_S = \frac{\mu_o len_f}{2\pi} log\left(\frac{2len_f}{r_f} - 1\right) \quad (2)$$

For mutual inductance estimation, drawing upon the principles of Faraday's law of electromagnetic induction and Lenz's law when an alternating current flows through two parallel wires separated by a distance d, a counteracting current is induced. This happens because the changing current gives rise to a changing magnetic field through the surrounding wires. Consequently, a current is generated that resists this magnetic field change. This phenomenon within the CFRP microstructure is depicted in Figure 3a. The direction of the magnetic field can be determined using the right-hand rule, and its magnitude can be expressed as below:

$$B = \frac{\mu_o I}{2\pi d} \quad (3)$$

where I is the electric current traversing the fiber and d denotes the radial distance from the fiber to the point of interest. Figure 3b depicts magnetic field vectors within a small region of the ply cross-section and illustrates the influence of surrounding fibers on the magnetic field intensity at any given fiber location. For illustration, the fibers are color-coded to correspond with their respective magnetic field vectors, with an 'X' symbol denoting the direction of current flow into the plane. It should be noted that the effective magnetic field at the arbitrary fiber experiences significant cancellation due to the influence of surrounding fibers. However, it remains nonzero owing to the nonuniform distribution of fibers within the microstructure. Finally, using the relation between magnetic flux φ and wire inductance L, the final form of mutual inductance denoted by $\left(L_f\right)_M$ can be expressed as follows:

$$\left(L_f\right)_M = \frac{\varphi}{I} = \frac{B.A}{I} = \frac{2Blen_f r_f}{I} \quad (4)$$

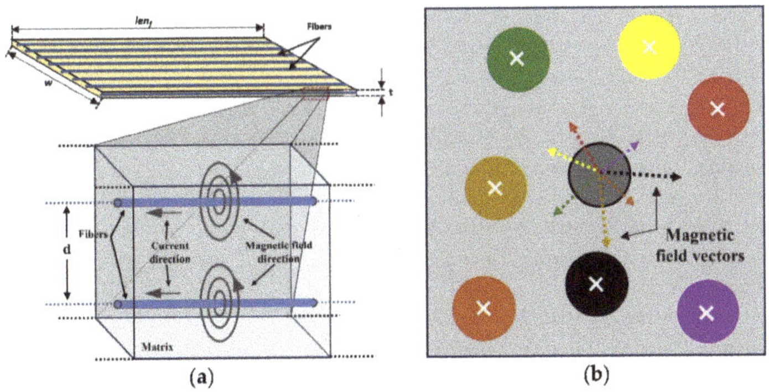

(a) (b)

Figure 3. (**a**) Magnetic fields due to parallel fibers separated by a distance d and (**b**) color-coded magnetic field vectors due to the current carrying neighboring fibers.

Utilizing a transverse CFRP RVE as depicted in Figure 4a, assuming that the fibers are connected in parallel, the inductance for each fiber and ultimately the overall inductance $\left(L_f\right)_{net}$ was estimated by summing $\left(L_f\right)_M$ and $\left(L_f\right)_S$.

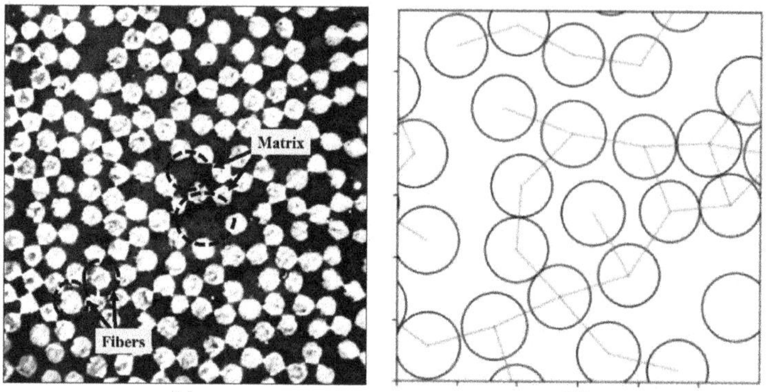

(a) (b)

Figure 4. (**a**) Transverse microstructure showing the fiber distribution used for estimating the overall inductance, (**b**) Interconnected RVE with straight line connections between pairs of fibers within the threshold distance.

The magnitude of total impedance in the longitudinal direction can be expressed in terms of fiber resistance R_f, total inductive reactance X_L, and the number of fibers N using the relation given below:

$$\left|Z_{long.}\right| = \sqrt{\left(\frac{R}{N}\right)^2 + X_L^2} \quad (5)$$

where X_L is further expressed in terms of the total inductance $\left(L_f\right)_{net}$ and signal frequency f as follows:

$$X_L = 2\pi f\left(L_f\right)_{net} \quad (6)$$

Finally, the AC conductivity in the longitudinal direction σ_L can be expressed using the following equation:

$$\sigma_L = \frac{len_f}{A\left|Z_{long.}\right|} \quad (7)$$

where A is the cross-sectional area of the CFRP lamina perpendicular to the fiber direction.

2.1.2. Transverse Conductivity

In the context of transverse conductivity, two primary mechanisms come into play. First, during the fabrication process, the likelihood of fiber-to-fiber contact can give rise to a conductive pathway within the sample. Second, when the gap between two fibers is within a threshold distance, quantum tunneling through the highly resistant epoxy barrier becomes a possibility. This section presents an in-depth explanation of a probabilistic conductivity model that encompasses both of these phenomena in conjunction with Monte Carlo simulations.

Figure 4a illustrates the transverse microstructure of a CFRP lamina. Based on the morphology, a maximum fiber–fiber separation of 100 nm was selected for generating the RVE. A square RVE with 23 fibers was generated, accounting for the mean and standard deviation of fiber radius, range of fiber spacing, and fiber volume fraction. For a 60% fiber volume fraction, the resulting RVE size was ~28.75 µm. To gain insights into the spatial relationships between fibers and potential conductive pathways, a connectivity matrix based on a threshold fiber pair distance (T) was established. It was observed that the tunneling probability was insignificant for fiber—fiber separation greater than 50 nm and was therefore chosen as the threshold for this study. This assumption is verified by the results presented in Section 3.2. Figure 4b illustrates the resulting connected RVE with doubly periodic boundaries, wherein all fiber pairs deemed potentially conductive are linked by straight lines.

Next, the connected RVE is analyzed to identify the shortest conductive pathway across the RVE. This pathway holds significance since electrical charges follow the route of least resistance. Subsequently, this shortest conductive path is subjected to a more in-depth analysis using principles of quantum tunneling to compute the transverse conductivity. The details of this methodology are presented in Figure 5. Finally, the Monte Carlo method was employed, which involved conducting 1000 simulations of the process outlined in Figure 5. The results were used to calculate the mean transverse conductivity.

To quantify the tunneling probability between each fiber pair within the conductive path, the one-dimensional, time-independent Schrödinger equation was solved, as given by Equation (8).

$$-\frac{\hbar^2}{2m_e}\frac{\partial^2\Psi(x)}{\partial x^2}+U(x)\Psi(x)=E\Psi(x) \tag{8}$$

where $\hbar$ is the reduced Planck's constant, m_e is the mass of the electron, and $\Psi(x)$ denotes the wave function of the particle at a given location x. The terms U and E represent the barrier potential energy and the particle energy, respectively. A schematic of fiber pairs with an enlarged view of the contact region is displayed in Figure 6. As illustrated in the figure, the fiber–fiber contact is divided into three regions. Regions A and C represent the fiber space, and region B represents the insulating matrix. By solving Equation (8) for each region individually, the unique wave functions Ψ_A, Ψ_B, and Ψ_C were computed.

The solutions to these wave functions, along with the derivation of the tunneling probability, is detailed in Appendix A. The final expression for the tunneling coefficient C_T is expressed as follows:

$$C_T=\left(1+\frac{\sinh^2(k_b a)}{4\eta(1-\eta)}\right)^{-1} \tag{9}$$

where η represents the ratio of the particle's energy (E) to the barrier potential energy (U) and k_b is a constant. Equation (9) reveals that the likelihood of tunneling is directly proportional to the energy of the electrons and inversely proportional to the barrier size a. The potential energy of the barrier is estimated by an equivalent circuit approach. As depicted in Figure 7a, the contact impedance can be translated into an equivalent circuit comprising a resistor-capacitor parallel circuit pair. The magnitude of this impedance $|Z_c|$ is expressed below:

$$|Z_c|=\frac{1}{\sqrt{\frac{1}{R_c^2}+\frac{1}{X_c^2}}} \tag{10}$$

where R_c and X_c are the resistance and capacitive reactance of the matrix region at the contact. For a curved matrix geometry, as illustrated in Figure 7b, the contact resistance and capacitive reactance is estimated by discretizing the matrix region and integrating. Their simplified expressions are as follows:

$$R_c=\frac{\rho_m}{len_f I} \tag{11}$$

$$X_c=\frac{1}{2\pi f C_c} \tag{12}$$

The contact capacitance C_c can be further expressed in terms of the material's relative permittivity ε_r and permittivity of free space ε_o as follows:

$$C_c = \varepsilon_0 \varepsilon_r l_c I \quad (13)$$

The term l_c denotes the fiber contact length, which is defined as the total length of the segment along the fiber direction where fibers consistently remain in close proximity. It's important to distinguish this term from the sample length along the fiber direction, as fibers exhibit a wavy, nonlinear nature, as illustrated in Figure 8. l_c was chosen as a percentage of the overall RVE length for which the wavy fibers maintain proximity. In Equations (11) and (13), the term I is a constant and can be expressed as shown below:

$$I = \int_0^{\frac{\pi}{2}} \frac{\cos\theta d\theta}{\left(\frac{T}{2r_f} - \cos\theta\right)} \quad (14)$$

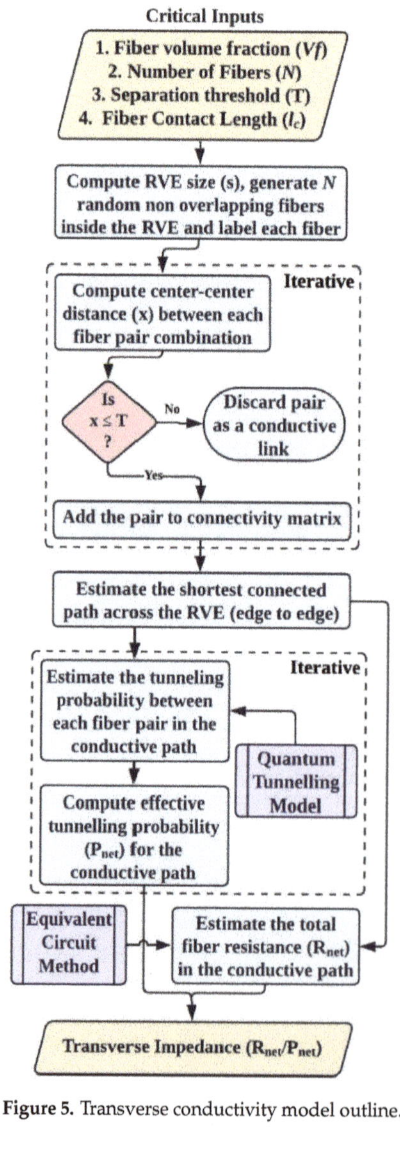

Figure 5. Transverse conductivity model outline.

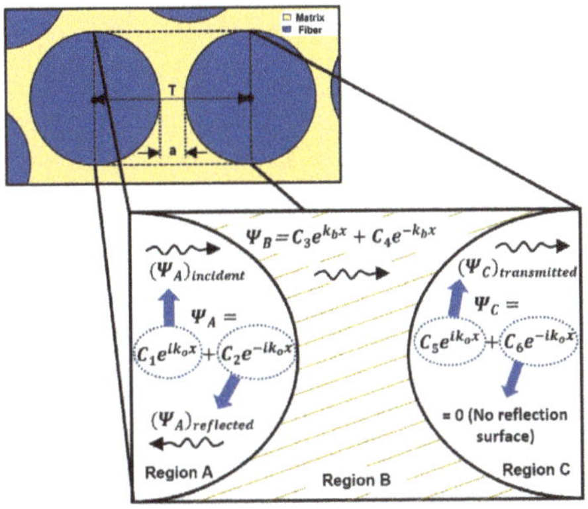

Figure 6. Fiber pair schematic with an enlarged view of the contact region showing the wave functions (Ψ) for different regions along the particle tunneling path.

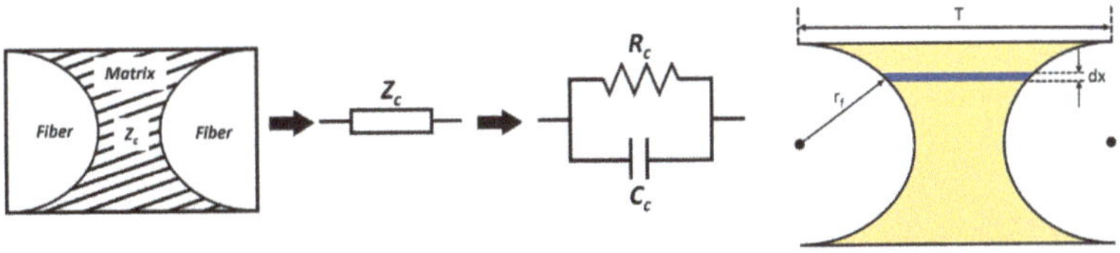

(a) (b)

Figure 7. (**a**) Equivalent circuit of a fiber–fiber contact pair and its equivalent contact impedance (Z_c) and (**b**) contact region discretization for determining effective contact resistance (R_c) and capacitance (C_c).

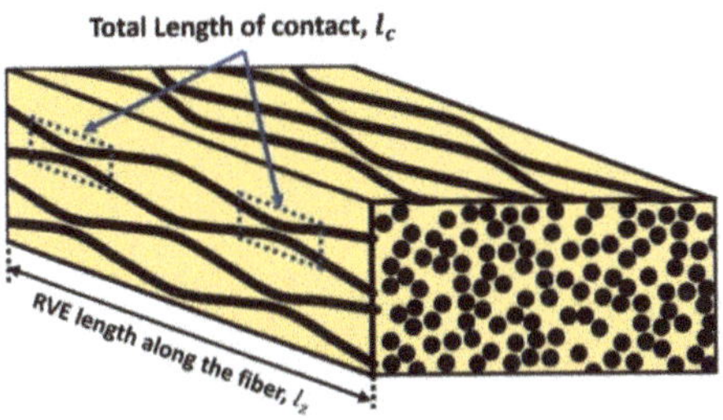

Figure 8. Schematic showing the fiber contact length (l_c) due to fiber waviness.

The energy required to transmit an electron through the matrix barrier is calculated from the energy, current, and voltage relationship as follows:

$$U = I_{rms}^2 \left| Z_c \right| \Delta t \tag{15}$$

where I_{rms} is the RMS current of the signal, and Δt is the time taken by the particle to pass through the barrier. Upon simplifying Equation (15) further, we obtain the following expression for the barrier potential energy U:

$$U = \frac{q_e{}^2 c \left| Z_c \right|}{a n_m} \tag{16}$$

where q_e, c, and n_m are the electron charge, speed of light, and refractive index of the matrix, respectively. The term c/n_m denotes the speed of electrons in the matrix medium. The energy of the particle E is expressed as follows:

$$E = 2\pi\hbar f \tag{17}$$

The shortest conductive path is analyzed next to determine the effective transverse impedance. Using Equation (9), the specific tunneling probabilities $(P_{link})_n$ associated with each nth fiber–fiber link within the conductive path can be calculated. Figure 9 visually represents the electron tunneling process along the shortest route. The cumulative tunneling probability across the RVE, denoted as P_{net}, can be expressed as the product of individual tunneling probabilities for all links as shown below:

$$P_{net} = \prod_{n=1}^{N_l} (P_{link})_n \tag{18}$$

where N_l corresponds to the total number of fiber–fiber links in the conductive path. In a hypothetical scenario where all the fibers along the conductive path are touching ($P_{net} = 1$), the overall RVE impedance can be determined by summing the resistance across each fiber cross-section within the conductive path. This cumulative impedance, denoted as R_{net}, can be mathematically represented as follows:

$$R_{net} = \sum_{n=1}^{N_f} \left(R_{ft}\right)_n \tag{19}$$

where N_f denotes the total number of fibers in the conductive path and $\left(R_{ft}\right)_n$ is the resistance across the nth fiber cross-section. Here, the subscript ft refers to the transverse fiber resistance. Finally, the RVE impedance $|Z_c|_{RVE}$ and transverse conductivity σ_T are determined by the following relations:

$$|Z_c|_{RVE} = \frac{R_{net}}{P_{net}} \tag{20}$$

$$\sigma_T = \frac{1}{l_z |Z_c|_{RVE}} \tag{21}$$

where l_Z is the length of the RVE along the fiber direction, as labeled in Figure 8.

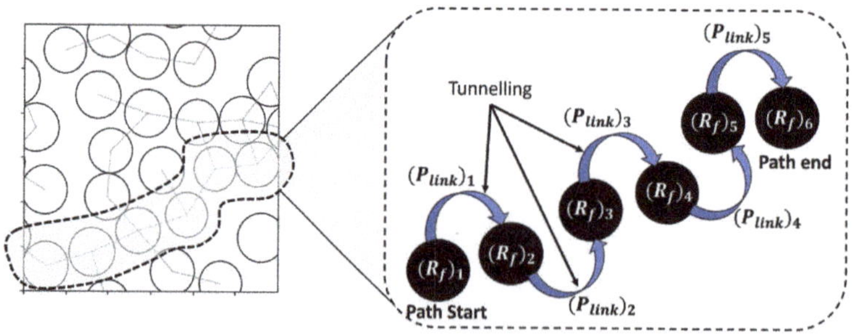

Figure 9. Quantum tunneling schematic including tunneling probabilities $(P_{link})_n$ and resistance across the nth fiber cross-section $\left(R_{ft}\right)_n$ for the shortest conductive path.

2.2. Electrical Conductivity of BP

The electrical conductivity of BP at the macroscopic level is generally considered transversely isotropic, primarily because CNTs within BP are randomly oriented and lack any inherent directional bias. Additionally, in this research, the BP conductivity is assumed to be independent of frequency. While this simplification is not entirely precise, it is reasonable for two main reasons. First, the CNTs in BP are short (~10 μm) and nonmagnetic and thus possess negligible inductance, and second, they are separated by only a few nanometers (~10 nm), as illustrated in Figure 1. At this length scale, the probability of quantum tunneling between nanotubes is notably high, facilitating the movement of electrons along many suitable conductive paths. In the through-thickness direction, due to the presence of a relatively thick matrix region, the electrical conductivity is considered similar to that of the DC transverse conductivity of the carbon fiber ply. The BP DC conductivity values used for this study are outlined in Table 1.

2.3. EMI Shielding Model

The longitudinal and transverse conductivity estimated from the equivalent circuit models explained above is used to predict the EMI SE of hybrid BP/CFRP laminates. The EMI shielding in materials is a consequence of three fundamental phenomena: reflection, absorption, and internal reflections. In the case of composites, the absorption-based shielding effect exceeds 10 dB; hence, the effect of shielding through multiple reflections is safely ignored. The EMI SE of the samples was determined using an ANSYS high-frequency structure simulator (HFSS), which uses a full-wave electromagnetic simulation technique to calculate the scattering parameters (S-parameters). The software employs the finite element boundary integral method (FEBIM) to solve Maxwell's equations that govern electromagnetic fields within each mesh element and account for interaction with exterior regions. Although the constituent properties employed for the EMI simulations are obtained from the micro-scale model explained in Section 2.1, the electromagnetic simulations themselves are conducted at the lamina scale and do not address the effects of multiple scattering at the constituent scale. For this reason, experimental testing using an appropriate setup and vector network analyzer is required to validate the ANSYS simulations.

In the modeling of each lamina, the material properties, including conductivity, permittivity, and permeability, are provided as inputs. The electrical conductivity for a lamina oriented at an arbitrary angle θ is represented in tensor form, as depicted in Equation (22) [24]. σ_L and σ_T denote the longitudinal and transverse conductivities. Due to the transversely isotropic nature of the lamina, the through-ply conductivity is assumed to be the same as the transverse conductivity. A semi-infinite representative shield element was simulated, and the scattering parameters (S-parameters) S_{11} and S_{21} were obtained for each sample under study, which allow for calculating the amplitude of the reflected and transmitted wave. A pictorial representation of a 10 mm × 10 mm rectangular waveguide used for the analysis is shown in Figure 10. To prevent boundary reflections/radiations and ensure precise results, periodic boundary conditions were applied along the walls, and Floquet ports [24] were employed at both the Inlet and Exit ports. The fundamental transverse electric mode of propagation TE_{10} was considered for the analysis due to the simplicity of the geometry and nonmagnetic nature of the specimen. This approach allowed for a comprehensive analysis of the electromagnetic response of the representative shield element under investigation.

$$\sigma_\theta = \begin{bmatrix} \sigma_L cos^2(\theta) + \sigma_T sin^2(\theta) & \frac{\sigma_L - \sigma_T}{2} sin(2\theta) & 0 \\ \frac{\sigma_L - \sigma_T}{2} sin(2\theta) & \sigma_L sin^2(\theta) + \sigma_T cos^2(\theta) & 0 \\ 0 & 0 & \sigma_T \end{bmatrix} \tag{22}$$

Using S-parameters obtained from the simulations, the shielding effectiveness due to reflection SE_R and absorption SE_A were calculated using the following relations:

$$SE_R = 10log_{10}\left(\frac{1}{1-|S_{11}^2|}\right) \text{ [dB]} \tag{23}$$

$$SE_A = 10log_{10}\left(\frac{1-|S_{11}^2|}{|S_{21}^2|}\right) \text{ [dB]} \tag{24}$$

Finally, the total shielding effectiveness SE can be expressed as follows:

$$SE = SE_A + SE_R \tag{25}$$

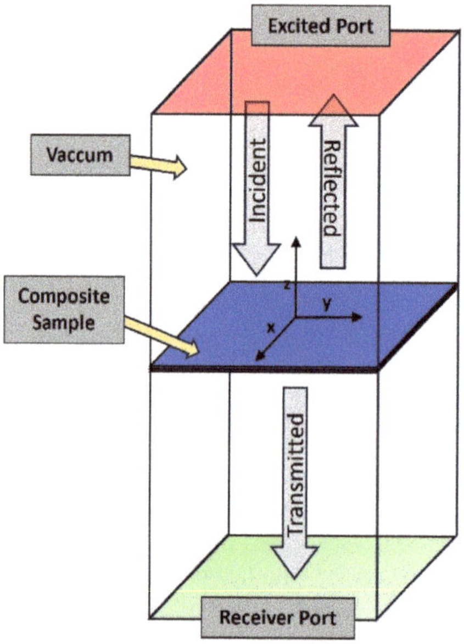

Figure 10. Waveguide model schematic showing the composite sample under study.

3. Results and Discussion

3.1. Longitudinal Electrical Conductivity

The CFRP plies exhibit high electrical conductivity in the longitudinal direction due to the long and continuous nature of fibers. Using the procedure detailed in Section 2.1, the frequency-dependent longitudinal conductivity (σ_L) of a unidirectional CFRP lamina in the 60 Hz to 50 GHz frequency range was calculated. The results are compared with experimental results obtained by Kim et al. [35] and are shown in Figure 11a. At frequencies above 0.5 GHz, the inductive reactance becomes notably pronounced, resulting in an overall reduction in conductivity. A slight variation in the location of the point where conductivity begins to reduce is observed with changes in ply size, as depicted in Figure 11b. Lamina with more fibers exhibit an earlier onset of the declining trend, while those with lesser fibers experience this decline at a later stage. This behavior can be attributed to the inverse relationship between the overall inductance of a ply in the longitudinal direction and the number of fibers within the ply, coupled with a direct proportionality to the ply length. Figure 12 depicts the overall inductance $\left(L_f\right)_{net}$ as a function of total number of fibers N, for a 1 cm long ply.

The decrease in overall inductance is a consequence of two primary factors. First, the effective inductance significantly reduces when more fibers are connected in parallel. Second, with an increasing number of fibers, the influence of mutual induction becomes negligible due to the cancelation of magnetic fields. Beyond ~5000 fibers, the net inductance saturates, which is attributed to the limited influence of distant fibers on the inductance of a given fiber.

For the case of a 10 mm × 10 mm × 0.25 mm ply with a 60% fiber volume fraction, comprising roughly 70,000 fibers, an overall inductance of ~2.5 pH was obtained. This value translated to an average fiber inductance of ~17.5 μH, which falls within the range published by other researchers [36]. Subsequently, employing this result, the frequency-dependent longitudinal conductivity σ_L can be calculated. The response obtained is cross-referenced with values documented in existing literature to validate the model [35,37]. Kim et al. [35] conducted experiments to investigate the AC conductivity of USN-150R prepregs with fiber conductivity between 600–660 S/cm. Their findings confirm the accuracy and technique of the model proposed in this study, as depicted in Figure 11a.

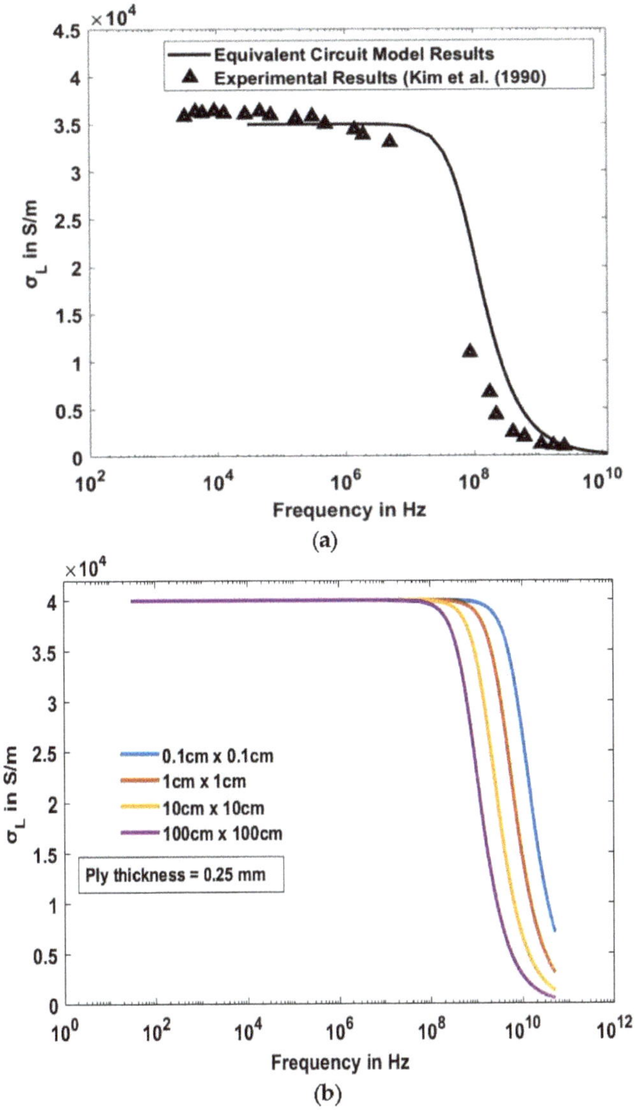

Figure 11. (**a**) Longitudinal CFRP conductivity declining at higher frequencies due to the rise of inductive reactance and (**b**) shift in conductivity decline point for different square ply sizes showing an early decline in conductivity in larger plies.

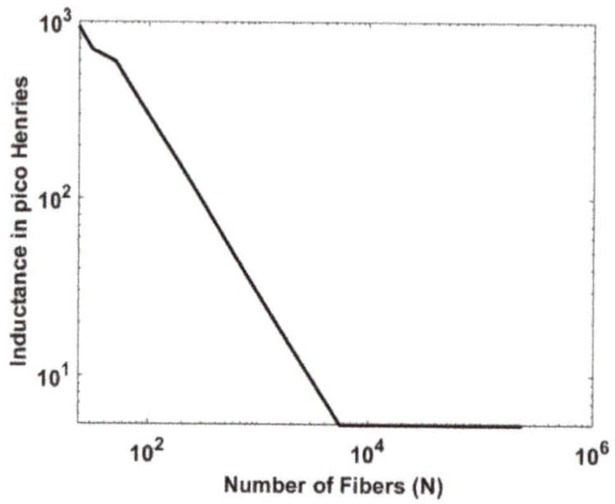

Figure 12. Overall inductance versus number of fibers on a log–log scale showing the saturation of the induction.

3.2. Transverse Electrical Conductivity

For fibers separated by a distance of less than 50 nm, the barrier potential energy is relatively low, resulting in a significant tunneling probability. Figure 13 depicts the trend of tunneling probability with signal frequency for various barrier sizes. It is observed that the tunneling probability rises with an increase in signal frequency. This holds true for all barrier sizes, as the particle's energy elevates with frequency according to Equation (17). As the barrier size diminishes, the charge can be conducted even at lower frequencies due to the reduced barrier potential energy required for passage. This observation underscores the closely packed fibers forming efficient conductive pathways along the transverse direction, leading to high electrical conductivity.

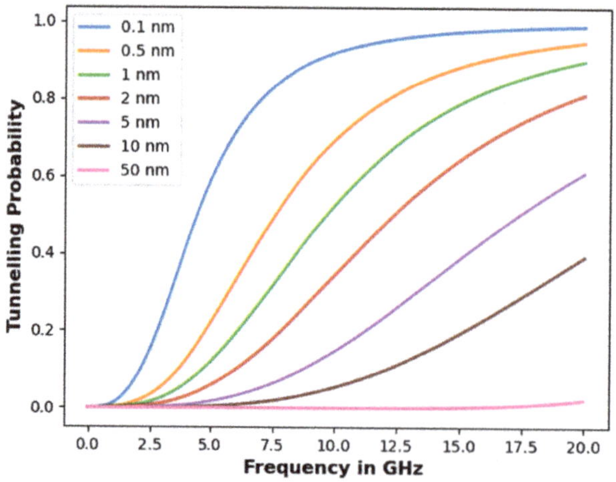

Figure 13. Tunneling probability versus signal frequency for different barrier sizes.

Another important parameter that dictates the transverse conductivity arises due to fiber waviness. The term l_c, as explained in Figure 8, is crucial in estimating the transverse conductivity of a lamina. In laminas with increased waviness, fibers tend to have multiple contacts with adjacent fibers along the longitudinal direction, thereby enhancing the probability of tunneling. The outcomes illustrated in Figure 13 correspond to a 20% fiber contact length along the fiber direction employed for this study. To understand the effect of fiber contact length l_c, three varying degrees of fiber contact along the longitudinal direction were considered, and the results are plotted as shown in Figure 14.

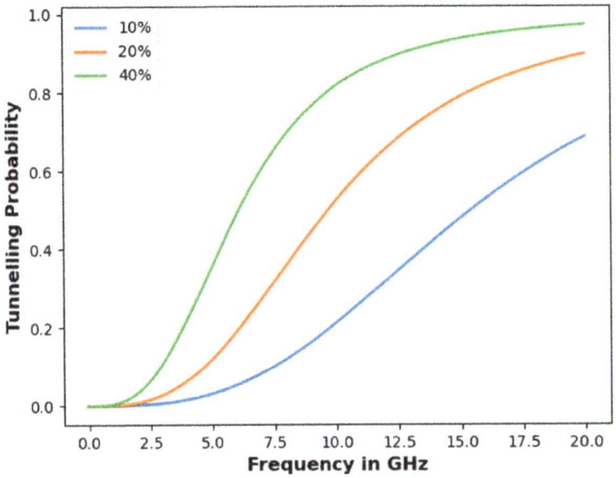

Figure 14. Tunneling probability versus signal frequency for different fiber contact lengths (contact length shown as a percentage of the entire fiber length).

The quantum tunneling-based Monte Carlo model described in Section 2.1.2 is used to generate 1000 datasets for the transverse conductivity (σ_T) in the frequency range 100 MHz to 1 GHz. Utilizing the mean and standard deviation from the dataset, σ_T was plotted with a 95% confidence interval, as shown in Figure 15. A consistent finding across all datasets is the negligible conductivity up until ~400 MHz. Beyond 450 MHz, an average transverse conductive of ~12.67 S/m and standard deviation ~1 S/m was observed, aligning with the broad spectrum of experimentally measured and simulated transverse conductivities as documented in existing literature. At frequencies above ~10 GHz, as the longitudinal conductivity rapidly declines, the transverse conductivity becomes comparatively significant and plays a crucial factor in dictating the EMI SE of CFRP composites.

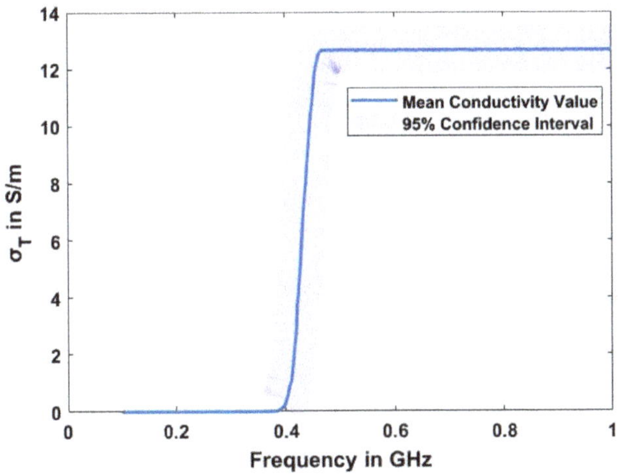

Figure 15. Transverse conductivity with a 95% confidence interval obtained from the quantum tunneling-based Monte Carlo simulations.

3.3. EMI Shielding of BP/CFRP Hybrid Composites

This study examines the effect of BP placement and relative CFRP ply orientation to understand the EMI SE of hybrid composites. All the specimen configurations used for the study are detailed in Table 4. Initially, the BP position is varied within a simple [90,0] baseline configuration to understand the effect of BP placement. Subsequently, to assess the influence of adjacent ply orientation, BP is introduced between two plies and one of the ply orientations is modified. Finally, the effects of BP in a composite with a [90, 0, ±45, 0, 90]$_s$ stacking sequence are investigated.

Table 4. Specimen configuration.

Label	Configuration
Case 1 (BP Placement)	[90, 0] [BP, 90, 0] [90, BP, 0] [90, 0, BP]
Case 2 (Adjacent Ply)	[0, BP, 0] [45, BP, 0] [90, BP, 0]
Case 3 (Standard Configuration)	[90, 0, ± 45, 0, 90]$_s$ [90, 0, BP, ± 45, 0, 90]$_s$

The total SE, along with its reflection and absorption components for a single BP film and a [90,0] baseline laminate, are shown in Figures 16a and 16b, respectively. These results show good agreement with the available literature [38–40]. At 2 GHz, the total shielding effectiveness (SE) for the BP is predicted to be approximately 43 dB, which aligns with the manufacturer's reported value of 40 dB [32]. Additionally, the total EMI SE for a [90,0] laminate with a thickness of 0.5 mm in the X-band (8 GHz to 12 GHz) is predicted to range from 62 dB to 57 dB, which is consistent with the experimental data reported in the literature. A dominant absorption effect was observed in the shielding effectiveness of the [90,0] laminate when contrasted with the BP film, owing to its thickness. It was observed that the presence of BP significantly enhances the overall SE of a composite. Particularly at frequencies exceeding 1 GHz, wherein the electrical conductivity of CFRP experiences a significant decline, the corresponding EMI SE also drops. The incorporation of BP interleaf proves effective in sustaining elevated electrical conductivity, even at higher frequencies, thereby preventing a reduction in the overall shielding effectiveness. The effect of BP location within the baseline [90,0] laminate is depicted in Figure 16c–e. By varying the position of BP within the composite, the SE also varies slightly. It was inferred that due to the complex reflection and absorption shielding mechanisms, the stacking sequence of the hybrid BP/CFRP composites greatly influences the overall shielding effectiveness.

Figure 17 illustrates the effect of ply orientations adjacent to the BP on its EMI SE. Optimal shielding against undesired radiation is attained by integrating two unidirectional plies with the greatest variation in orientation angles. The inclusion of more uniformly oriented plies aids in transitioning towards a quasi-isotropic laminate, resulting in a notable decrease in the anisotropy of electrical conductivity and enhancing the overall SE. Moreover, the isotropic nature of BP further enhances the EMI shielding performance by reducing the anisotropy in electrical properties. In the case of thicker balanced symmetric laminates, as the frequency increases, the effect of BP is more pronounced due to the drop in the CFRP conductivity. The use of BP interleaves can help significantly improve the EMI SE of thick CFRP laminates. This behavior can be observed in Figure 18, which shows the EMI SE of a 12-ply sample with and without the BP.

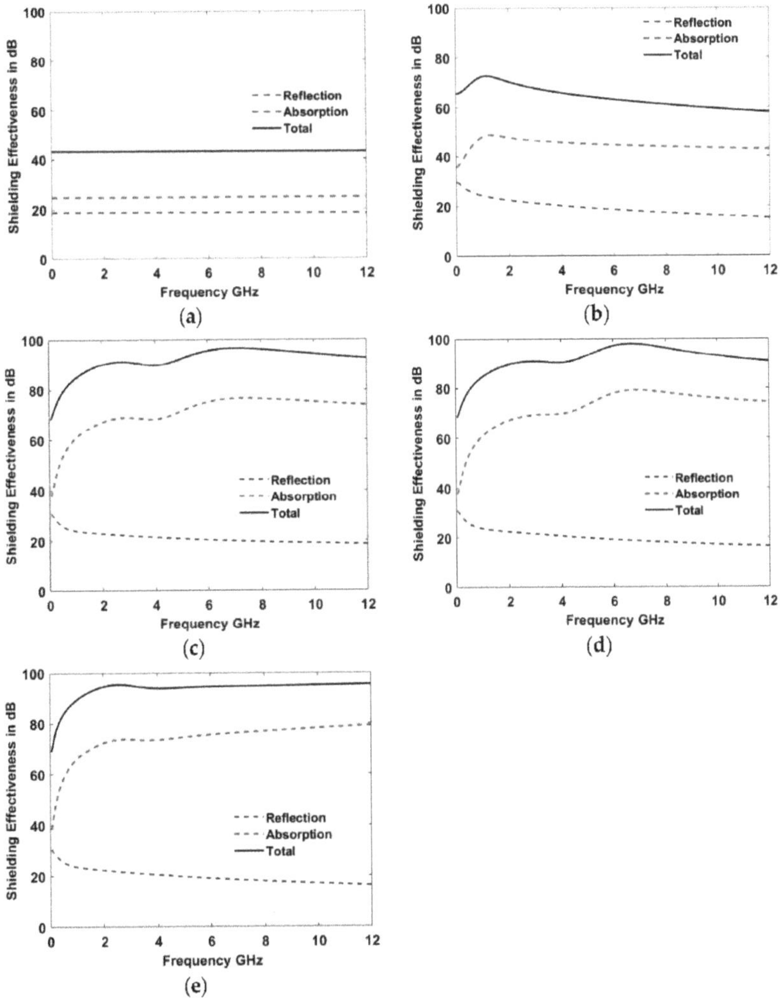

Figure 16. Shielding effectiveness in the 50 MHz–12 GHz frequency range for (**a**) buckypaper (BP), (**b**) [90,0], (**c**) [BP,90,0], (**d**) [90,BP,0], and (**e**) [90,0,BP].

Finally, the shielding effectiveness of BP is compared to other techniques, such as incorporating copper mesh [24]. Figure 19 illustrates that the EMI shielding effectiveness of BP composites remains significantly high beyond 1 GHz. This is credited to the increased electrical conductivity of the BP layer at high frequencies, in contrast to copper, which experiences a gradual decline in conductivity with frequency due to the skin effect. At elevated frequencies, instead of flowing uniformly, the current tends to concentrate on the surface of copper, thereby restricting the penetration of electromagnetic waves. The current density decreases exponentially as the wave penetrates the material, and the depth at which it falls to approximately 37% of its surface value is known as the skin depth. The skin depth decreases with increasing frequency. As a result, the effective area available for current flow is reduced, leading to a decrease in EMI SE. Additionally, BP offers a crucial advantage over metallic layers by seamlessly bonding with carbon fiber lamina without compromising mechanical performance, thereby rendering it suitable for high performance in various aspects [17].

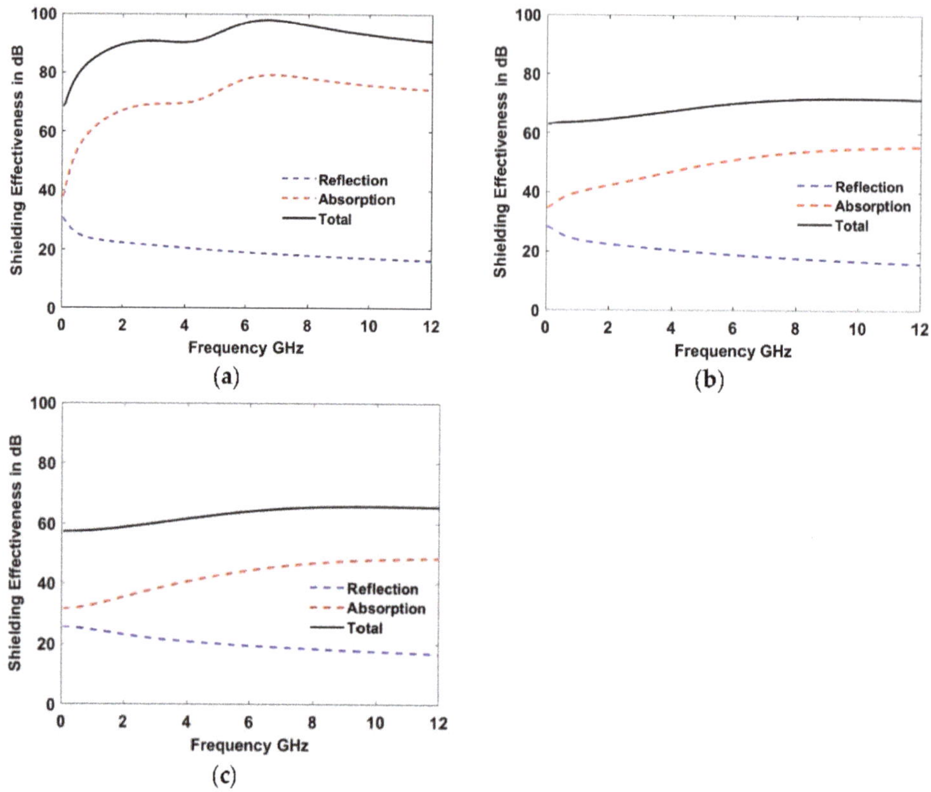

Figure 17. Effect of ply orientation on the shielding effectiveness: (**a**) [90,BP,0], (**b**) [45,BP,0], and (**c**) [0,BP,0].

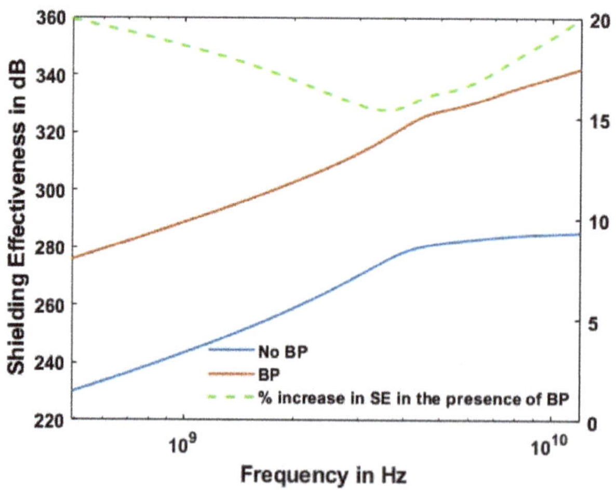

Figure 18. Effect of buckypaper (BP) on the shielding effectiveness of a 12-ply laminate showing the percentage enhancement in SE by incorporating BP.

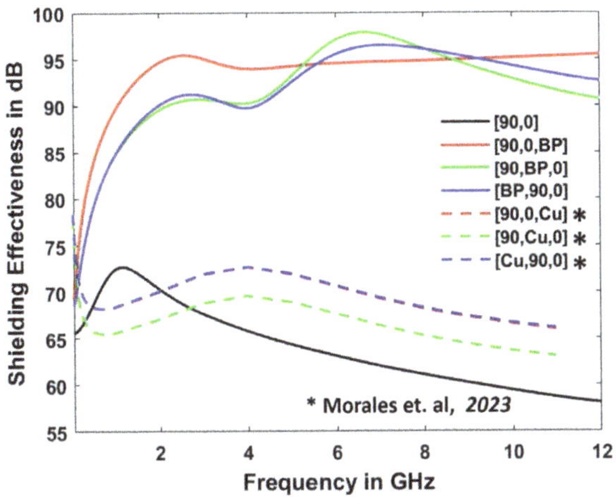

Figure 19. Comparison between the shielding effectiveness exhibited by composites with buckypaper (BP) and copper mesh interleaves [24].

4. Conclusions

This study investigates the electromagnetic interference shielding properties of hybrid buckypaper/carbon fiber-reinforced composites. A novel quantum tunneling-based equivalent circuit approach combined with Monte Carlo simulations was introduced to predict the frequency-dependent electrical conductivity of unidirectional carbon fiber lamina. The model provides an accurate approach for estimating both longitudinal and transverse electrical conductivity of unidirectional plies of arbitrary size, morphology, and constituent electrical properties. Subsequently, effective electrical properties were employed to evaluate the shielding performance of buckypaper against external electromagnetic radiation in the 50 MHz–12 GHz frequency range. The findings demonstrate a significant improvement in the overall shielding effectiveness of carbon fiber-reinforced polymer composites with the inclusion of buckypaper interleaves at high frequencies. This enhancement is attributed to the consistent electrical conductivity of buckypaper, even as the conductivity of carbon fiber lamina declines beyond 1 GHz. The integration of buckypaper into fiber-reinforced composites leads to a comprehensive improvement in both the electrical and mechanical properties, making them an ideal choice for use in composite structures.

Author Contributions: Conceptualization, A.C., T.C.H., A.H. and M.A.M.; methodology, K.T.; software, K.T.; validation, K.T.; formal analysis, K.T. and M.H.H.; investigation, K.T.; resources, A.C.; data curation, K.T.; writing—original draft preparation, K.T.; writing—review and editing, A.C. and M.H.H.; visualization, K.T.; supervision, A.C.; project administration, T.C.H., A.H. and M.A.M.; funding acquisition, A.C. All authors have read and agreed to the published version of the manuscript.

Funding: This research was funded by Army Research Laboratory grant number W911NF-17-2-0207.

Institutional Review Board Statement: Not applicable.

Informed Consent Statement: Not applicable.

Data Availability Statement: The raw data required to reproduce these findings will be made available upon request.

Acknowledgments: This research was sponsored by the Army Research Laboratory (Cooperative Agreement W911NF-17-2-0207). The views and conclusions contained in this document are those of the authors and should not be interpreted as representing the official policies, either expressed or implied, of the Army Research Laboratory or the U.S. Government. The U.S. Government is authorized to reproduce and distribute reprints for Government purposes, notwithstanding any copyright notation herein. This article is a revised and expanded version of a paper entitled "Electrical Characterization and Electromagnetic Interference Shielding Properties of Hybrid Buckypaper-Reinforced Polymer Matrix Composites" [41], which was presented at the Aerospace Structures, Structural Dynamics, and Materials Conference at Renton, Washington on 29 April 2024.

Conflicts of Interest: The authors declare no conflicts of interest.

Nomenclature

ρ_f	Fiber resistivity
ρ_m	Matrix resistivity
ε_o	Vacuum permittivity
ε_r	Relative permittivity of medium
μ_o	Vacuum permeability
μ_r	Relative permeability of medium
n_m	Refractive Index of matrix
R_f	Fiber resistance
len_f	Fiber length
r_f	Mean fiber radius
X_L	Fiber inductive reactance
L_f	Fiber inductance
f	Signal frequency
T_{avg}	Average fiber separation
$\lvert Z_{long.}\rvert$	Longitudinal CFRP impedance
N	Number of fibers
σ_{11}	Longitudinal AC conductivity
$\hbar$	Reduced Planck's constant
m_e	Electron mass
U_o	Barrier potential energy
E	Particle's energy
a	Barrier size
C_T	Tunneling coefficient
$\lvert Z_c\rvert$	Contact impedance
R_c	Contact resistance
C_c	Contact capacitance
T	Separation between neighboring fibers
X_C	Contact capacitive reactance
l_c	Total fiber contact length
S	RVE length transverse to the fiber
l_z	RVE length along the fiber
c	Speed of light
q_e	Electron charge
σ_{22}	Transverse AC conductivity
SE_A	Absorption shielding effectiveness
SE_R	Reflection shielding effectiveness
SE	Total shielding effectiveness

Appendix A

The general solutions for wave functions Ψ_A, Ψ_B, and Ψ_C obtained from Equation (8) are given below:

$$\Psi_A(x) = C_1 e^{ik_o x} + C_2 e^{-ik_o x} \tag{A1}$$

$$\Psi_B(x) = C_3 e^{k_b x} + C_4 e^{-k_b x} \tag{A2}$$

$$\Psi_C(x) = C_5 e^{ik_o x} + C_6 e^{-ik_o x} \tag{A3}$$

where C_1, C_2, C_3, C_4, C_5, and C_6 are solution coefficients. Constants k_o and k_b can be further expressed as follows:

$$k_o = \frac{\sqrt{2m_e E}}{\hbar} \quad \text{(A4)}$$

$$k_b = \frac{\sqrt{2m_e(U-E)}}{\hbar} \quad \text{(A5)}$$

In the present case, the second term in $\Psi_C(x)$ (Equation (A3)) can be ignored because there is no interface to induce reflection; the particle is moving unidirectionally to the right. Additionally, the solution coefficient C_1 associated with the probability density of particle incident on the barrier is one. In the matrix region, denoted by B, the wave function exhibits a decaying behavior, as represented in Figure A1. Upon enforcing C^1 continuity at the two fiber matrix interfaces (x = 0 and x = a where 'a' represents the barrier size), a deterministic system of equations involving four unknowns is obtained.

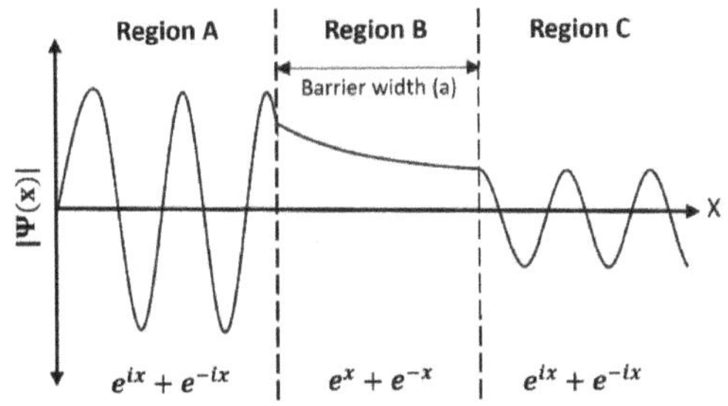

Figure A1. The wave function in the three different regions.

At the incident interface (x = 0):

$$1 + C_2 = C_3 + C_4 \quad \text{(A6)}$$

$$ik_o(1 - C_2) = k_b(C_3 - C_4) \quad \text{(A7)}$$

At the transmitted interface (x = a):

$$C_3 e^{k_b a} + C_4 e^{-k_b a} = C_5 e^{ik_o a} \quad \text{(A8)}$$

$$k_b\left(C_3 e^{k_b a} - C_4 e^{-k_b a}\right) = ik_o C_5 e^{ik_o a} \quad \text{(A9)}$$

Solving Equations (A6)–(A9), the coefficient C_5, which is linked to the tunneling probability, can be obtained. Subsequently, the transmission coefficient C_T is computed using the following equation:

$$C_T = \frac{P_t}{P_i} \quad \text{(A10)}$$

where P_t and P_i are the probability density of finding the particle as it exits the barrier in region C and the particle incident on the barrier in region A, respectively. P_t can be further expressed as a product of the wave function Ψ_C and its complex conjugate Ψ_C^* using the following expression:

$$P_t = \Psi_C \Psi_C^* \quad \text{(A11)}$$

Upon further simplification, P_t is found equivalent to C_5^2. Given that C_1 equals one, P_i also equals one, resulting in a streamlined tunneling probability formulation and yielding the ultimate tunneling coefficient as shown below:

$$C_T = \left(1 + \frac{\sinh^2(k_b a)}{4\eta(1-\eta)}\right)^{-1} \quad \text{(A12)}$$

where η represents the ratio of the particle's energy (E) to the barrier potential energy (U).

References

1. Iijima, S. Helical microtubules of graphitic carbon. *Nature* **1991**, *354*, 56–58. [CrossRef]
2. Eatemadi, A.; Daraee, H.; Karimkhanloo, H.; Kouhi, M.; Zarghami, N.; Akbarzadeh, A.; Abasi, M.; Hanifehpour, Y.; Joo, S.W. Carbon nanotubes: Properties, synthesis, purification, and medical applications. *Nanoscale Res. Lett.* **2014**, *9*, 393. [CrossRef] [PubMed]
3. Li, J.; Zhang, Z.; Fu, J.; Liang, Z.; Ramakrishnan, K.R. Mechanical properties and structural health monitoring performance of carbon nanotube modified FRP composites: A review. *Nanotechnol. Rev.* **2021**, *10*, 1438–1468. [CrossRef]
4. Yokozeki, T.; Iwahori, Y.; Ishiwata, S. Matrix cracking behaviors in carbon fiber/epoxy laminates filled with cup-stacked carbon nanotubes (CSCNTs). *Compos. Part A Appl. Sci. Manuf.* **2007**, *38*, 917–924. [CrossRef]
5. Du, F.; Scogna, R.C.; Zhou, W.; Brand, S.; Fischer, J.E.; Winey, K.I. Nanotube Networks in Polymer Nanocomposites: Rheology and Electrical Conductivity. *Macromolecules* **2004**, *37*, 9048–9055. [CrossRef]
6. El Sawi, I.; Olivier, P.A.; Demont, P.; Bougherara, H. Processing and electrical characterization of a unidirectional CFRP composite filled with double walled carbon nanotubes. *Compos. Sci. Technol.* **2012**, *73*, 19–26. [CrossRef]
7. Xia, X.; Wang, Y.; Zhong, Z.; Weng, G.J. A frequency-dependent theory of electrical conductivity and dielectric permittivity for graphene-polymer nanocomposites. *Carbon* **2017**, *111*, 221–230. [CrossRef]
8. Mohan, L.; Kumar, P.N.; Karakkad, S.; Krishnan, S.T. Determination of electrical percolation threshold of carbon nanotube-based epoxy nanocomposites and its experimental validation. *IET Sci. Meas. Technol.* **2019**, *13*, 1299–1304. [CrossRef]
9. Guadagno, L.; Vertuccio, L.; Naddeo, C.; Raimondo, M.; Barra, G.; De Nicola, F.; Volponi, R.; Lamberti, P.; Spinelli, G.; Tucci, V. Electrical Current Map and Bulk Conductivity of Carbon Fiber-Reinforced Nanocomposites. *Polymers* **2019**, *11*, 1865. [CrossRef] [PubMed] [PubMed Central]
10. Ma, P.-C.; Siddiqui, N.A.; Marom, G.; Kim, J.-K. Dispersion and functionalization of carbon nanotubes for polymer-based nanocomposites: A review. *Compos. Part A Appl. Sci. Manuf.* **2010**, *41*, 1345–1367. [CrossRef]
11. Zhao, Q.; Zhang, K.; Zhu, S.; Xu, H.; Cao, D.; Zhao, L.; Zhang, R.; Yin, W. Review on the Electrical Resistance/Conductivity of Carbon Fiber Reinforced Polymer. *Appl. Sci.* **2019**, *9*, 2390. [CrossRef]
12. Gojny, F.H.; Wichmann, M.H.G.; Fielder, B.; Bauhofer, W.; Schulte, K. Influence of nano-modification on the mechanical and electrical properties of conventional fibre-reinforced composites. *Compos. Part A Appl. Sci. Manuf.* **2005**, *36*, 1525–1535. [CrossRef]
13. Schulz, S.C.; Schlutter, J.; Bauhofer, W. Influence of Initial High Shearing on Electrical and Rheological Properties and Formation of Percolating Agglomerates for MWCNT/Epoxy Suspensions. *Macromol. Mater. Eng.* **2010**, *295*, 613–617. [CrossRef]
14. Song, Y.S.; Youn, J.R. Influence of dispersion states of carbon nanotubes on physical properties of epoxy nanocomposites. *Carbon* **2005**, *43*, 1378–1385. [CrossRef]
15. Datta, S.; Fard, M.Y.; Chattopadhyay, A. High-Speed Surfactant-Free Fabrication of Large Carbon Nanotube Membranes for Multifunctional Composites. *J. Aerosp. Eng.* **2015**, *29*, 04015060. [CrossRef]
16. Li, Z.; Liang, Z. Optimization of Buckypaper-enhanced Multifunctional Thermoplastic Composites. *Sci. Rep.* **2017**, *7*, 42. [CrossRef]
17. Tripathi, K.; Hamza, M.H.; Chattopadhyay, A.; Henry, T.C.; Hall, A. Impact of buckypaper on the mechanical properties and failure modes of composites. In Proceedings of the American Society for Composites-Thirty-Eighth Technical Conference, Woburn, MA, USA, 18–20 September 2023.
18. Munalli, D.; Dimitrakis, G.; Chronopoulos, D.; Greedy, S.; Long, A. Electromagnetic shielding effectiveness of carbon fibre reinforced composites. *Compos. Part B Eng.* **2019**, *173*, 106906. [CrossRef]
19. Banerjee, P.; Schmidt, J.L. Electrical conductivity modeling and validation in unidirectional carbon fiber reinforced polymer composites. In Proceedings of the COMSOL Conference, Rotterdam, The Netherlands, 23–25 October 2013.
20. Wasselynck, G.; Trichet, D.; Fouladgar, J. Determination of the Electrical Conductivity Tensor of a CFRP Composite Using a 3-D Percolation Model. *IEEE Trans. Magn.* **2013**, *49*, 1825–1828. [CrossRef]
21. Wasselynck, G.; Trichet, D.; Ramdane, B.; Fouladgar, J. Microscopic and Macroscopic Electromagnetic and Thermal Modeling of Carbon Fiber Reinforced Polymer Composites. *IEEE Trans. Magn.* **2011**, *47*, 1114–1117. [CrossRef]
22. Hirano, Y.; Yamane, T.; Todoroki, A. Through-thickness electric conductivity of toughened carbon-fibre-reinforced polymer laminates with resin-rich layers. *Compos. Sci. Technol.* **2016**, *122*, 67–72. [CrossRef]
23. Zhang, R.; Bin, Y.; Chen, R.; Matsuo, M. Evaluation by tunneling effect for the temperature-dependent electric conductivity of polymer-carbon fiber composites with visco-elastic properties. *Polym. J.* **2013**, *45*, 1120–1134. [CrossRef]
24. Morales, M.A.; Henry, T.C.; Salamanca-Riba, L.G. Model of electromagnetic interference shielding effectiveness for a multifunctional composite containing carbon-fiber-reinforced polymer and copper mesh layers. *Carbon* **2023**, *212*, 118179. [CrossRef]
25. Mostafavi Yazdi, S.J.; Lisitski, A.; Pack, S.; Hiziroglu, H.R.; Baqersad, J. Analysis of Shielding Effectiveness against Electromagnetic Interference (EMI) for Metal-Coated Polymeric Materials. *Polymers* **2023**, *15*, 1911. [CrossRef] [PubMed]
26. Hu, S.; Wang, D.; Kyosev, Y.; Kremenakova, D.; Militky, J. The novel approach of EMI shielding simulation for metal coated nonwoven textiles with optimized textile module. *Polym. Test.* **2022**, *114*, 107706. [CrossRef]
27. Liu, Y.; Song, D.; Wu, C.; Leng, J. EMI shielding performance of nanocomposites with MWCNTs, nanosized Fe_3O_4 and Fe. *Compos. Part B Eng.* **2014**, *63*, 34–40. [CrossRef]

28. Karimi, P.; Ostoja-Starzewski, M.; Jasiuk, I. Experimental and computational study of shielding effectiveness of polycarbonate carbon nanocomposites. *J. Appl. Phys.* **2016**, *120*, 145103. [CrossRef]
29. Zhang, M.; Cao, M.-S. Assembly of Multi-Dimensional Microstructures of MXene towards Wearable Electromagnetic Attenuating Devices. *Crystals* **2023**, *13*, 1612. [CrossRef]
30. Holloway, C.L.; Sarto, M.S.; Johansson, M. Analyzing carbon-fiber composite materials with equivalent-Layer models. *IEEE Trans. Electromagn. Compat.* **2005**, *47*, 833–844. [CrossRef]
31. Available online: https://3937524.app.netsuite.com/core/media/media.nl?id=172313&c=3937524&h=dab869f57f81854c6a97&_xt=.pdf (accessed on 8 September 2024).
32. Available online: https://www.huntsman.com/docs/Documents/MIRALON%20Sheet%20Technical%20Datasheet.pdf (accessed on 8 September 2024).
33. Matsuo, S.; Sottos, N.R. Single carbon fiber transverse electrical resistivity measurement via the van der Pauw method. *J. Appl. Phys.* **2021**, *130*, 115105. [CrossRef]
34. Maradei, F.; Caniggia, S. Appendix A: Formulae for partial inductance calculation. In *Signal Integrity and Radiated Emission of High-Speed Digital Systems*; Francescaromana, M., Caniggia, S., Eds.; John Wiley & Sons, Ltd.: Hoboken, NJ, USA, 2008; pp. 481–486.
35. Kim, H.C.; See, S.K. Electrical properties of unidirectional carbon-epoxy composites in wide frequency band. *J. Phys. D Appl. Phys.* **1990**, *23*, 916. [CrossRef]
36. Sun, J.; Li, X.; Tian, X.; Chen, J.; Yao, X. Dynamic electrical characteristics of carbon fiber-reinforced polymer composite under low intensity lightning current impulse. *Adv. Compos. Lett.* **2020**, *29*, 2633366X20942775. [CrossRef]
37. Galehdar, A.; Rowe, W.S.T.; Ghorbani, K.; Callus, P.J.; John, S.; Wang, C.H. The Effect of Ply Orientation on the Performance of Antennas in or on Carbon Fiber Composites. *Prog. Electromagn. Res.* **2011**, *116*, 123–136. [CrossRef]
38. Rea, S.; Linton, D.; Orr, E.; McConnell, J. Electromagnetic shielding properties of carbon fibre composites in avionic systems. *Microw. Rev.* **2005**, *29*, 29–32.
39. Mikinka, E.; Siwak, M. Recent advances in electromagnetic interference shielding properties of carbon-fibre-reinforced polymer composites—A topical review. *J. Mater. Sci. Mater. Electron.* **2021**, *32*, 24585–24643. [CrossRef]
40. Guan, H.; Chung, D. Effect of the planar coil and linear arrangements of continuous carbon fiber tow on the electromagnetic interference shielding effectiveness, with comparison of carbon fibers with and without nickel coating. *Carbon* **2019**, *152*, 898–908. [CrossRef]
41. Tripathi, K.; Morales, M.; Hamza, M.H.; Henry, T.C.; Hall, A.; Chattopadhyay, A. Electrical Characterization and Electromagnetic Interference Shielding Properties of Hybrid Buckypaper Reinforced Polymer Matrix Composites. In Proceedings of the ASME 2024 Aerospace Structures, Structural Dynamics, and Materials Conference, Renton, WA, USA, 29 April–1 May 2024; ASME: New York, NY, USA, 2024; p. V001T01A006. [CrossRef]

applied sciences

MDPI

Article

Observations about Propagation, Attenuation, and Radiation of EMI from HVDC Converter Stations

John B. Schneider *,† and Robert G. Olsen *,†

School of Electrical Engineering and Computer Science, Washington State University, Pullman, WA 99164, USA
* Correspondence: john_schneider@wsu.edu (J.B.S.); bgolsen@wsu.edu (R.G.O.); Tel.: +1-509-334-6457 (J.B.S.); Tel.: +1-509-335-4950 (R.G.O.)
† These authors contributed equally to this work.

Featured Application: Given the increased use of high-voltage DC (HVDC) systems and the conversion of AC to DC using power electronics with faster switching times, the issue of electromagnetic interference (EMI) is receiving more attention. In this paper, the origin and evolution of EMI to the outside of HVDC converter stations is examined along with measures for managing this EMI.

Abstract: The electromagnetic interference (EMI) generated by switching operations within high-voltage DC (HVDC) converter stations is an issue that has been addressed by CIGRE, and methods to manage EMI should be considered in the design phase using electromagnetic modeling techniques. It is shown that the methods of the moments-based techniques that have been used in many previous studies may not be sufficient. Here, a finite-difference time-domain method is used to study the properties of single switching events using a realistic model of an HVDC converter station with a special emphasis on determining the impact of the valve hall on shielding EMI. It is shown that the valve hall does not confine high-frequency electromagnetic fields to the valve hall; rather, it delays them from exiting through bushings in the wall and spreads them out in time. Further, it is shown that incorporating electromagnetic absorbing material into the valve hall's design can significantly reduce EMI outside the converter station.

Keywords: HVDC; converter station; valve hall; EMI; RFI; FDTD; MoM

1. Introduction

Electromagnetic interference (EMI) due to high-voltage direct current (HVDC) converter stations or flexible alternating current transmission system (FACTS) installations has been a subject of interest for many years. There are several distinct sources of EMI associated with these systems, including corona-initiated noise from conductors and other hardware, spark discharges due to arcs between hardware at different space potentials, and commutation noise due to the current-switching operations of power electronic devices. Here, the emphasis will be on the EMI that is the result of switching operations in HVDC converter stations. Limits on the EMI generated by these stations have been proposed by CIGRE [1]. Further, these limits have often been adopted by companies that employ HVDC converter stations within their electrical transmission network [2]. Hence, without measures to suppress EMI, these stations may not comply with the limits required by the companies that purchase them. One possible consequence is a significant economic impact in terms of delays and contractual penalties.

Over the years, the technology used in power electronics has evolved. For example, many modern converter stations are based on modular multilevel converter technology that uses insulated gate bipolar transistors (IGBTs) to switch currents. These IGBTs have switching times in the order of 100 ns; much shorter than those of the thyristors they have replaced. Although the CIGRE EMI limits are defined up to 18 GHz, the IGBT switching speed will limit the frequencies of interest to below approximately 30 MHz. This assumption has been validated by a statement in a later CIGRE paper [3]. "Above 30 MHz it is not possible to distinguish the level generated by the installation from the background noise level".

Along with proposing EMI limits, the CIGRE Guide has described the methods to be used to measure EMI (e.g., antenna and detector types, including their peak and quasi-peak) [1]. In addition, guidance is provided for interpreting measurements, since EMI from other sources generally exists and must be separated from that generated by HVDC converter stations [1].

Generally, when HVDC converter station vendors propose equipment for these facilities, they are required to predict its EMI and will want to ensure that the EMI meets CIGRE limits at this stage. Further, since these stations are expensive and time-consuming to construct or modify, modeling EMI is required to reduce the cost and time needed to meet EMI limits.

Given this background, a considerable amount of research has been published in the area of the electromagnetic modeling of HVDC converter stations. As a necessity, these models utilize numerical modeling techniques, given the complexity of the stations. In fact, in Sec. 6.1 of the CIGRE Guide, it is noted that [1], "Describing an electromagnetic complex environment like the one of an electric substation is very complicated not only for the model itself, but mainly for the difficulty in knowing, even with rough precision, all the sources of electromagnetic field contained inside the fence circumscribing the area relevant to the substation ..."

There are three components to constructing a model. The first is to define the fundamental source(s) of the EMI. The second is to construct a model that accounts for the propagation of EMI currents throughout the HVDC converter station. This model will account for various components such as transformers, the smoothing reactor, and the filters (if any). Appropriate techniques (e.g., circuit theory, transmission line theory, and full-wave theory) are used depending on the distances involved relative to the appropriate wavelength. Finally, the resulting currents can be used for calculating the electric and magnetic fields in the vicinity of the HVDC converter station. This model should include the effect of shielding provided by the valve hall.

The specific characteristics of EMI sources within a station are not the subject of this paper because they are usually proprietary to a given manufacturer. Rather, the disposition of the electromagnetic energy generated by these sources will be carefully examined. (For a transient source voltage with no DC component, energy inserted into the non-radiating near field is reabsorbed by the source by the time the transient voltage ends. What remains is the radiated field.) One specific issue of interest is the impact of the valve hall's construction. Part of the reason for this comes from the comment in Sec. 5.2 of the CIGRE Guide [1]: "If the power electronic equipment is installed in a metallic building, or a building with metallic structures, the building will screen the radiation from the power electronic equipment by the induced compensating currents in the conducting metallic parts. The level of RFI [radio frequency interference] is reduced". In this statement, there is no discussion about the attenuation or frequency variation of the shielding, the effect of openings in the shield for connecting the AC and DC sides to the outside, or the ultimate disposition of the electromagnetic energy confined within the valve hall. Yet, these topics are critical issues in determining the measured EMI at specified distances from the installation. They are also the major issues discussed in this paper.

Previous Research

Given the purpose of this paper, the following discussion of previous research identifies the relevance (or not) of each paper in determining the impact of valve hall shielding or other aspects of reducing EMI. The earliest reported research on this topic is that of Hylten-Cavallius and Olsson in 1962 [4]. The next was the work reported by Annestrand in

1972 [5]. There, measurements were reported of EMI in the switchyard and along transmission lines. In addition, the attenuation of a valve hall was measured before equipment was installed. From [5], "The effectiveness of the valve hall screen at Celilo was verified before equipment was installed in the valve hall. A noise generator, essentially a vertical electric dipole, was erected in the valve hall and the noise level measured as a function of distance from the source. The dip in the profile at the wall screen was taken as a measure of the attenuation... In the frequency range from 0.15 to 5 MHz an attenuation of 40 to 45 dB was measured". Soon after this, Sarma and Gilsig reported their work on calculating the EMI from HVDC converter stations [6]. (Maruvada P. Sarma also publishes under the names P. Sarma Maruvada and P. S. Maruvada.) They discussed the source of EMI as the transient change in voltage across the terminals of a valve. Then, they calculated the high-frequency transient currents in the converter station using the appropriate high-frequency equivalent circuits for different elements of the network. Comments were made about the effectiveness of inductive filters (15 to 20 dB reduction). These currents were then used to calculate the EMI fields in the vicinity of the station.

In the 1980s, a large EPRI project was conducted to better understand the EMI from HVDC converter stations [7–10]. In this project, a scale model of an HVDC converter station was developed and used for measurements. Further, an EMI prediction program was developed using wideband circuit equivalents as components [8]. One relevant result was that [9] "the study showed that if the smoothing reactor is in the HVDC bus, it does attenuate the RF signal on the HVDC bus significantly". There was no attempt to model the shielding of the valve hall.

In 1989, Dallaire and Maruvada [11], rather than producing a full analysis of a converter station, applied full-wave analysis using the method of moments to evaluate the effectiveness of traditional shielding and filtering techniques. This is a useful reference for at least two reasons: (1) specific high-frequency equivalent circuits of components are identified and (2) the results could potentially be used to identify the relative effectiveness of shielding and filtering. It was shown that there are situations in which the radiated field in a particular direction may actually be enhanced because of a shielding enclosure [11]. This stands in contrast, for example, to the blanket statements in [5,12] that indicate a valve hall will provide at least 40 dB of reduced radiation. The statements in [5,12] are based on measurements and the authors do not provide a full-wave analysis of an electromagnetic environment representative of an actual valve hall. Later, Maruvada, Malewski, and Wong reported measurements of the electromagnetic environment, but no modeling [13].

In 1999, Murata et al. built a 1/400 scale model of an HVDC converter station and analyzed it using the method of moments in the frequency range 200–1000 MHz [14]. Included in the model were busses, the equivalent circuits of devices such as DC reactors, the valve hall, and other buildings. No details were provided about the meshing or about how the openings in the valve hall through which the busses passed were (if they were) modeled; however, the calculated and measured fields were reasonably close.

In 2006, Juhlin et al. provided useful information about background noise as well as the characteristics of different sources of EMI attributable to HVDC converter stations [3]. These include coronas and sparking, as well as commutation noise. Several measurement results are given for points near HVDC converter stations, but no modeling was performed.

In two papers that appeared in 2007, and another in 2009, Yu and various colleagues [15–17] described a complete circuit of a converter station including wideband equivalent circuits for components such as the smoothing reactor and parasitic elements. The entire circuit was evaluated using lumped circuit theory. Hence, the results are not likely to be valid at higher frequencies, since at 30 MHz the wavelength is 10 m and distances are no longer small compared to a wavelength. No attempt was made to model the shielding of the valve hall. Similarly, no attempt was made to model the valve hall shielding in a 2016 paper by Sun, Cui, and Du [18] or in the 2020 work of Wei, Wang, and Zou [19].

In a 2019 ABB review article, it was reported that "ABB's smart simulation models, or digital twins, reproduce the entire converter stations including valves, valve hall, converter reactors, wall bushings, converter transformers, high frequency (HF) filters and the entire wiring in the AC- and DC-yards" [20]. While details of the modeling are not given in this article, some information about their models has been published in [2,21,22]. Their simulations were performed in the CST Studio Suite using the transmission line matrix (TLM) time-domain solver with broadband circuit models for the smoothing reactors and transformers. It was noted in [2] that the highest frequency of interest was 10 MHz. However, while the comparison between the simulation and measurements reported in their Figure 17 was generally good, there are significant differences at some frequencies below 10 MHz and the measurements in the range above 10 MHz suggest that some attention being paid to modeling in this range is warranted. Further, in [20] it is noted that bushings, doors, vents, and possible deficiencies in wall construction will reduce shielding efficiency because they create openings in the Faraday cage. These issues will be addressed in this paper.

Most recently, Ning et al. developed a model for calculating the electromagnetic radiation of the converter valve based on the method of moments [23]. This model has been used to study the interaction between the converter valve and its operating system. It does not report any calculations of the EMI outside the station.

2. Materials and Methods

Simulations were performed using the method of moments (MoM) and the finite-difference time-domain (FDTD) method. The MoM simulations employed the NEC-5 software [24,25]. The FDTD code was developed in-house. The software that was used is a variant of that described in [26] and is available from GitHub [27].

3. Results

3.1. The Method of Moments: Attributes and Limitations

One approach to modeling a substation, as suggested by CIGRE, is to use a large number of dipoles [1]. However, surfaces, such as the metallic shielding surfaces that surround a valve hall, are not readily modeled merely by dipoles. Given this, several authors have modeled substations using the method of moments (MoM), which provides a method for describing the electromagnetic fields of a collection of conductors with specified sources [12,14–18,28–30]. The conductors are divided into small segments (effectively acting like dipoles) or surface patches (thus permitting the modeling of conducting enclosures such as those associated with a valve hall). The MoM is inherently a frequency-domain technique and thus the transient nature of the sources in an HVDC station must be accounted for by specifying the correct magnitude and phase of the spectral components of the excitation and transforming them to the time domain. The MoM does have several compelling attributes in terms of its modeling abilities. First, it permits the simple and accurate modeling of thin conductors such as those one would find throughout an HVDC converter station. Additionally, one can easily incorporate lumped elements at a point on a line or distribute a load over some portion of a line. However, caution must be exercised when modeling using a lumped circuit, since the modeling may require that the physical elements of the structure responsible for its lumped circuit behavior be eliminated. Finally, one need only determine the currents that flow on wires and surfaces. Once these are obtained, the fields can be obtained throughout the space.

MoM was used by Dallaire and Maruvada [11] to show that it is critically important to create a full-wave model of the enclosure with the associated conductors passing through the shielding. As mentioned, they showed that there are situations in which the radiated field in a particular direction may be *enhanced* because of the shielding enclosure [11]. As also mentioned, this contrasts, for example, with the statements in [5,12] that indicate a valve hall provides at least 40 dB of reduced radiation. Again, the statements in [5,12] are based on measurements and the authors do not provide a full-wave analysis of an electromagnetic environment representative of an actual valve hall.

Because MoM has frequently been used to model emissions from converter stations, the authors decided to explore the degree to which the most recent version of the MoM-based NEC code, i.e., NEC-5, could be expected to yield meaningful results for scenarios related to HVDC converter stations [24,25]. More specifically, a simple test was created where a dipole source was enclosed in a simplified valve hall represented by a hollow metallic cube with 10 m sides. The dipole was aligned with the axes of the cube and centered in two of the dimensions but offset from the center by 3 m in one of the dimensions. The valve hall was "perfect", in that there were no holes in its walls nor conductors passing from the inside to the outside. The magnitude of the electric field was recorded over a circle that was 200 m in radius and centered about the z axis. If the numeric code was performing perfectly, the exterior fields would be identically zero at all points throughout the space exterior to the cube. However, that was not the result that was obtained.

Since the MoM is a numerical method and the currents on the patches that represent the walls of the enclosure are discretized, the currents are always approximations and one would not expect perfect cancellation beyond the enclosure. Realistically, the best that one could hope for is that the exterior fields would be "small", but what constitutes "small", and the degree to which MoM can achieve that, appears open to question. In many applications, when comparing numerical results to either theoretical or measured results, an error of one percent is quite acceptable. Another way of thinking of one percent is as two orders of magnitude or 40 dB. But in the scenario described above, with a source in a perfect valve hall, where the exterior field should be zero, the difference between any non-zero field and the correct value is infinite on a dB scale. Alternatively, one might compare the "numerically leaked" field at the exterior to the field when no valve hall is present. This was carried out for excitation frequencies of 200 kHz, 1 MHz, 5 MHz, and 25 MHz. Table 1 shows the reduction of the measured field caused by the valve hall compared to that when the valve hall is not present (a finitely conducting ground beneath a perfect valve hall is present for all calculations with $\epsilon_r = 15$ and $\sigma = 10$ mS/m).

Table 1. The shielding effectiveness of a "perfect" valve hall as determined by NEC-5. The measurement point was 200 m from the center of a $10 \times 10 \times 10$ m cubic hall.

Frequency	Change Caused by Valve Hall
200 kHz	−41.94 dB
1 MHz	−49.12 dB
5 MHz	−58.84 dB
25 MHz	−7.96 dB

Because the hall has no apertures, the correct change is $-\infty$ on a dB scale, i.e., the lower the value the better. At 200 kHz, we note that the reduction is in the order of 40 dB, which is the value often given as the effect of the valve hall. Recall that a 40 dB reduction appeared in [5,12] and was based on measurement. However, the approximate 40 dB reduction observed at 200 kHz with this NEC-5 simulation is purely a numeric artifact. The fact that this may agree with the measurements should be considered coincidental and not indicative of this software capturing the full physics of the effect of the valve hall. In [25], the authors also report a "shielding effectiveness" at 15 MHz of −59.6 dB. That matches the value of 5 MHz in Table 1 remarkably closely. But, again, this is mere coincidence. If one continues to higher frequencies (up to 100 MHz), ref. [12] reports that the shielding effectiveness further increases to −74.54 dB, whereas the MoM results mysteriously drop to −7.96 dB at 25 MHz (the highest frequency that was considered). In contrast to the behavior at these frequencies, the results obtained at 25 MHz are unacceptably high (predicting only about −8 dB of shielding from the perfect valve hall) and apparently nearly immune to an increase in the discretization of the valve hall walls. This is a cause for concern and at present we have no suitable explanation for this behavior.

The results in Table 1 were obtained using a discretization that adheres to the recommendations in the NEC-5 user's manual [25]. Nevertheless, keeping in mind that any non-zero field represents an error and observing that the results ran into difficulties at 25 MHz, the test was run using a finer discretization. The results in Table 1 were obtained using 400 patches per face of the cube (i.e., 20 patches along a side of the cube for $20 \times 20 = 400$ patches). There was no significant difference in the results.

3.2. The Finite-Difference Time-Domain Method: Initial Insights from Simple Geometries

The finite-difference time-domain (FDTD) method [31,32] employs a fundamentally different approach. One must discretize not only the sources themselves, but also the space in the vicinity of the sources and the space that surrounds any physical features that might affect the field of interest. In addition to discretizing space (where the discretization assigns the appropriate material properties to each point in the discretization), the FDTD method also discretizes time. In this way, all the derivatives in Maxwell's equations (specifically the curl equations) are replaced with central differences. This yields a set of equations where future (unknown) fields are expressed in terms of past (known) fields. One then advances by incremental steps in time, revealing the time-domain behavior of the fields. While the MoM is a frequency-domain technique that allows one to obtain fields throughout space at a single frequency, the FDTD method is a time-domain technique that can potentially yield results over a broad spectrum with a single simulation. However, the FDTD method will only directly yield the fields within the space that has been discretized. If one is interested in fields outside of the discretized space, one must carry out a transformation of the known fields to the point of interest. The FDTD method is generally considered to be quite computationally demanding. Furthermore, special consideration is needed to handle any number of features of a particular scenario, such as a relevant physical structure that is small compared to the discretization that is used.

Several FDTD simulations were run using code developed in-house that explored various aspects of HVDC converter station modeling with an emphasis on the effect of the valve hall. The first considered was a "perfect" valve hall. As with the initial MoM testing, this hall had no openings or any conductors running from the interior to the exterior. However, instead of using a cube with 10 m sides, the dimensions were based on those of an actual converter station. The valve hall dimensions were (x) 42 m by (y) 68 m by (z) 19 m. The hall and surrounding space were discretized with a step size Δs of 0.5 m. Initial simulations, which incorporated part of the AC yard and the DC hall (reactor hall), used a computational domain of $340 \times 211 \times 107$ cells in the x, y, and z directions, respectively.

The computational domain is depicted in Figure 1, where the scale is in terms of FDTD cells (one can divide by two to convert to meters). Red lines correspond to conductors and the rectangular structure near the center of the figures corresponds to the valve hall. Conductors are modeled as a line of electric field nodes set to zero. Several researchers have reported that the "effective" radius r_0 of such a line of cells is approximately $0.2\Delta s$ [33–37], ranging from $0.135\Delta s$ [33] to $0.23\Delta s$ [35]. In [33,35–37], the researchers proposed ways to modify the FDTD update equations to model wires that had radii that were either smaller or larger than this effective radius. In [38], Taniguchi et al. reexamined the effective radius in the context of surge impedance and compared FDTD results to the theoretical work of Chen [39]. They obtained an effective radius of $r_0 = 0.199\Delta s$. This is taken to be the most accurate value of the effective radius. Applying that to the modeling performed here yields an effective conductor radius of roughly 10 cm.

The green plane on which the valve hall sits is a perfectly conducting ground plane. Figure 1a shows its projection onto the xy plane (the green ground plane has been removed from this view). The AC yard is to the left (west) of the valve hall. The three lines in the AC yard ultimately are shorted together at what would be, in practice, the location of a transformer. Although this is a crude approximation of the electromagnetic behavior of the transformer at these frequencies, it is not an unreasonable one given the lack of a validated model of such a transformer in the open literature. Nevertheless, because of the shorting of the lines at this location, the model presented here could not predict the high-frequency currents on the AC lines exiting the AC yard.

The conductors to the right (east) of the valve hall correspond to several of the lines that exist in the DC hall. The walls of the DC hall are not modeled. The 18 rectangular conductors seen within the valve hall serve as coarse approximations of the valves themselves. No attempt was made to model the detailed physical structure of the valves, but rather we modeled their overall size and shape to approximate how electromagnetic energy might propagate in their presence. A horizontal line passes through the center of each of the six sets of three "valves", moving from the AC to the DC side of the hall. Figure 1b shows a projection of the computational domain onto the xz plane while Figure 1c shows an oblique perspective projection of the computational domain.

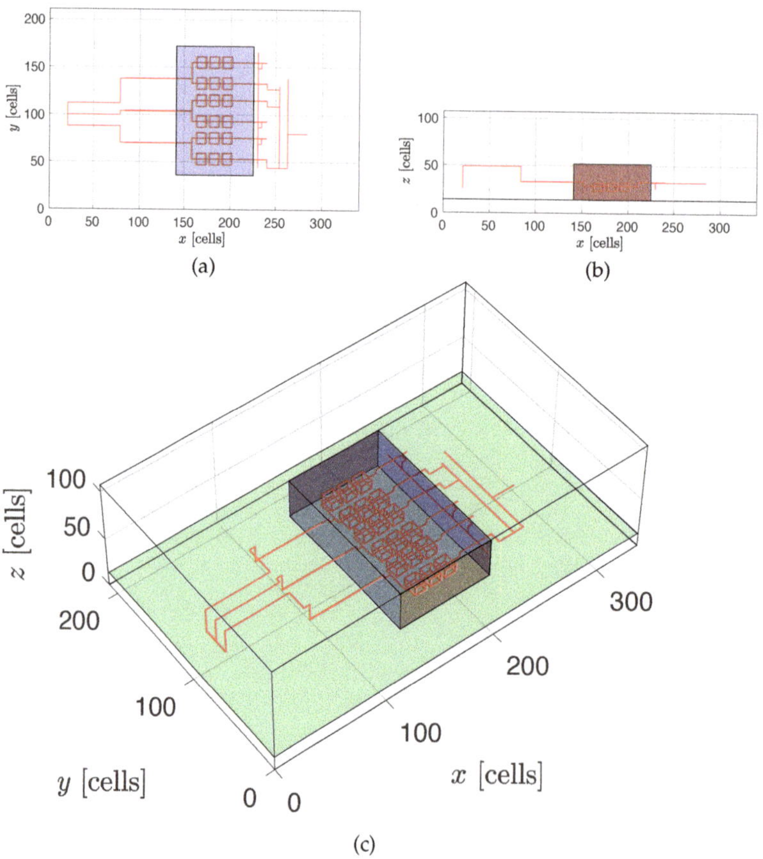

Figure 1. Computational domain showing (**a**) a projection onto the xy plane, (**b**) projection onto the xz plane, and (**c**) an oblique perspective view. The red lines represent conductors, the shaded box corresponds to the valve hall, and the green-shaded plane on which the hall sits is a perfectly conducting ground plane.

Energy is introduced into the computational domain by specifying the electric field at a particular FDTD node along one of the conductors. Specifically, the value of an E_x node along a line that transits from the AC to the DC side of the hall is specified by a source function at each time step of the simulation. This is the equivalent of specifying the voltage at that location. The source function was a unit-amplitude Ricker wavelet. (A Ricker wavelet is obtained by taking the second derivative of a Gaussian. In this way, the Ricker wavelet has somewhat similar spectral properties to a Gaussian pulse, but no DC component.) The Ricker wavelet had its peak spectral content at 10 MHz. This source can be thought of as introducing energy over a range of frequencies that can be associated with commutation events. But, the simulations are performed with a single "firing" of the Ricker pulse at a single location. For this analysis, the source location is near the right edge of the center "valve" at the northern-most (greatest y) side of the valve hall, as shown in Figure 1a.

Although the computational domain is finite, an infinite space is approximated by surrounding the computational domain with a perfectly matched convolutional layer [40]. Simulations were typically run for 65,536 time steps. Using the computational domain mentioned above ($340 \times 211 \times 107$ cells), simulations took approximately one hour and 49 min to complete on Apple Mac Studio with an M1 Ultra processor and 128 GB of memory. For the discretization that was used, where the time step Δt was 962.25 ps, 65,536 time steps correspond to a total of 63.06 μs. (The Courant number $c\Delta t/\Delta s$ was the 3D limit of $1/\sqrt{3}$.)

In the FDTD method, boundary conditions at material interfaces "take care of themselves", as one merely specifies the material properties at all points throughout the discretized space, i.e., throughout the FDTD grid, and then advances in time via the discretized version of Maxwell's equations. However, perfect electric conductors (or nearly perfect conductors), such as those that pertain to the metal walls of a valve hall, are realized not by specifying material values per se, but rather by setting the electric fields tangential to such conductors to zero and leaving them at zero. (This is similar to the way in which conductors are realized, where a line of electric field nodes is set to zero.)

Simulations were performed of a perfectly intact valve hall where the conductors (red lines) shown in Figure 1 were present but terminated by unbroken conducting walls. The FDTD model of such an intact valve hall yields results consistent with our expectations: (1) In contrast to the MoM solution, the fields exterior to the hall are identically zero. (2) Once energy is introduced into the valve hall, either in the form of fields that radiate away from the source or currents induced on the line in which the source is embedded (and the fields associated with such currents), this energy simply oscillates inside the hall until the simulation is terminated. (3) Strong resonances occur owing to the structure of the hall and its associated interior conductors. When the model consists solely of perfectly conducting wires and walls, there is no loss mechanism and the fields resonate indefinitely in this rectangular cuboid. (The simulation does allow for "loss" in the form of radiation and subsequent absorption in the perfectly matched layer that surrounds the computational domain. However, such a loss can only take place if fields escape the valve hall. And, with the hall intact, no fields escape.)

Nine holes were then introduced into the valve hall walls corresponding to the locations of each of the lines seen in Figure 1 that transitioned from the interior to the exterior of the valve hall (three on the AC side of the hall and six on the DC side). The size of the holes was not the minimum possible, but rather the minimum possible that would subsequently let us model a conductor passing through the hole. For the somewhat coarse discretization that was used ($\Delta s = 0.5$ m), the hole could be said to be approximately 1 m in diameter. Given that the peak spectrum of the pulsed excitation was at 10 MHz, i.e., a wavelength of 30 m, the holes were small compared to the wavelength of the bulk of the excitation.

With the holes thus created, multiple simulations were performed with various modifications. First, in addition to introducing the holes, a 2 m section of each conductor passing through the walls was removed which was centered about the valve hall walls, i.e., a 1 m section of each conductor was removed to either side of the wall it passed through. Second, the conductors were left intact and merely passed through the holes. For both simulations, the magnitude of the electric field at two points was recorded, one point inside the hall and one point outside. These points were selected somewhat arbitrarily. Relative to the hole that was introduced to accommodate the northern-most line in Figure 1a, which passes from the valve hall to the DC hall, the "inside point" was 4 m away in x (i.e., 4 m further into the hall relative to the wall with the hole), 3 m away from the hole in y (3 m closer to the center of the hall), and 2 m below the hole in the z direction. The outside point was 4 m away in x (i.e., 4 m away from the wall), but, as with the inside point, 3 m away in y and 2 m away (below) in z. Thus, each observation point was approximately 5.39 m away from the center of the hole. This northern-most line is also the one along which the commutation event is simulated.

Figure 2a shows the field measured (for the remaining discussion of the results obtained via the FDTD method, measured and measurement should be considered synonyms for calculated and calculation) at the inside and outside points when the 2 m segments are removed from each line. The fields are plotted on a log scale over the first 5 μs of the simulation. Note that the field inside the hall becomes non-zero prior to the field outside the hall due to its closer proximity to the source. The field inside the hall varies in the vicinity of roughly −45 db, while the field outside the hall varies in the vicinity of roughly −95 db, i.e., a reduction in the field of approximately 50 db.

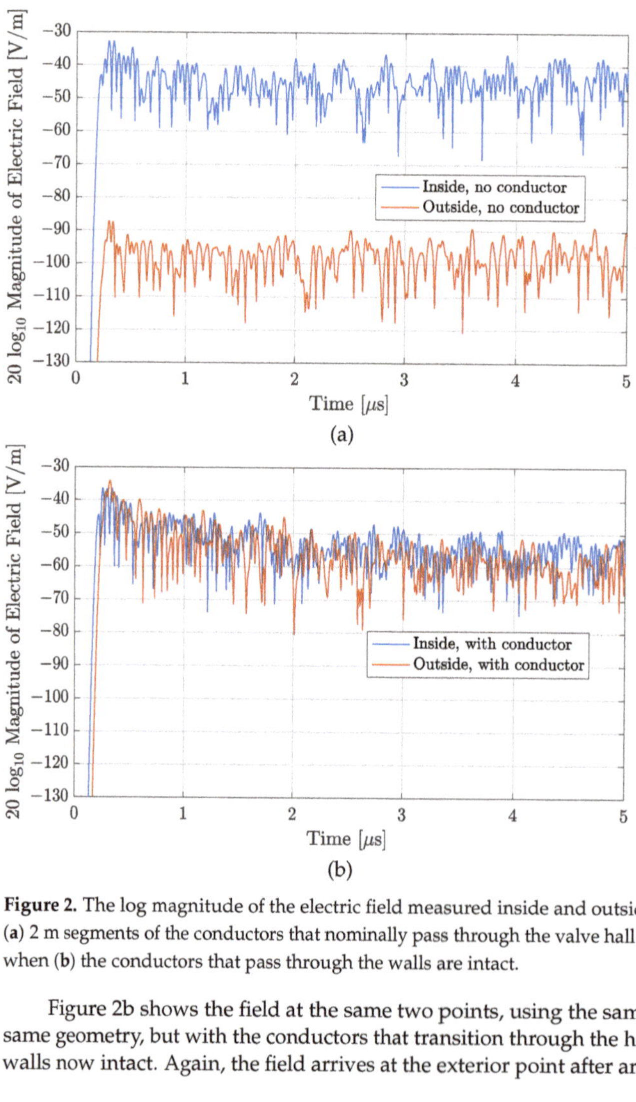

Figure 2. The log magnitude of the electric field measured inside and outside the valve hall when (**a**) 2 m segments of the conductors that nominally pass through the valve hall walls are removed and when (**b**) the conductors that pass through the walls are intact.

Figure 2b shows the field at the same two points, using the same excitation and the same geometry, but with the conductors that transition through the holes in the valve hall walls now intact. Again, the field arrives at the exterior point after arriving at the interior point. However, in this case, the fields are almost comparable in magnitude. The inside field is generally larger, as would be anticipated, but only slightly so, and there are instances where the field at the point outside the hall is greater than inside the hall. Another way to think about the contrast between Figure 2a,b is that having intact conductors raises the field exterior to the hall by approximately 50 dB.

The simulation in Figure 2b merely had the conductors passing through holes surrounded by free space. Another simulation was performed that modeled the bushings associated with each hole. These bushing were assumed to be cylindrical, extend 1 m to either side of the wall, and have a relative permittivity, ϵ_r, of 3. The radius of the bushings was such that they filled the holes through which the conductors passed. The bushings made essentially no difference to the overall observed behavior seen without the bushings. Another simulation was performed where, in addition to having the bushings present, lumped-element inductors were incorporated into the FDTD grid to represent the smoothing reactors found in the DC hall. The inductance used was 0.98 mH (in keeping with the inductance provided in [21]; its implementation followed the work described in [41]). Again, this made effectively no difference to the roughly similar field observed inside and outside the valve hall.

In Figure 2a, note that after about 1 μs, the fields decrease very little. Contrast this to the slight, but clearly evident, decrease in the field as time progresses in Figure 2b. In both simulations there are nine holes in the valve hall walls, but only in the simulation for Figure 2b do conductors pass through the holes. In the case of Figure 2a, given the relatively small size of the holes compared to the wavelength, the fields escaping the hall are primarily evanescent. These fields can couple onto the conductors that are present outside the hall, but this represents a nearly vanishingly small radiation of the fields, which explains why there is seemingly no decay (with time) of the fields in Figure 2a. (Figure 2 only displays the first 5 μs of the simulation, but this persistence of the field is evident for the entire 66.03 μs simulation.) On the other hand, when the conductors are intact, they serve to create what is effectively a transverse electromagnetic (TEM) transmission line for the fields to pass from the interior to the exterior of the valve hall. There is no cut-off frequency for this transmission line (which needs to be the case to move DC and low-frequency energy in and out of the hall).

A conductor above the ground plane serves as a TEM transmission line. The lines in the AC yard can be thought of as segments of such lines. As a line passes through the valve hall wall, one might envision this transit as a very short co-axial line, where the conductor is the inner core of the co-axial line and the wall is the outer conductor. There is an impedance mismatch going from one form of transmission line to another, but given the short extent of the intervening "co-axial" line, this mismatch does not cause significant reflections, i.e., fields directed out of the hall largely continue to exit the hall, a point which will be returned to when the currents in the conductors are considered.

The E_z component of the electric field at these observation points when the conductors are intact will now be considered. Figure 3a shows the log magnitude (which is twenty times the log base 10 of the magnitude) of the Fourier transform of the E_z component of the electric field, measured at the points inside and outside the hall over a range of frequencies from 0 to 35 MHz. For reference, the Fourier transform of the source function is also provided. One can see that peak energy occurs at 10 MHz. Note that below approximately 14 MHz, there is very little difference between the fields inside and outside the hall and, at several frequencies, the field outside the hall is greater than that inside the hall. For frequencies greater than 14 MHz, the field inside the hall is generally larger than that outside the hall. This appears to indicate that the valve hall is serving as a filter. However, such an assertion can be called into question, as discussed below. Note that the lowest-order resonant frequency for a valve hall of this size is 4.20 MHz. The strongest resonances generally occur below this frequency. Thus, these resonances must be associated with the overall structure of the station, including the conductors in the AC yard and DC hall, and not merely with the valve hall itself. As will be seen, these "low-frequency" resonances remain rather persistent regardless of what is done to the valve hall.

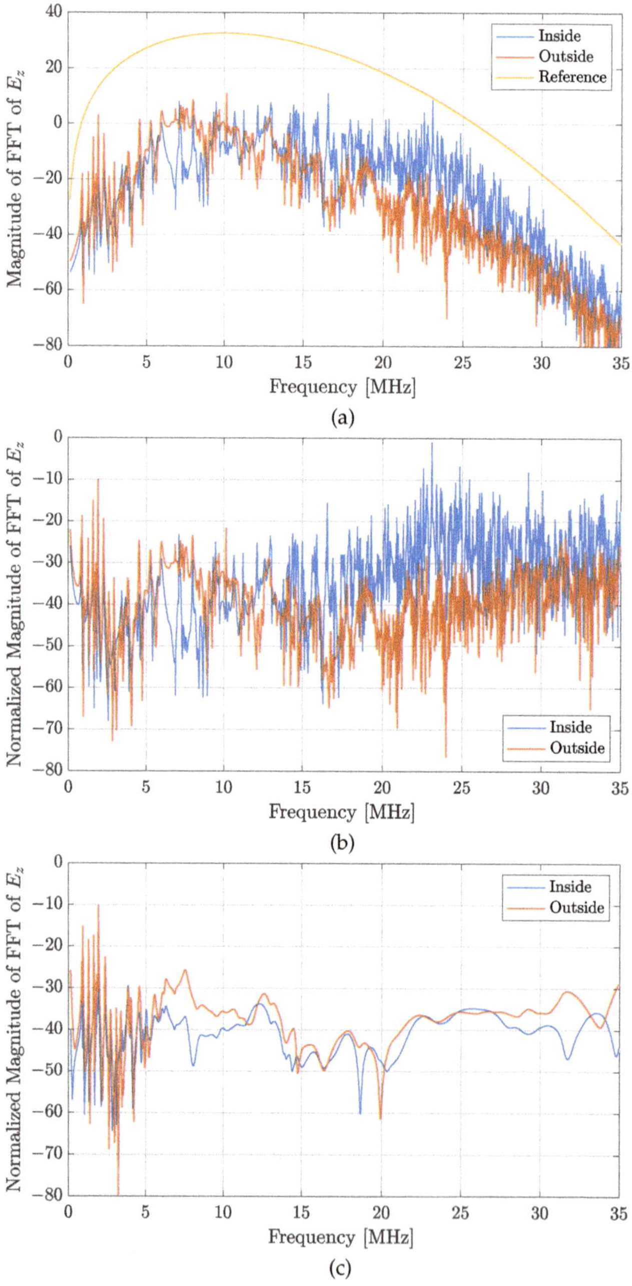

Figure 3. The log magnitude of the spectral content of the E_z component of the field, measured at two points when (**a**) the observation points are inside and outside the valve hall. The reference curve shows the Fourier transform of the source function. (**b**) Fields at the observations points normalized by the source function. (**c**) The valve hall has been removed, with no other changes to the simulation.

Consider the fields normalized by the spectral content of the source function, as shown in Figure 3b. Here, the similarity of the fields below 14 MHz and their apparent filtering above 14 MHz are somewhat easier to discern. However, also consider the results shown in Figure 3c, which are identical to those of Figure 3b, except that now the valve hall has been removed. (These curves will be labeled "Inside" and "Outside", although in Figure 3c there is no valve hall present to distinguish inside from outside while the location of the observation points have not changed.) Several of the peaks, especially those below about 7 MHz, remain nearly unchanged. The persistence of these resonances, independent of the existence of the valve hall, indicates that they are associated with the conductors throughout the model. For frequencies above this, the results in Figure 3c are much "quieter" than they are when the valve hall is present. Consider the "outside" curves in Figure 3b,c; although the curve in Figure 3b is much noisier than that of Figure 3c, the fluctuations in Figure 3b appear to be roughly about the "baseline" shown in Figure 3c. The same cannot be said of the results for the "inside" curves. It is certainly true that the results for the inside point in Figure 3c are quieter than those for Figure 3b, but one would not say that the fluctuations in Figure 3b are roughly about the curve present in Figure 3c. Instead, when the valve hall is present, rather than filtering the outside field, its presence tends to elevate the interior field.

Putting these observations together, we can say that the valve hall does not diminish the exterior field but rather enhances the interior field. Note, however, that this claim pertains to these two particular observation points, and both are in relatively close proximity to the line on which the source pulse exists. Were one to consider an exterior point located close to one of the two walls with no conductors passing through them, the field measured with the presence of the valve hall would certainly be reduced relative to having no hall present. Nevertheless, the interior field is enhanced by the presence of the valve hall (independent of where the field is observed in the interior). This enhancement may not be evident at a particular time or at a particular frequency. Rather, if one integrates the square of the field over time, the overall value will be higher when the hall is present due to the way in which it hinders the radiation of energy in all directions except via the transmission lines that pass out of the hall. Because there is no loss present within the valve hall, the only place energy can go after it is introduced into the valve hall is out along the conductors that pass through the valve hall walls, i.e., along the transmission lines out of the hall. At these frequencies, these lines serve as antennas which then radiate fields into the surrounding environment (as will be considered below).

Instead of a component of the electric field, consider the current on a conductor measured both inside and outside the valve hall. The current is obtained by taking the line integral of the magnetic field around the conductor at the desired measurement point. The current was calculated on the same line on which the source exists (the northern-most line exiting the valve hall to the right in Figure 1a). The inside and outside measurement points are each 2 m from the wall (with one being inside the valve hall and the other being in the DC hall). Temporal plots of the currents at these points reveal a difference in the arrival time of fluctuations, which is consistent with the different locations of the observation points. There are some other differences in the currents, but these are relatively minor. Rather than showing a temporal plot, Figure 4a shows the log magnitude of the normalized spectrum of the currents when the valve hall is present. Figure 4b is the same scenario as Figure 4a, except the valve hall has been removed. Although there are some slight differences in the features observed in Figures 3 and 4, overall, the results in Figure 4 serve to confirm the statement made in connection to the measurement of the electric field: the valve hall does practically nothing in terms of confining high-frequency energy to the interior of the valve hall. One could say that the hall serves to redirect energy while not eliminating it (i.e., delaying energy's eventual escape from the hall). Once a current has been established on a line within the valve hall, if that line exits the valve hall, the current found outside the hall will be nearly the same as that inside.

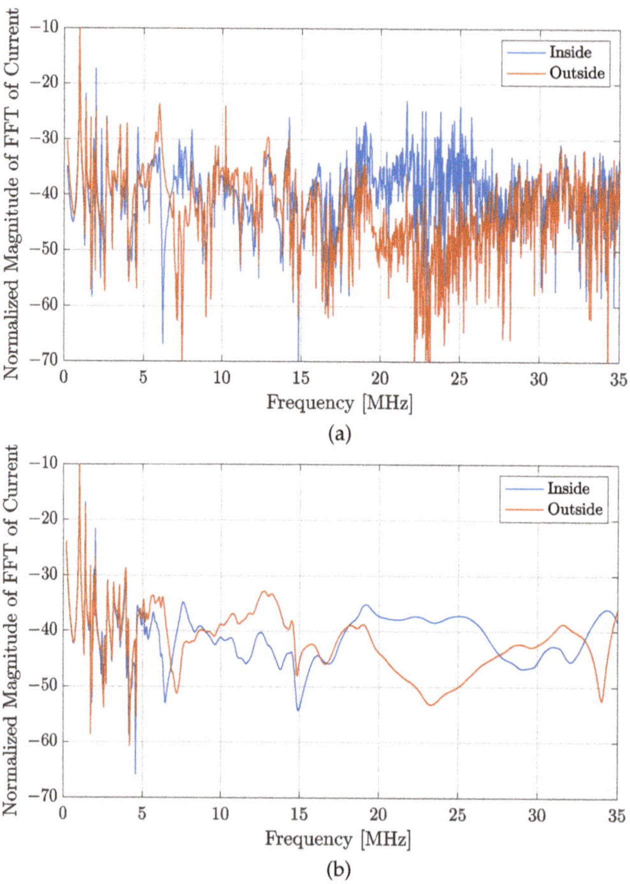

Figure 4. The log magnitude of the normalized spectrum of the current measured at two points along the top-most line from the valve hall to the DC hall. (**a**) The hall is present. (**b**) The hall has been removed.

Having established that the valve hall does little in terms of reducing the ultimate escape of high-frequency currents and fields, attention is turned to measurements more aligned with those related to compliance. CIGRE [1] is frequently concerned with the fields that exist "200 m from the closest active part of the station". The computational domain was extended 200 m (400 cells) in the y direction (simulations of 65,536 time steps now required approximately five and half hours to complete on the Apple Mac Studio mentioned before) and the field was measured at several observation points that were 200 m "south" of the southern-most valve hall wall depicted in Figure 1a (i.e., 400 cells away from the valve hall wall in the y direction). Recall that the entire computational domain sits atop a perfectly conducting ground plane. All components of the electric and magnetic fields were measured at 15 different locations with varying x and z values. The observation points were separated 15 m in x and had vertical displacements (z values) either 2, 3, or 4 m above the ground plane. In the discussion that follows, any, or all, of these points could have been selected and would lead to the same observations. For the sake of concreteness, however, an observation point was selected that is 20.5 m to the "west" of the western-most valve hall wall (i.e., the one leading to the AC yard), 200 m south of the southern-most wall, and 2 m above the ground plane.

Figure 5a shows the electric field magnitude over the initial 5 μs of two simulations, one when the valve hall is present and the other where all the conductors are unchanged but the valve hall has been removed. Unsurprisingly, when the valve hall is not present, the initial fields are stronger than when the hall is present (please refer to the interval between 1 and 2 μs). This is a consequence of the fields being free to directly radiate in all directions (other than "down" into the perfectly conducting ground) from the commutation event. However, after this initial burst, the fields without the valve hall drop below those observed when the hall is present. This is because without the hall present much of the energy is quickly lost to radiation throughout the surrounding environment. On the other hand, with the hall present (and there being no loss mechanisms contained within the hall) and the energy unable to radiate in all directions, the energy confined to the hall can only escape slowly via the conductors which, in turn, serve as antennas to radiate the field to the surrounding environment. As previously mentioned, including inductances (which model the smoothing reactors) and bushings in the simulation has little effect on the radiated field (affecting mostly its phase but not its magnitude).

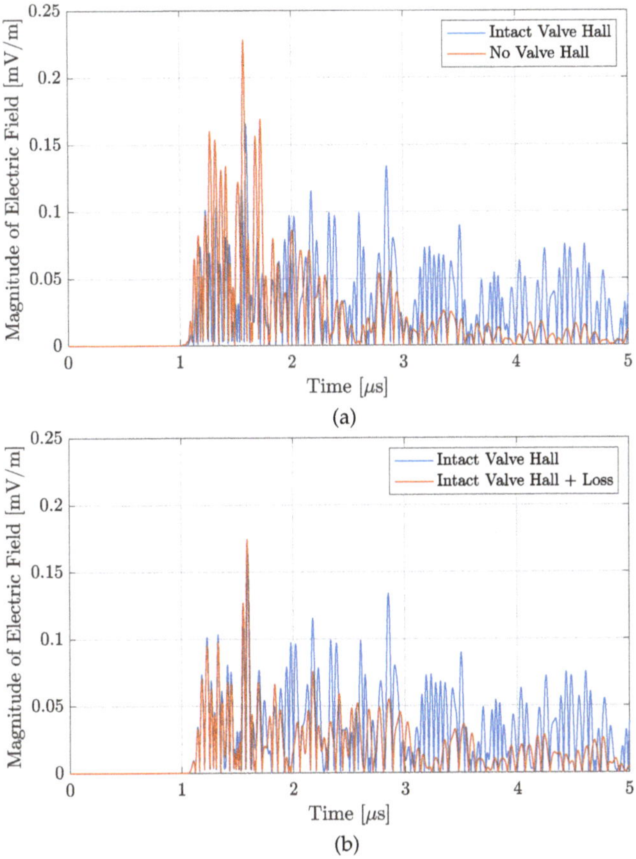

Figure 5. The electric field magnitude measured approximately 200 m from the converter station. (**a**) Field observed when the valve hall is and is not present. (**b**) Fields observed when there is and is not a lossy dielectric slab placed in the valve hall.

Two other scenarios were considered. In one, the walls of the valve hall remained in place, but the roof was removed. Potentially, this could have directed significant amounts of energy "up" so that this energy would not be present for terrestrial measurements, such as at the observation point being considered. However, at the frequencies being considered, it appears that the height of the valve hall walls was not sufficient to direct the energy up. Instead, the fields largely diffract over the walls so that the observed fields, in this case, are similar to having no valve hall present at all, and thus these results are not shown here.

The other scenario we considered introduced absorption loss into the valve hall. This took the form of a 2 m thick (in the z direction) dielectric slab adjacent to the ceiling of the intact valve hall. The slab nearly spanned the x and y extents of the hall and was displaced from the walls and roof by 0.5 m. The slab had a relative permittivity, ϵ_r, of 1.5 and a conductivity, σ, of 13.8 mS/m. Figure 5b again shows the electric field magnitude at the observation point over the initial 5 µs of two simulations. The results are again shown for an intact valve hall (the same as in Figure 5a) but also for the case in which the lossy dielectric slab is in the valve hall. In this case, the initial fields in the interval from 1 to 1.5 µs are nearly identical, but with the lossless case having slightly higher peaks. However, beyond that, the field for the lossy hall decreases significantly relative to the lossless case.

Figure 6 shows the log magnitude of the normalized spectrum of the x component of the magnetic field for the three scenarios considered in Figure 5. (If H_x is scaled by the impedance of free space, nearly identical results are obtained compared to plotting the spectrum of E_z. In practice, the horizontal component of the magnetic field is measured using a loop antenna, which is what motivated the plotting of H_x in Figure 6). Across much of the spectrum, when the lossy slab is present, the field is smaller than in the other two scenarios and, over certain ranges of frequencies, significantly so. Above approximately 14 MHz, the lossy results are below those of the intact hall by roughly 10 dB. Between 7 MHz and 14 MHz, the difference between these results varies, but the lossy hall consistently out-performs the intact hall in terms of having lower fields. Below 5 MHz, the results are all quite similar. Again, this range of frequencies is largely below the lowest-order resonance for a cuboid the size of the valve hall.

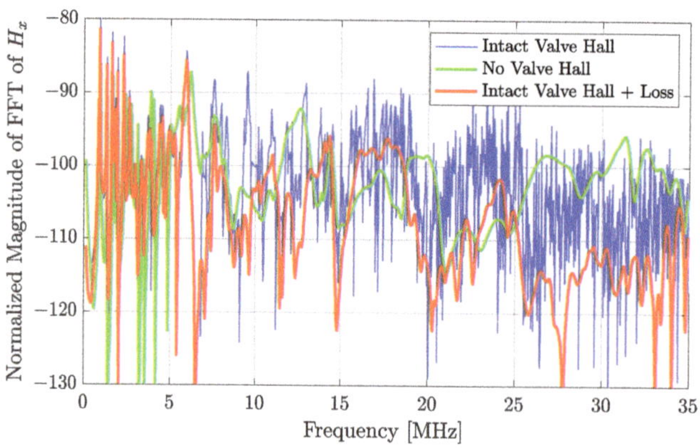

Figure 6. The normalized spectrum of the H_x field (in dB), measured approximately 200 m from the valve hall. The results are shown for an intact valve hall, when no valve hall is present, and when the valve hall is intact but contains a lossy dielectric slab.

4. Discussion

The electromagnetic interference (EMI) generated by switching operations within high-voltage DC (HVDC) converter stations is an issue that has been addressed by CIGRE, and methods to manage EMI should be considered in the design phase using electromagnetic modeling techniques. More specifically, CIGRE has proposed guidelines for the maximum values of EMI fields.

Investigations into the modeling of high-frequency (e.g., 9 kHz–30 MHz) EMI emissions from an HVDC converter station leads to the conclusion that one must use a full-wave model to be able to capture the complicated physics of the problem over the broad spectrum of excitation that exists. However, work with MoM and the FDTD method points to the fact that there are numerous ways in which such a full-wave model might lead one astray or be rather limited in the information it can provide. For example, it is shown here that MoM calculations of the small fields outside a shielded valve hall may be numerical artifacts rather than accurate results. Although there are various papers that purport to characterize the radiation from or into a shielding structure (e.g., [42,43]), to our knowledge there does not appear, in the open literature, to be a thorough analysis in which one can be confident that the effects of a valve hall are fully captured.

Here, a realistic model of an HVDC converter station has been developed and analyzed using the FDTD method. The model consists of a (shielded or unshielded) valve hall as well as the conductors within it, the AC yard, the DC hall, and the territory surrounding the station so that the EMI can be predicted at distant points.

It is often claimed that a valve hall provides perfect (or near-perfect) shielding because it is assumed to be a Faraday cage. This is shown here to be true even if there are holes (small compared to a wavelength) in the shield, despite the claims that small (compared to a wavelength) seams from valve hall construction can cause significant leakage [20]. It is also shown that the spectrum of the field inside the valve hall is greatly influenced by the cavity resonances associated with the cuboid geometry of the valve hall shield.

If, however, conductors penetrate the valve hall wall, significant leakage of the electromagnetic fields within the valve hall occurs. And, in practice, there must be conductors that pass through bushings uninterrupted, from the interior to the exterior, to carry the AC and DC currents into and out of the valve hall. Hence, the valve hall must be considered "leaky". It is shown here that the difference between the fields inside and outside the valve hall with and without conductors that make holes in the shield can be to the order of 50 dB.

To understand what happens to the high-frequency electromagnetic energy generated by commutation events in a valve hall, it must first be noted that this energy is dissipated only by its absorption into lossy materials or radiation to infinity. A shielded valve hall by itself (i.e., without absorbing material or conductors penetrating its walls) merely serves to trap radiated fields; it is shown here that they reflect around the valve hall indefinitely. If conductors inside the valve hall that penetrate its walls are introduced into the model, the interior fields can and will eventually couple onto these conductors as a means of escape. Once a current is induced on a line, the valve hall does little to filter the current passing out of the hall. The result of this process is that the original pulse of EMI energy is spread over a significant amount of time. After it exits the valve hall, it is radiated to infinity. It is this radiation that is the subject of CIGRE guidelines.

A converter station can be envisioned as a source of high-frequency energy (owing to these commutations), where this source is surrounded by a lossless shield (the valve hall) that redirects the flow of this energy. Rather than radiating directly away from the source, the valve hall walls redirect any radiated energy back into the interior until it eventually manages to escape via the conductors that pass through the walls. The energy could be said to be temporarily trapped within the hall. However, if a loss mechanism exists within the hall, it becomes possible to absorb this trapped energy before it has a chance to escape and radiate to the surrounding environment.

It is shown here that adding electromagnetic absorbing material only in the vicinity of the ceiling of the valve hall can considerably reduce the EMI energy present in the valve hall before it exits via the conductors that penetrate its walls. As a result, the radiated EMI outside the valve hall is significantly reduced. It should be noted that this calculation is simply an illustration of what can be done. No attempt has been made to optimize either the electrical properties of the absorbing material, its geometry, or its location within the valve hall. This would be a fruitful endeavor.

Note that the results presented here concerned a single "commutation event", and the focus was on the high-frequency radiation associated with that. No definitive claims were made about the precise field that might be measured for such an event. Rather, the goal was to better understand the degree to which the valve hall does or does not prevent the fields generated from such an event escaping to locations where their presence may be problematic. In practice, in a modern voltage source converter (VSC) station, thousands of commutations may be occurring in ways that have overlapping and interacting fields. The issue of multiple and interacting commutations will be the subject of future work.

Given all the complexity of an HVDC station, although full-wave modeling is necessary, it is by no means sufficient. It can provide clues and insights into the behavior of various aspects of a converter station, but it cannot be trusted, by itself, to tell the full story. The base-level modeling of sufficiently complicated structures that mimic the relevant interacting components of an HVDC converter station needs to be validated by actual measurements. Having validated the model, some level of changing the components of the models to guide design is certainly reasonable. However, it would be completely unreasonable, as appears to have been the case in some previous works, to obtain results from a model and then assume that the introduction of a valve hall that was not previously included in the model will provide a further reduction of, for instance, 40 dB in the radiated fields.

Author Contributions: Conceptualization, R.G.O.; Methodology, J.B.S.; Software, J.B.S.; Validation, J.B.S. and R.G.O.; Investigation, R.G.O.; Writing—original draft, J.B.S. and R.G.O.; Writing—review and editing, J.B.S. and R.G.O.; Project administration, R.G.O. All authors have read and agreed to the published version of the manuscript.

Funding: This research was funded by the Electric Power Research Institute, Contract 10016675.

Institutional Review Board Statement: Not applicable.

Informed Consent Statement: Not applicable.

Data Availability Statement: The data supporting the conclusions of this article will be made available by the authors on request.

Conflicts of Interest: The authors declare no conflicts of interest. The funders had no role in the design of the study; in the collection, analyses, or interpretation of data; in the writing of the manuscript; or in the decision to publish the results.

Abbreviations

The following abbreviations are used in this manuscript:

AC	Alternating current
CIGRE	Conseil International des Grands Réseaux Electriques (Council on Large Electric Systems)
DC	Direct current
EMI	Electromagnetic interference
EPRI	Electric Power Research Institute
FACTS	Flexible AC transmission systems
FDTD	Finite-difference time domain
HVDC	High-voltage direct current
IGBT	Insulated gate bipolar transistors
MoM	Method of moments
NEC	Numerical Electromagnetics Code
RFI	Radio frequency interference
TEM	Transverse electromagnetic

References

1. CIGRE. *Guide for Measurement of Radio Frequency Interference from HV and MV Substations*; CIGRE: Paris, France, 2009. ISBN 978-2-85873-078-0.
2. Cottet, D.; Eriksson, G.; Ostrogorska, M.; Skansens, J.; Wunsch, M.; Grecki, F.; Piasecki, W.; Andersson, O. Electromagnetic modeling of high voltage multi-level converter substations. In Proceedings of the Asia-Pacific international Electromagnetic Compatibility (APEMC) Symposium, Singapore, 14–17 May 2018; pp. 1001–1006.
3. Juhlin, L.-E.; Larsson, T.; Skansen, J.; Petersson, E. *Considerations Regarding RI Limits for High Voltage HVDC or FACTS Stations*; CIGRE Paper B4-207; CIGRE: Paris, France, 2006.

4. Hylten-Cavallius, N.; Olsson, E. Radio Noise from Terminal Equipment of High Voltage Direct Current Systems. *Direct Curr.* **1962**, *4*, 172–181.
5. Annestrand, S.A. Radio Interference from HVDC Converter Stations. *IEEE Trans. Power Appar. Syst.* **1972**, *PAS-91*, 874–882.
6. Sarma, M.P.; Gilsig, T. A Method of Calculating the RI from HVDC Converter Stations. *IEEE Trans. Power Appar. Syst.* **1973**, *PAS-92*, 1009–1018.
7. EPRI. *Radio Interference from HVDC Converter Stations: Modeling and Characterization*; Electric Power Research Institute Report EL 4956; EPRI: Washington, DC, USA, 1986.
8. Kasten, D.G.; Caldecott, R.; Sebo, S.A.; Liu, Y. A Computer Program for HVDC Converter Station RF Noise Calculations. *IEEE Trans. Power Deliv.* **1994**, *9*, 1129–1136. [CrossRef]
9. Kasten, D.G.; Liu, Y.; Caldecott, R.; Sebo, S.A. Radio Frequency Performance Analysis of High Voltage DC Converter Stations. In Proceedings of the IEEE Porto Power Tech Conference, Porto, Portugal, 10–13 September 2001.
10. Caldecott, R.; DeVore, R.V.; Kasten, D.G.; Sebo, S.A.; Wright, S.E. HVDC Converter Station Tests in the 0.1 to 5 MHz Frequency Range. *IEEE Trans. Power Deliv.* **1988**, *3*, 971–977. [CrossRef]
11. Dallaire, R.D.; Maruvada, P.S. Evaluation of the Effectiveness of Shielding and Filtering of HVDC Converter Stations. *IEEE Trans. Power Deliv.* **1989**, *4*, 1469–1475. [CrossRef]
12. Ding, Z.F.; Liu, L.; Liu, Y.Y.; Liu, X.F.; He, R.; Hu, Y.F.; Jiao, C.Q. Measurement and Analysis on Electromagnetic Shielding Effectiveness of a 500kV UPFC Valve Hall. In Proceedings of the IEEE 5th International Symposium on Electromagnetic Compatibility (EMC-Beijing), Beijing, China, 28–31 October 2017; pp. 1–4.
13. Sarma Maruvada, P.; Malewski, R.; Wong, P.S. Measurement of the Electromagnetic Environment of HVDC Converter Stations. *IEEE Trans. Power Deliv.* **1989**, *4*, 1129–1136. [CrossRef]
14. Murata, Y.; Tanabe, S.; Tadokoro, M.; Ukawa, Y.; Fukunaga, E.; Nakao, H. 3D-MoM Analysis of Radio Frequency Noise Radiation from HVDC Converter Station. In Proceedings of the IEEE International Symposium on Electromagnetic Compatibility, Symposium Record (Cat. No. 99CH36261), Tokyo, Japan, 17–21 May 1999.
15. Yu, Z.; He, J.L.; Zeng, R.; Zhang, B.; Chen, S.M.; Rao, H.; Zhao, J.; Li, X.L.; Wang, Q. Simulation Analysis on EMI of +-800-kV UHVDC Converter Station. In Proceedings of the IEEE International Symposium on Electromagnetic Compatibility, Honolulu, HI, USA, 8–13 July 2007; pp. 1–5.
16. Yu, Z.; He, J.; Zhang, B.; Zeng, R.; Chen, S.; Zhao, J.; Wang, Q. Simulation Analysis on Radiated EMI in ±500-kV HVDC Converter Stations. In Proceedings of the 18th International Zurich Symposium on EMC, Munich, Germany, 24–28 September 2007.
17. Yu, Z.; He, J.; Zeng, R.; Rao, H.; Li, X.; Wang, Q.; Zhang, B.; Chen, S. Simulation Analysis on Conducted EMD Caused by Valves in ±800-kV UHVDC Converter Station. *IEEE Trans. Electromagn. Compat.* **2009**, *51*, 236–244. [CrossRef]
18. Sun, H.; Cui, X.; Du, L. Electromagnetic Interference Prediction of ±800 kV UHVDC Converter Station. *IEEE Trans. Magn.* **2016**, *52*, 9400404. [CrossRef]
19. Wei, L.; Wang, X.; Zou, J. Analysis of Electromagnetic Characteristics of Controllable Devices in HVDC Converter Station. In Proceedings of the IEEE Int. Conference on High Voltage Engineering and Application (ICHVE), Beijing, China, 6–10 September 2020; pp. 1–4.
20. ABB. *HVDC Light® Digital Twin Facilitates EMC Design*; ABB Review: Vasteras, Sweden, February 2019.
21. Cottet, D.; Wunsch, B.; Grecki, F.; Piasecki, W.; Ostrogorska, M. Hybrid Model for Air Core Reactors in EMC Simulations of High Voltage Converter Stations. In Proceedings of the Joint International Symposium on Electromagnetic Compatibility, Sapporo and Asia-Pacific International Symposium on Electromagnetic Compatibility (EMC Sapporo/APEMC), Sapporo, Japan, 3–7 June 2019; pp. 429–432.
22. Wunsch, B.; Cottet, D.; Eriksson, G. Broadband Models of High Voltage Power Transformers and Their Use in EMC System Simulations of High Voltage Substations. In Proceedings of the IEEE International Symposium on Electromagnetic Compatibility and IEEE Asia-Pacific Symposium on Electromagnetic Compatibility (EMC/APEMC), Singapore, 14–17 May 2018.
23. Ning, J.; Li, J.; Hao, B.; Zhao, Z. Calculation and Analysis of Electromagnetic Radiation in UHV High-capacity Flexible DC Converter Valve. In Proceedings of the 3rd International Conference on Electrical Engineering and Control Science (IC2ECS), Hangzhou, China, 29–31 December 2023.
24. NEC 5. Available online: https://softwarelicensing.llnl.gov/product/nec-v50 (accessed on 31 July 2024).
25. Burke, G.J.; Poggio, A.J. *Numerical Electromagnetics Code—NEC-5 Method of Moments, User's Manual*; Lawrence Livermore National Laboratory, LLNL-SM-742937; Lawrence Livermore National Laboratory: Livermore, CA, USA, 2017.
26. Schneider, J.B. Understanding the Finite-Difference Time-Domain Method. 2010. Available online: https://eecs.wsu.edu/~schneidj/ufdtd (accessed on 31 July 2024).
27. uFDTD. Available online: https://github.com/john-b-schneider/uFDTD (accessed on 31 July 2024).
28. Sun, H.F.; Du, L.S.; Liang, G.S. Calculation of Electromagnetic Radiation of VSC-HVdc Converter System. *IEEE Trans. Magn.* **2016**, *52*, 9400504. [CrossRef]
29. Zhang, W.D.; Wan, Q.; Qi, L.; Zhao, D.L.; Zhao, G.L.; Cai, L.H. The Prediction of Radio Frequency Interference from HVDC-flexible converter valve. In Proceedings of the International Symposium on Electromagnetic Compatibility—EMC EUROPE, Wroclaw, Poland, 5–9 September 2016; pp. 790–793.

30. Zhang, J.; Lu, T.; Zhang, W.; Zhang, X.; Xu, B. Simulation of electromagnetic environment on ±320kV VSC-HVDC converter valve. In Proceedings of the IEEE International Symposium on Electromagnetic Compatibility & Signal/Power Integrity (EMCSI), Washington, DC, USA, 7–11 August 2017; pp. 103–107.
31. Yee, K.S. Numerical Solution of Initial Boundary Value Problems Involving Maxwell's Equations in Isotropic Media. *IEEE Trans. Antennas Propag.* **1966**, *AP-14*, 302–307.
32. Taflove, A.; Hagness, S. *Computational Electrodynamics: The Finite-Difference Time-Domain Method*, 3rd ed.; Artech House: London, UK, 2005.
33. Umashankar, K.; Taflove, A.; Beker, B. Calculation and Experimental Validation of Induced Currents on Coupled Wires in an Arbitrary Shaped Cavity. *IEEE Trans. Antennas Propag.* **1987**, *35*, 1248–1257. [CrossRef]
34. Hockanson, D.M.; Drewniak, J.L.; Hubing, T.H.; Van Doren, T.P. FDTD Modeling of Thin Wires for Simulating Common-Mode Radiation from Structures with Attached Cables. In Proceedings of the International Symposium on Electromagnetic Compatibility, Atlanta, GA, USA, 14–18 August 1995; pp. 168–173.
35. Noda, T.; Yokoyama, S. Thin Wire Representation in Finite Difference Time Domain Surge Simulation. *IEEE Trans. Power Deliv.* **2002**, *17*, 840–847. [CrossRef]
36. Railton, C.J.; Paul, D.L.; Dumanli, S. The Treatment of Thin Wire and Coaxial Structures in Lossless and Lossy Media in FDTD by the Modification of Assigned Material Parameters. *IEEE Trans. Electromagn. Compat.* **2006**, *48*, 654–660. [CrossRef]
37. Asada, T.; Baba, Y.; Nagaoka, N.; Ametani, A. An Improved Thin Wire Representation for FDTD Transient Simulations. *IEEE Trans. Electromagn. Compat.* **2015**, *57*, 484–487. [CrossRef]
38. Taniguchi, Y.; Baba, Y.; Nagaoka, N.; Ametani, A. An Improved Thin Wire Representation for FDTD Computations. *IEEE Trans. Antennas Propag.* **2008**, *56*, 3248–3252. [CrossRef]
39. Chen, K. Transient Response of an Infinite Cylindrical Antenna. *IEEE Trans. Antennas Propag.* **1983**, *31*, 170–172. [CrossRef]
40. Roden, J.A.; Gedney, S.D. Convolutional PML (CPML): An Efficient FDTD Implementation of the CFS-PML for Arbitrary Media. *Microw. Opt. Technol. Lett.* **2000**, *27*, 334–339. [CrossRef]
41. Piket-May, M.; Taflove, A.; Baron, J. FD-TD Modeling of Digital Signal Propagation in 3-D Circuits with Passive and Active Loads. *IEEE Trans. Microw. Theory Tech.* **1994**, *42*, 1514–1523. [CrossRef]
42. Zhao, Z.B.; Cui, X.; Li, L.; Zhang, B. Analysis of the Shielding Effectiveness of Rectangular Enclosure of Metal Structures with Apertures above Ground Plane. *IEEE Trans. Magn.* **2005**, *41*, 1892–1895. [CrossRef]
43. Nie, X.C.; Yuan, N.; Li, L.W.; Gan, Y.B. Accurate Modeling of Monopole Antennas in Shielded Enclosures with Apertures. *Prog. Electromagn. Res.* **2008**, *PIER 79*, 251–262. [CrossRef]

applied sciences

MDPI

Article

Modeling Absorbed Energy in Microwave Range for Nanocomposite Hot Melts Containing Metallic Additives

Radu F. Damian and Romeo Cristian Ciobanu *

Department of Electrical Measurements and Materials, Gheorghe Asachi Technical University, 700050 Iasi, Romania; rdamian@etti.tuiasi.ro
* Correspondence: r.c.ciobanu@tuiasi.ro

Abstract: This paper describes the simulation of the S (scattering)-parameters and absorbed energy for polymeric nanocomposites with metallic insertions (iron and aluminum, with two particle dimensions). The considered frequency domain, 0.1–3 GHz, is specific for a wide range of applications in microwave technologies. The actual limitations of electromagnetic simulations are described, along with an application related to the Nicolson–Ross–Weir procedure in infinite media, which opens new perspectives in nanocomposite electromagnetic modeling. It was demonstrated that, due to a higher conductivity, nanocomposites with Al particles can absorb a larger amount of energy compared to Fe particles within composite materials, at both insertion dimensions. At higher frequencies, the power loss density increases for both metals. The significant reduction in transmission, linked to a minimal reflection, verifies that for composites with added metallic powder, microwave energy is extensively absorbed by the materials, particularly at frequencies above 1.5 GHz, confirming their potential functionality as hot melts for advanced reversible bonding technologies.

Keywords: microwave domain; Nicolson–Ross–Weir procedure in infinite media; S-parameters; absorbed energy; polymeric nanocomposites; metallic insertions

1. Introduction

Currently, hot melt bonding methods are significantly influential in the market, with the hot melt adhesives industry expected to increase from USD 10.07 billion in 2024 to USD 13.42 billion by 2032, reflecting a 4.33% annual growth rate during the forecast period [1]. Hot melt adhesives, glues made of polymers that are melted before application, transition from liquid to solid over a narrow temperature range. After cooling, the hot melt solidifies and forms a sticky bond between two chosen surfaces [2]. Hot melt adhesives are advantageous for bulk production at a low cost because of their easy application, fast bonding, and the ability to bond with just one component, eliminating the need for extra materials or catalysts. Hot melts can adhere to various materials including plastics, paper, wood, metal, and fabrics, leading to a variety of application possibilities. Conventional bonding techniques with hot melts may include thermal, electrical, or laser methods, either individually or in combination [3]. The electromagnetic field-activated adhesive bonding technology is a unique and modern alternative to the current thermal bonding methods utilized in various manufacturing industries. The primary benefit of this is the pre-heating of the hot melt prior to heating the surfaces being bonded, ensuring the thorough heating of the hot melt. A more recent technique involves utilizing microwave radiation at frequencies over 1 GHz [4–6], but this addresses mainly fluid hot melts, with limited applicability. The difficulty with microwave technologies lies in creating new solid hot melt materials and related bonding processes that are multifunctional, highly reliable, and can create a strong, energy-efficient joint. On the other hand, the joints should be able to be disassembled by a specific electromagnetic field trigger too, allowing for a reversible bonding process in line with the circular economy concept [7,8]. Developing new hot melt adhesives with specialized micro-/nanosized fillers can enhance the mechanical strength of adhesives, improve joint performance, increase adhesion to various substrates, and accelerate heating through electromagnetic fields. Different materials have been extensively researched as inserts in dielectrics due to their current application in microwave absorber technology [9–12], or for their impact on electromagnetic energy [13–15], particularly by enhancing absorption in the microwave frequency range. But there is no current research on using MW-activatable adhesives with metallic functional particles, and the present study demonstrates, by modeling with specialized software, that adding metal particles to thermo-plastic formulations may create MW-activatable adhesives with lower energy usage. The main focus of this paper is on developing the concept of polyethylene-based hot melt adhesives with metallic inserts of Al and Fe, which may offer advantages like long-term stability, improved heat resistance, a broader range of bonding capabilities, and no potential for charring or producing unpleasant odors.

Analytical methods were initially used to determine the parameters of a material with inserts from another material, Lord Rayleigh and Lewin, cited in [16,17], being among the first to investigate these issues. The interest in theoretical approaches has remained constant as new dielectric/insert combinations have been found to obtain adequate behavior at microwave frequencies [18–21]. However, analytical approaches have only been successful for simple structures and are generally ineffective in the microwave frequency range where the wavelength starts to become comparable to the interparticle distances and/or particle size.

To estimate the electrical parameters, some methods for estimating the relative permittivity of a composite material, e.g., the Kraszewski, Landau, Lifshitz, Looyenga, and Lichtenecker equations are available [22]. A quick method of determination, especially valid from an energetic point of view, starting from the observation that the energy is dissipated in the dielectric with inserts which affects the material density, is the weighted method (1), where ε—equivalent dielectric permittivity of a composite material, ε_i—dielectric permittivity of each component, and v_i—mass ratio of the respective component.

$$\varepsilon = \sum_i v_i \cdot \varepsilon_i \tag{1}$$

Figure 1 shows that in the case of composite materials with a polymer matrix (here, polystyrene), the weighted relation (1) can be used with minimal error because, typically, such materials have a low content of functional ingredients. However, if nanocomposites contain materials that interact with electromagnetic fields (mainly metal-based insertions) all results, shown in Figure 1, with their different macroscopic methods of estimating losses in materials are found to be inaccurate.

The electromagnetic simulation is based on the working procedure originally proposed by Nicolson, Ross, and Weir (NRW procedure), improved, as presented, e.g., in [23–26], in order to eliminate the instability of the NRW procedure in the case of low-loss materials, in which situation measurement errors decisively affect the accuracy of the results. The NRW method provides an accurate direct calculation of both the permittivity and S-parameters (mainly reflection and transmission parameters, S_{11} and S_{21}, respectively). This procedure aims to replace a dielectric material with inserts with an equivalent material in terms of behavior in the electromagnetic field, in both cases, the excitation being in the form of a plane wave. Volumetric energy losses (W/m^3) can be determined directly in the case of the analysis of polymeric materials with metal inserts, as the only source of heat is the dissipation of electromagnetic energy inside the structure.

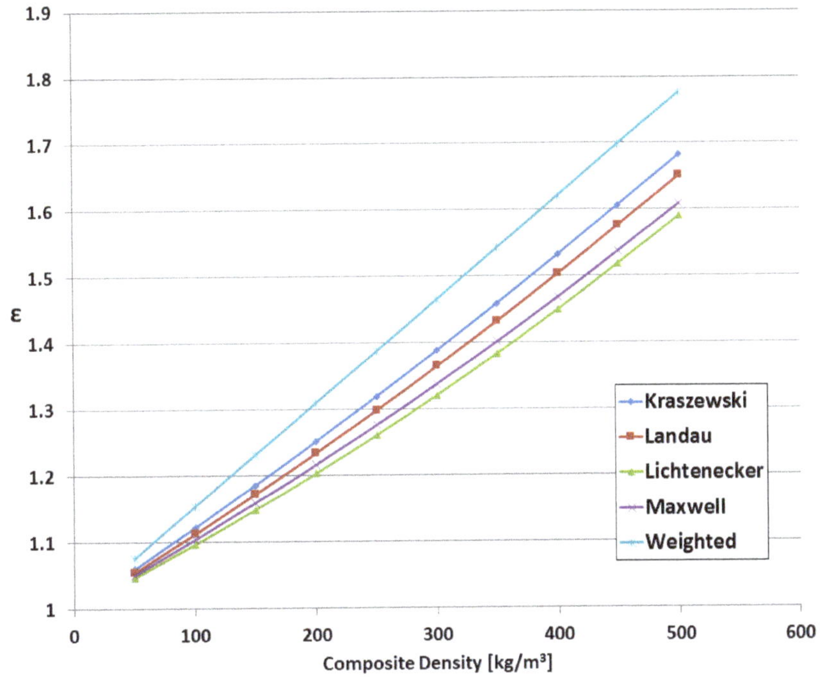

Figure 1. Equivalent permittivity of a composite material (with polystyrene matrix).

The software for electromagnetic simulation was chosen as CST Studio Suite [27]. The primary electromagnetic analysis is based on the calculated electromagnetic field values within the composite structure, as described, e.g., in [28–30]. The simulation illustrated below for metallic additive-containing nanocomposite hot melts is innovative and aims to determine the optimal selection regarding particle size and power loss across various frequencies, as a prerequisite for developing bonding technology in the microwave spectrum.

2. Electromagnetic Simulation

The boundary conditions (Figure 2) aimed to investigate the interaction of a plane wave with the considered sample.

The generation of plane wave characteristics of the electromagnetic field was determined by choosing the conditions of the electric wall (x_{min} and x_{max}) and magnetic wall (y_{min} and y_{max}). The conditions for the input and output ports are labeled "open". The generation of the interaction of the analyzed structure with a plane type wave is presented in Figure 3. Details about the procedure and related conditions can be found in [29].

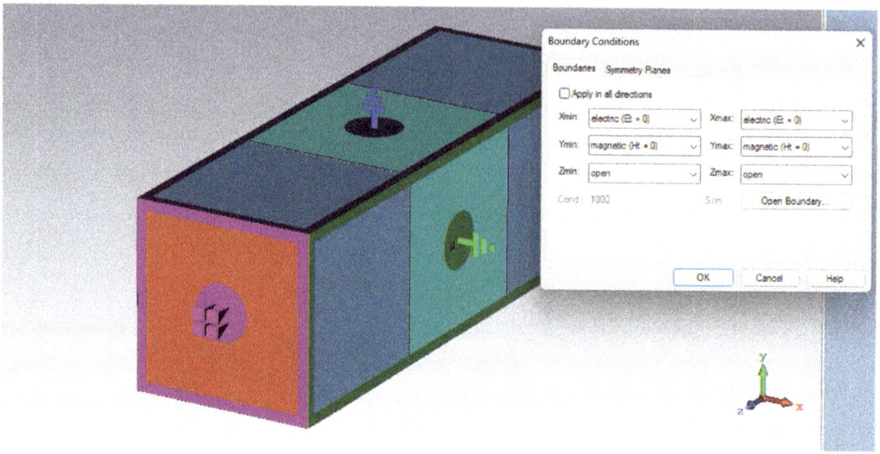

Figure 2. Boundary conditions.

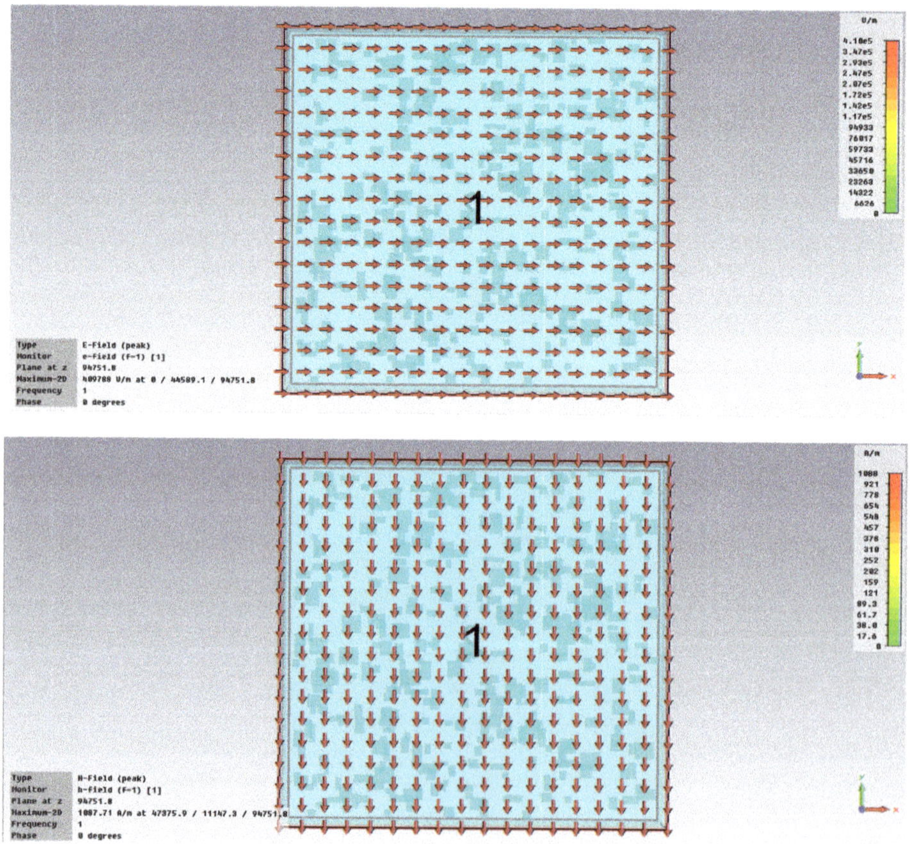

Figure 3. Plane wave configuration: Electric and magnetic field lines at the entrance port.

The polymer matrix that will host the inserts was low-density polyethylene (LDPE), for which CST Studio Suite software [27] does not have models in its own library. As a result, a new model was made based on 1st-order Debye Equation (2) (where ω represents the angular velocity), and on the physical characteristics of LDPE, listed in Table 1.

$$\varepsilon^*(\omega) = \varepsilon_\infty + \frac{\varepsilon_S - \varepsilon_\infty}{1 + j \cdot \omega \cdot \tau} \tag{2}$$

Table 1. LDPE matrix properties.

Matrix	LDPE
Model type	Normal
Thermal conductivity	0.48 [W/K/m]
Density	945 kg/m^3
Electrical permittivity ε_∞	2.2
Electrical permittivity ε_S	4
Relaxation time	0.2 s
Magnetic permeability	1
Specific heat	1.9 kJ/K/kg

As regards the metal inserts (iron and aluminum) the software has predefined models in libraries, as presented in Table 2.

Table 2. Metal insert properties.

Material	Fe	Al
Model	Lossy metal	Lossy metal
Electrical conductivity	1.04×10^7 [S/m]	3.56×10^7 [S/m]
Thermal conductivity	79.5 [W/K/m]	237.0 [W/K/m]
Density	7870.0 kg/m^3	2700.0 kg/m^3
Specific heat	0.45 kJ/K/kg	0.9 kJ/K/kg

This study analyzed the electromagnetic energy loss enhanced by the presence of inserts inside the polymer matrix. For this, the SAR (specific absorption rate) was calculated by integrating the entire volume of the considered domain (matrix and particles), at various frequencies in the microwave domain. The modeling is more accurate if considering quasi-cubic/parallelepipedal particles, equating the volume of particles of different shapes, because, in the case of metallic materials, we expect the electromagnetic energy losses to be superficial rather than volumetric. As a result, in all the following stages of the simulation, parallelepipedal particles equivalent to spheres with radii of 50 nm and 800 nm, respectively, were taken into account, the mathematical equating methodology being largely described in [29,30]. The edge of the reference cubic particle with the same volume as the ideal spherical particle will be $R_{ref} = R \cdot (4\pi/3)^{1/3}$. To account for detected irregularities in the size of the nanoparticles, we allow every side of every cube to variate between half and twice the reference cubical edge, resulting in a randomly shaped parallelepipedal particle, also randomly placed.

By decreasing the dimensions of the inserts, a limitation of the capacity of the computing systems may be reached relatively quickly if working with larger simulated volumes. A reasonable solution adopted was to reduce the total volume of the structure as the particle size decreases, so that the total number of particles is approximately constant for both particle dimensions, at around 500 (variations appear due to their random distribution).

For the random generation (spatial/size) of the particles, a VB macro was implemented, for which CST has an interpreter, counting the total generated volume and the number of particles. The generation of particles continues until the volume reaches the predefined limits. As a result, the typical simulated structures are similar to those shown, e.g., in Figures 4 and 5 (MR—mass ratio; VR—volume ratio).

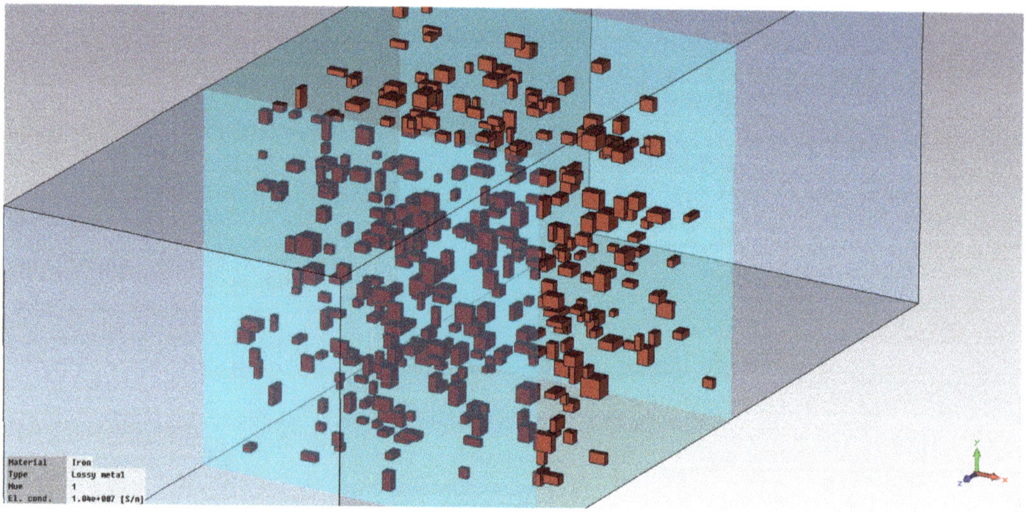

Figure 4. Example of simulated structure, 405 parallelepipeds, Fe, L = 800 nm, MR = 8%, VR = 1.00024%.

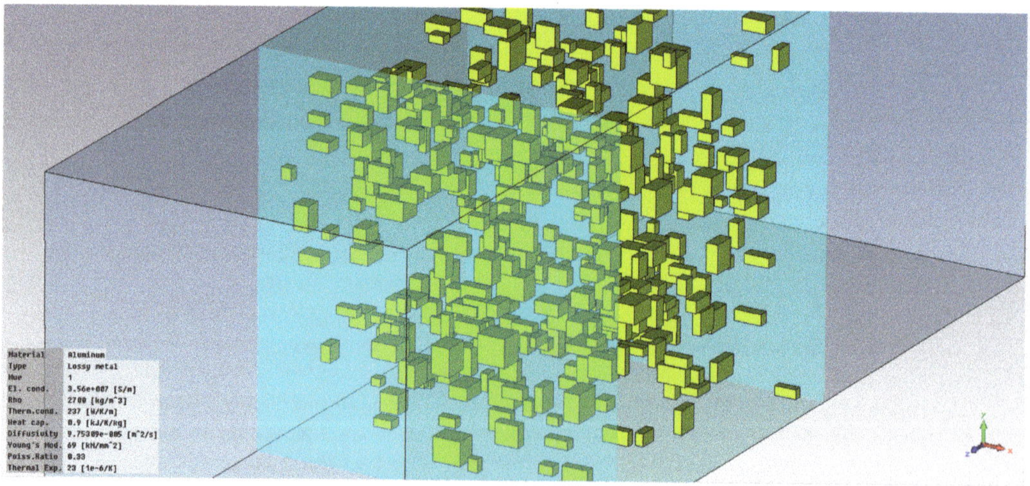

Figure 5. Example of simulated structure, 394 parallelepipeds, Al, L = 50 nm, MR = 8%, VR = 2.8048%.

The particle sizes in the nanometer area involving simulations in smaller cubic structures, associated with the frequency domain used (0.1–3 GHz), were determined to display difficult convergence. The frequency domain solver (FEM) was employed despite the limitation of being unable to create cells within the lossy metals. However, this drawback is not significant because previous research [30] has found that the losses in the metals are minimal, with the inserts mainly affecting the losses by altering the surrounding fields.

3. Results and Discussion

3.1. Reference Structure

A reference analysis was carried out upon a cubic layer of LDPE without inserts, with the size of the composite similar to that to be further used for LDPE with Fe and Al inserts. The results are presented in Figure 6 and show, as expected, a uniform power loss density across the layer. The parameters S_{11} (reflection coefficient) and S_{21} (direct transmission coefficient and power loss density) were simulated and analyzed.

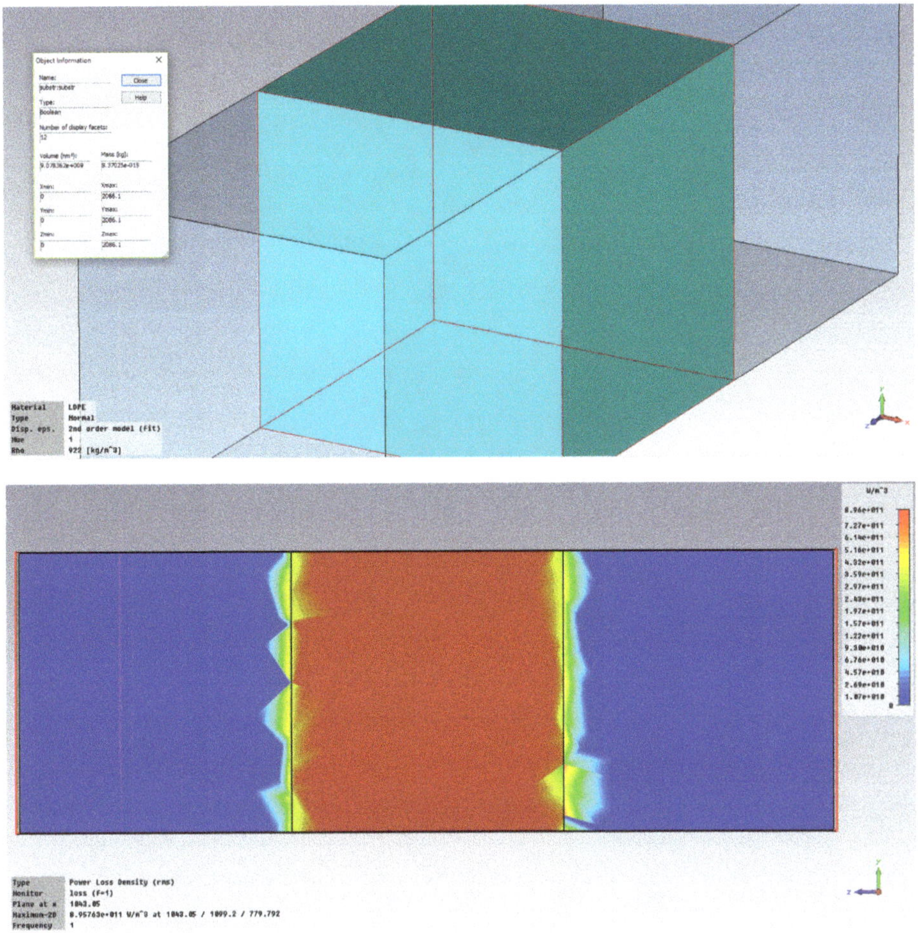

Figure 6. Reference structure and related volumetric distribution of power loss density (uniform).

3.2. Analysis of Composites with 50 nm Al Inserts

For the composite with 50 nm aluminum (equivalent) inserts, the edge of the cubic layer of the LDPE matrix was 2086.1 nm, as shown in Figure 7. The simulations of the volumetric distribution of the power loss density, as the absorption intensity inside the material, are plotted in Figures 8 and 9 at 0.1 GHz and 3 GHz, respectively, showing the effect of the randomly placed metallic inserts as a nonuniform power loss distribution inside the material. Significantly higher values for the exposure at 3 GHz were emphasized, with an average of 4.145×10^{13} W/m^3, compared to 2.204×10^{11} W/m^3 for 0.1 GHz.

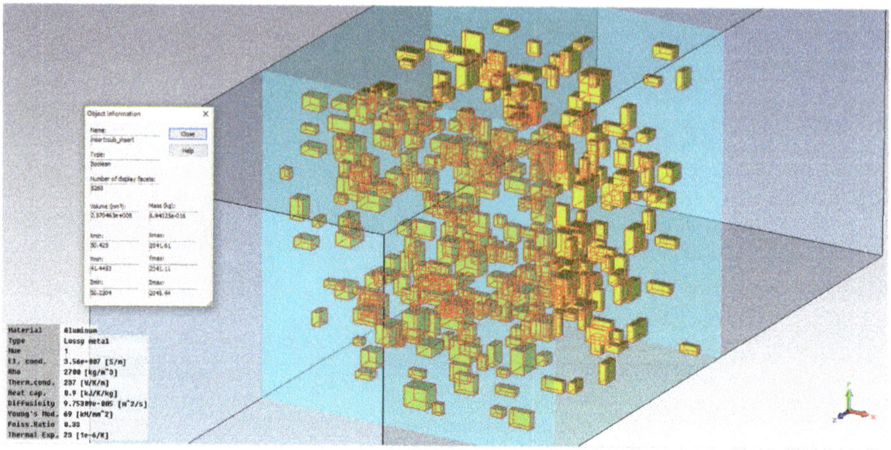

Figure 7. Analyzed structure: LDPE composite with 50 nm Al inserts.

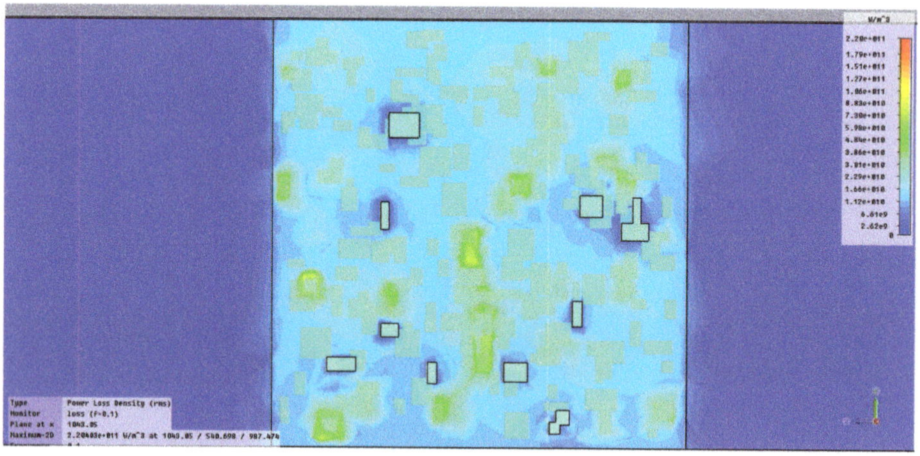

Figure 8. Volumetric distribution of power loss density at 0.1 GHz: LDPE composite with 50 nm Al inserts.

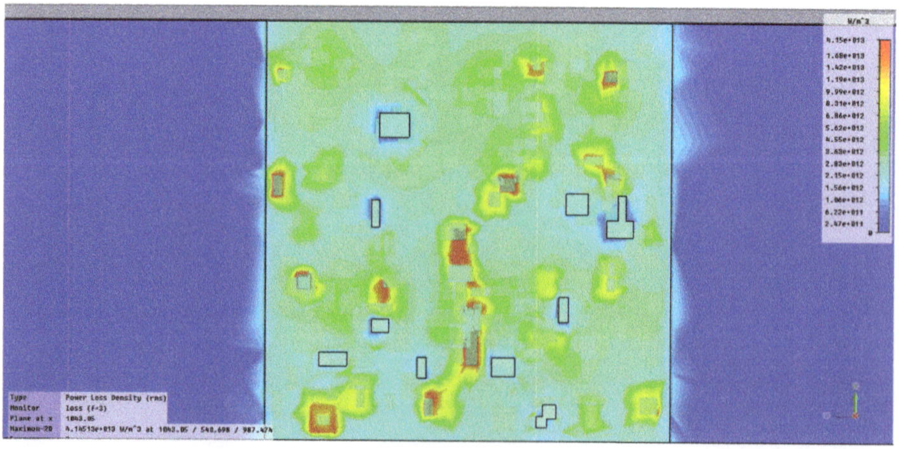

Figure 9. Volumetric distribution of power loss density at 3 GHz: LDPE composite with 50 nm Al inserts.

3.3. Analysis of Composites with 800 nm Al Inserts

For the composites with 800 nm aluminum (equivalent) inserts, the edge of the cubic layer of the LDPE matrix was 33,377.7 nm. The structure is plotted in Figure 10, and the volumetric distribution of the power loss density is displayed in Figures 11 and 12, showing a similar nonuniform absorption inside the material. Here, we reconfirmed the significantly higher values for the exposure at 3 GHz, with an average of 4.717×10^{11} W/m^3, compared to 2.507×10^9 W/m^3 for 0.1 GHz, but the values are about 100 times lower compared to the composite with aluminum 50 nm (equivalent) inserts.

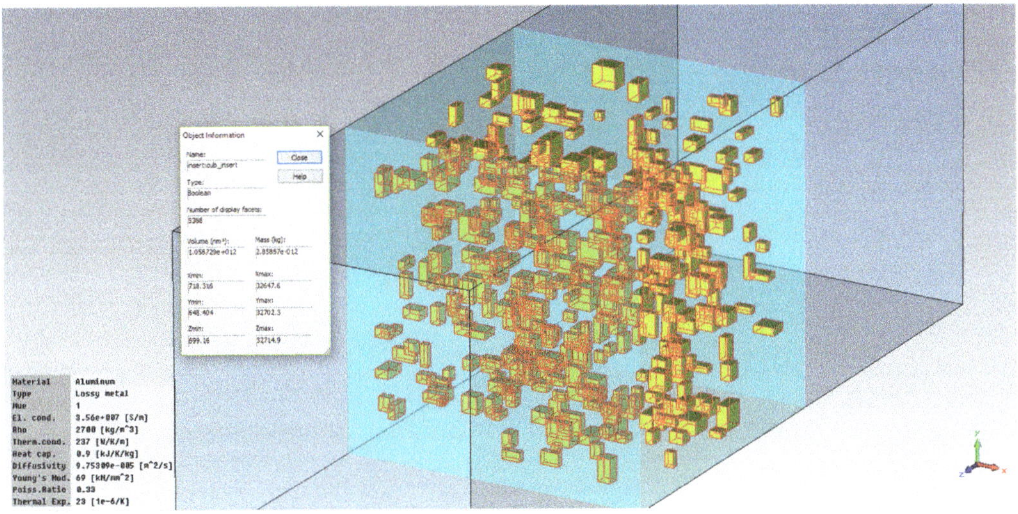

Figure 10. Analyzed structure: LDPE composite with 800 nm Al inserts.

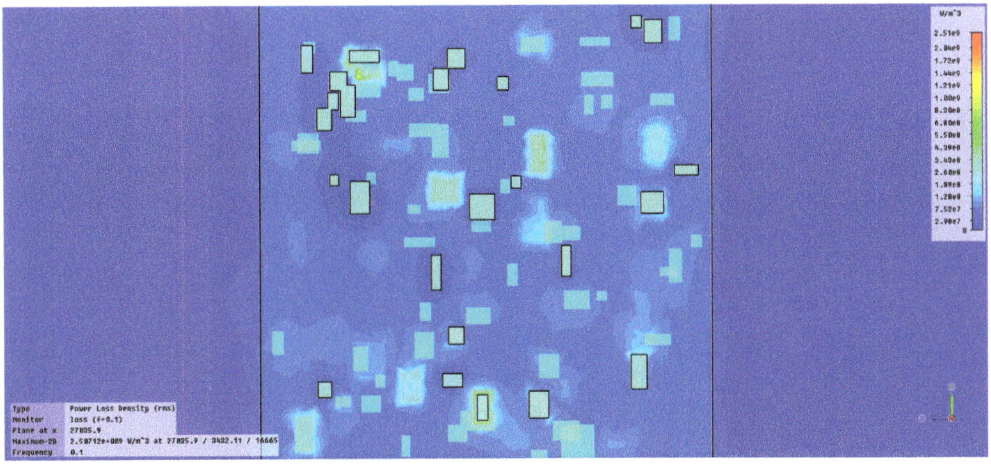

Figure 11. Volumetric distribution of power loss density at 0.1 GHz: LDPE composite with 800 nm Al inserts.

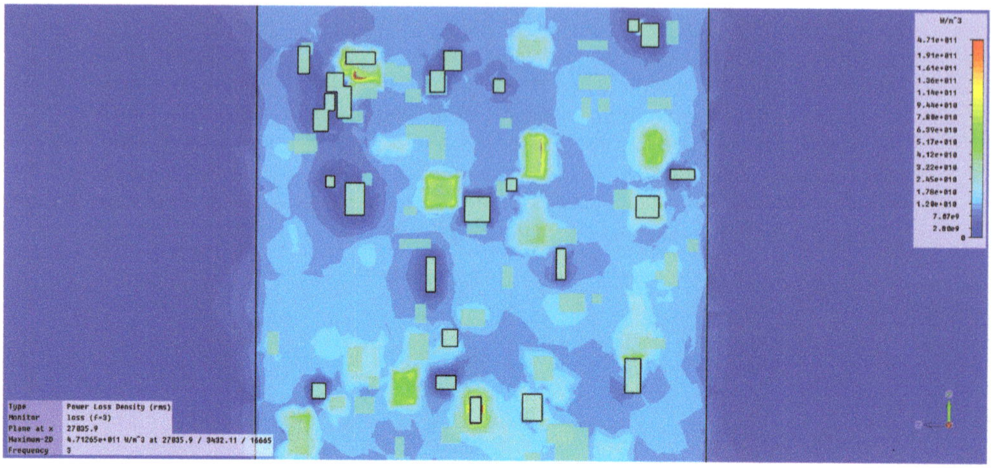

Figure 12. Volumetric distribution of power loss density at 3 GHz: LDPE composite with 800 nm Al inserts.

3.4. Analysis of Composites with 50 nm Fe Inserts

As regards the composites with 50 nm iron (equivalent) inserts, the edge of the cubic layer of the LDPE matrix was 2961.0 nm, and the model is plotted in Figure 13. The volumetric distribution of the power loss density is presented in Figures 14 and 15 at 0.1 GHz and 3 GHz, respectively. Referring to the power loss density values, significantly higher values for the exposure at 3 GHz were found, with an average of 2.475×10^{13} W/m^3, compared to 1.318 10^{11} W/m^3 for 0.1 GHz. Compared to the composite with aluminum 50 nm (equivalent) inserts, the values for the composite with iron 50 nm (equivalent) inserts are about 50% lower, showing a higher technological efficiency in using Al inserts.

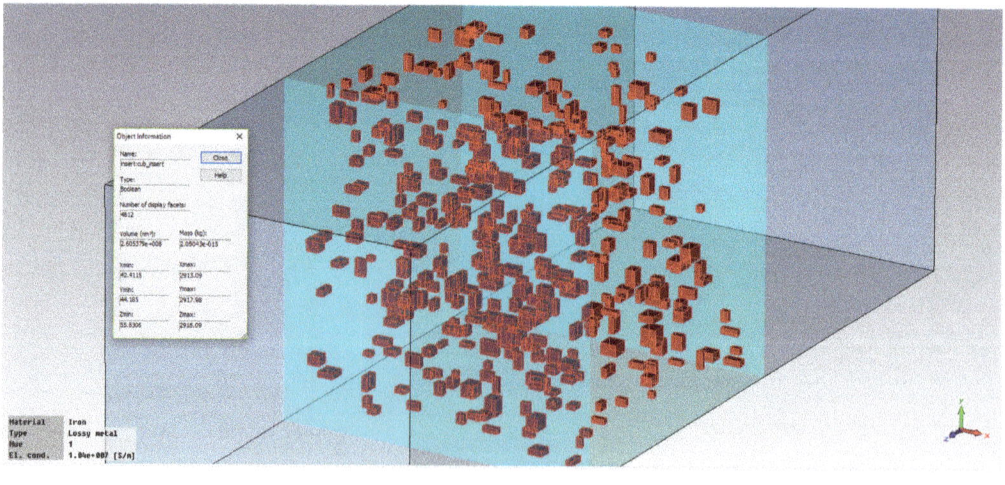

Figure 13. Analyzed structure: LDPE composite with 50 nm Fe inserts.

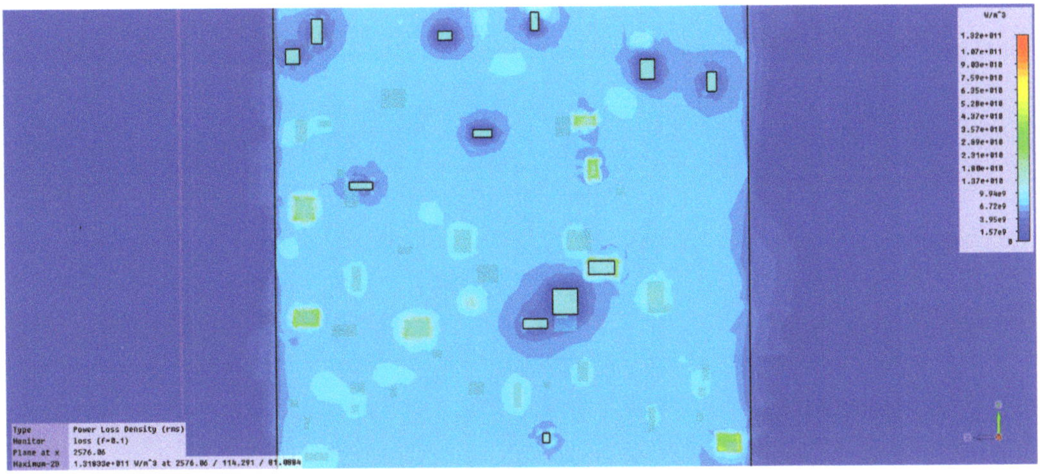

Figure 14. Volumetric distribution of power loss density at 0.1 GHz: LDPE composite with 50 nm Fe inserts.

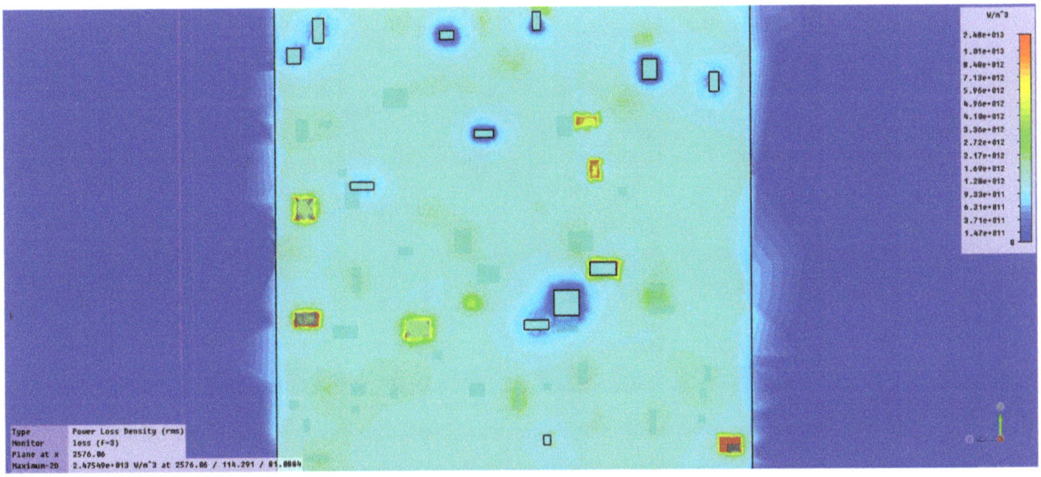

Figure 15. Volumetric distribution of power loss density at 3 GHz: LDPE composite with 50 nm Fe inserts.

3.5. Analysis of Composites with 800 nm Fe Inserts

Finally, the composites with 800 nm iron (equivalent) inserts are presented, with the edge of the cubic layer of the LDPE matrix being 47,518.0 nm, and the model is plotted in Figure 16. The volumetric distribution of the power loss density is presented in Figures 17 and 18 at 0.1 GHz and 3 GHz, respectively. Regarding the power loss density values, significantly higher values for the exposure at 3 GHz were found, with an average of 2.949×10^{11} W/m^3, compared to 1.568×10^9 W/m^3 for 0.1 GHz. Compared to the composite with 50 nm iron (equivalent) inserts, the values are about 100 times lower at both frequencies. In comparison to the composite containing 800 nm aluminum (equivalent) inserts, the values for the composite with 800 nm iron (equivalent) inserts are approximately 30% lower, indicating greater technological efficiency when utilizing Al inserts.

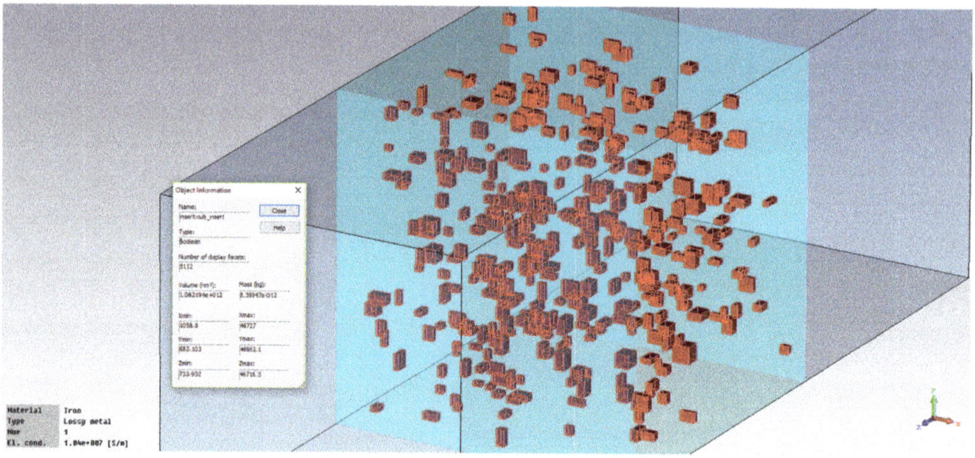

Figure 16. Analyzed structure: LDPE composite with 800 nm Fe inserts.

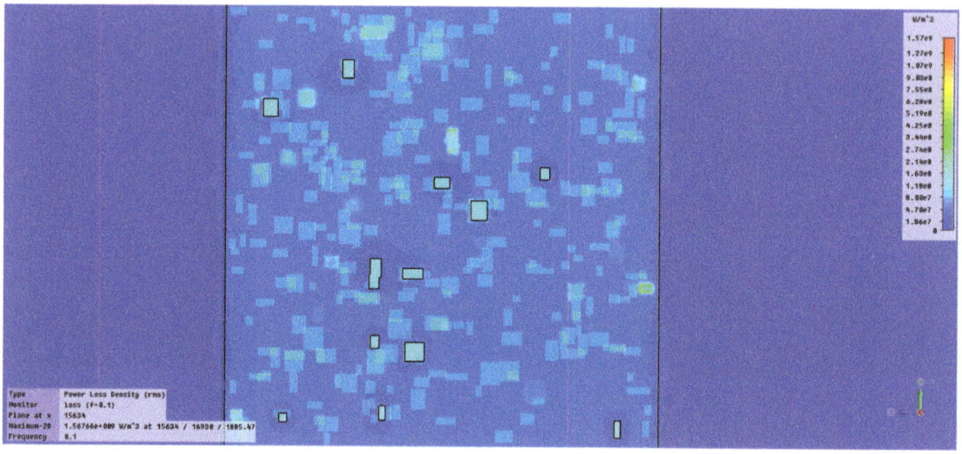

Figure 17. Volumetric distribution of power loss density at 0.1 GHz: LDPE composite with 800 nm Fe inserts.

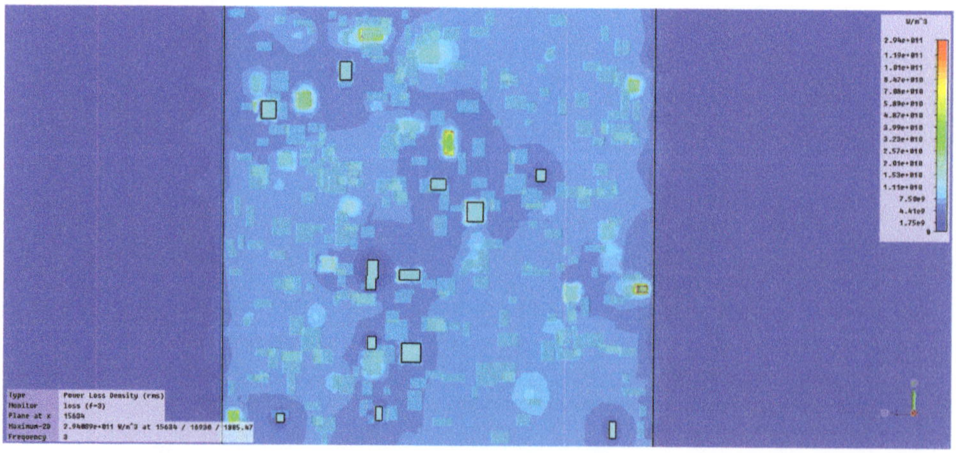

Figure 18. Volumetric distribution of power loss density at 3 GHz: LDPE composite with 800 nm Fe inserts.

4. Final Remarks

Based on the simulations presented above, at first glance, it appears that the composites with metallic ingredients with a lower dimension are more adequate for usage. At higher frequencies, the power loss density is larger for both metals, so the usage of higher frequencies is recommended for technological applications. The particles with a larger dimension (800 nm) absorb a much lower energy compared to the ones with a small dimension (50 nm), at the same mass ratio, because the equivalent external surface exposed to microwave radiation is lower. On the other hand, due to a higher conductivity, Al particles within composite materials can absorb a larger amount of energy compared to Fe particles, at both insertion dimensions. However, the impact of the metallic type might lose significance when considering the energy consumption that is just 30% more for Fe, and/or the technological productivity, alongside the expense of powders, as Al powders are considerably more expensive than Fe powders.

Figures 19 and 20 compare the S11 reflection coefficient parameters, and Table 3 presents some data considering the most relevant frequencies for microwave technology.

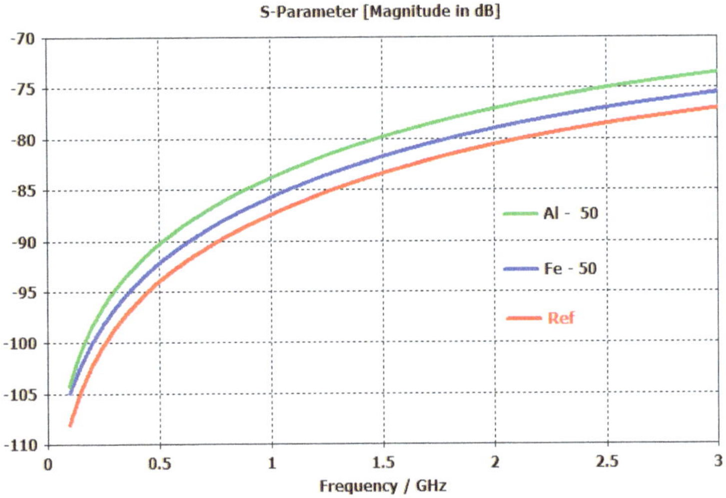

Figure 19. S_{11} reflection coefficient comparison for composites with 50 nm inserts.

The reflection coefficient is influenced by the dimension of the insertion, decreasing for higher-dimensional cases. It also relies on frequency, decreasing as the frequency increases. Concerning the effect of the metal type, the Al-containing composite shows a marginally lower reflection at both sizes or analyzed frequencies (a less than 5% difference). Overall, based on the data, the reflection phenomena are minimal; however, it is important to also examine these phenomena when taking into account the types of insertion and frequency dependence, as it confirms that the key factor for bonding technology is the absorption of radiation energy.

Analyzing the S_{21} transmission parameters, the findings are shown in Figures 21 and 22, which illustrate that incorporating nanometric inserts will reduce transmission through the material. It is clear that the transmission significantly declines as the frequency rises in the GHz range, yet a saturation effect was observed after 2–2.5 GHz. The significant reduction in transmission, linked to minimal reflection, verifies that, for composites with added metallic powder, microwave energy is extensively absorbed by the materials, particularly at frequencies above 1.5 GHz, confirming their potential functionality as hot melts.

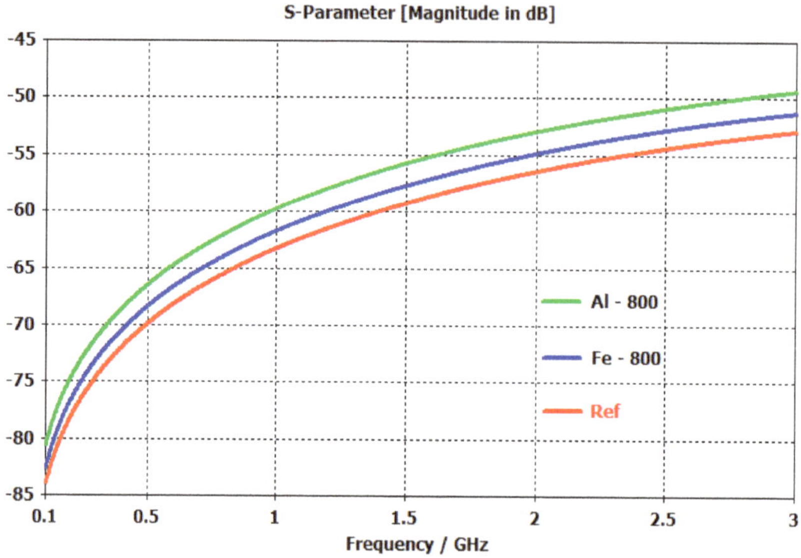

Figure 20. S_{11} reflection coefficient comparison for composites with 800 nm inserts.

Table 3. S_{11} magnitude [dB] vs. insertion type.

Frequency	Al—50 nm	Al—800 nm	Fe—50 nm	Fe—800 nm
1 GHz	83.5	60	86	62
3 GHz	73	49	76	51.5

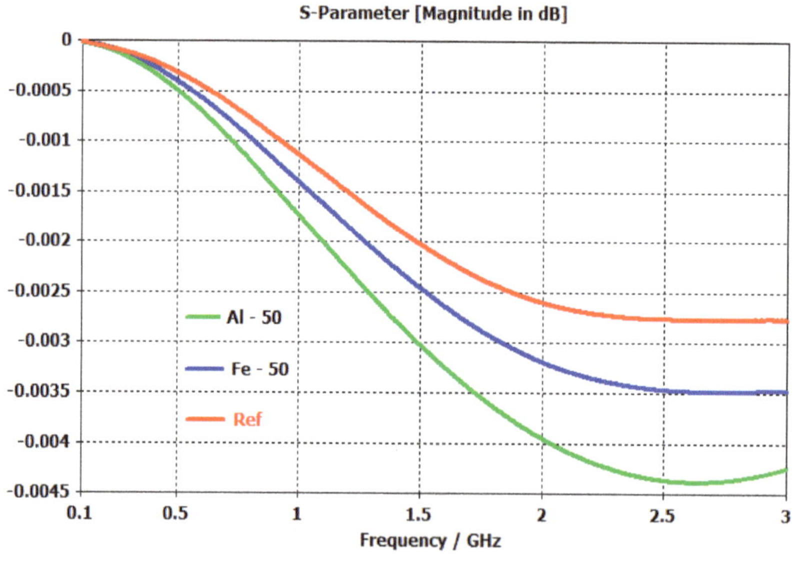

Figure 21. S_{21} transmission coefficient comparison for composites with 50 nm inserts.

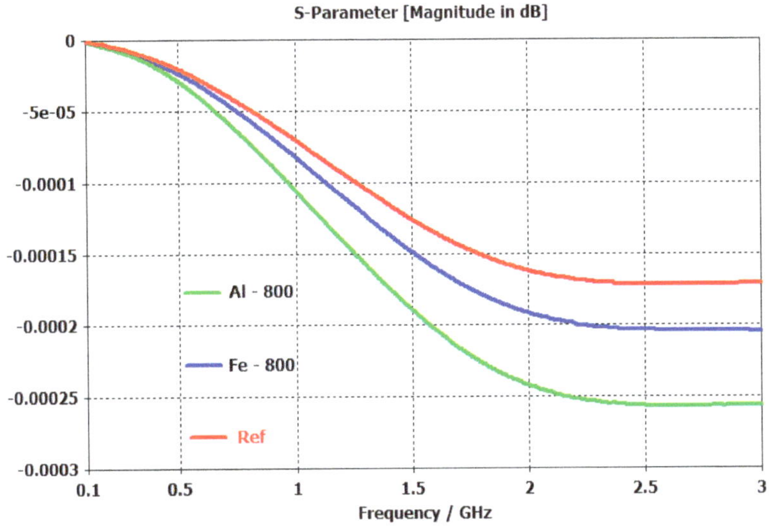

Figure 22. S_{21} transmission coefficient comparison for composites with 800 nm inserts.

The reduction in transmission shown in Figures 21 and 22 may not appear significant, but it is important to note that the simulated layers are here merely 2 µm thick for 50 nm insertions, and 33 µm thick for 800 nm insertions, respectively. However, when transforming the absorption into dB/unit length, notably high values for composite layers, such as those 1 or 2 mm in thickness, are achieved, which is the typical dimension in bonding applications utilizing hot melt technology. The figures in Table 4 are pertinent in this manner.

Table 4. Attenuation [dB] vs. frequency for 1mm layer, 8% mass ratio, Al and Fe inserts.

Insertion Type	1 GHz	2 GHz	3 GHz
Fe—800 nm	14.8	38.7	40.5
Al—800 nm	25.7	61.2	64.4
Fe—50 nm	47.3	81.8	86.6
Al—50 nm	51.5	98.2	104.1

Overall, a value exceeding 80 dB is regarded as adequate for a hot melt layer of 1 mm thickness; however, it can be noted that the peak values are attained for Al insertions and, respectively, for both metallic particles at a size of 50 nm, reaffirming the results obtained from the power loss simulations.

Furthermore, the volumetric distribution of the power loss density within the material containing inserts continues to exhibit a significantly nonuniform loss distribution. One could attempt to manage the spatial arrangement of the inserts during manufacturing to direct the waves toward an area with superior absorbent qualities; however, this could lead to greater technological complexity and higher material expenses.

5. Conclusions

This article outlines a simulation for S-parameters and absorbed energy in polymeric nanocomposites containing metallic inserts (iron and aluminum, featuring two particle sizes). The examined frequency range of 0.1–3 GHz is tailored for numerous applications in microwave technologies. The electromagnetic simulation software selected was CST

Studio Suite, analyzing the power loss at different frequencies, serving as a necessary step in advanced bonding technology within the microwave range. The real constraints in the electromagnetic simulation are detailed, including a related application of the Nicolson–Ross–Weir method in infinite media, which unveils new perspectives in modeling the behavior of nanocomposites.

It was shown that, because of their greater conductivity, Al particles can capture more energy than Fe particles in composite materials, at both insertion scales. At elevated frequencies, the power loss density increases for both metals. Overall, based on the data, the reflection phenomena are minimal. It was demonstrated that adding nanometric inserts will decrease transmission across the material. It is evident that the transmission drops considerably as the frequency increases in the GHz spectrum, although a saturation effect was noted after 2–2.5 GHz. In general, a value over 80 dB is considered sufficient for a 1 mm thick hot melt layer; nonetheless, it is observed that the highest values are reached for Al insertions and, correspondingly, for both metallic particles sized at 50 nm, reinforcing the findings from the power loss simulations.

The substantial decrease in transmission, associated with a slight reflection, confirms that composites containing metallic powder absorb microwave energy effectively, especially at frequencies exceeding 1.5 GHz, validating their possible use as hot melts for innovative reversible bonding applications.

Author Contributions: Conceptualization, R.F.D. and R.C.C.; methodology, R.F.D. and R.C.C.; validation, R.C.C.; formal analysis, R.F.D. and R.C.C.; investigation, R.F.D. and R.C.C.; data curation, R.F.D. and R.C.C.; writing—original draft preparation, R.F.D. and R.C.C.; writing—review and editing, R.C.C.; visualization, R.F.D. and R.C.C.; supervision, R.C.C. All authors have read and agreed to the published version of the manuscript.

Funding: This research received no external funding.

Institutional Review Board Statement: Not applicable.

Informed Consent Statement: Not applicable.

Data Availability Statement: The data are presented in this study.

Conflicts of Interest: The authors declare no conflicts of interest.

References

1. Global Hot Melt Adhesives Market Overview. Available online: https://www.marketresearchfuture.com/reports/hot-melt-adhesives-market-4640 (accessed on 28 October 2024).
2. What Is Hot Melt Made of? Available online: https://www.hotmelt.com/blogs/blog/what-is-hot-melt (accessed on 28 October 2024).
3. Gharde, S.; Sharma, G.; Kandasubramanian, B. Hot-Melt Adhesives: Fundamentals, Formulations, and Applications: A Critical Review. In *Progress in Adhesion and Adhesives*; John Wiley & Sons: Hoboken, NJ, USA. [CrossRef]
4. Liebezeit, S.; Mueller, A.; Leydolph, B.; Palzer, U. Microwave-induced interfacial failure to enable debonding of composite materials for recycling. *Sustain. Mater. Technol.* **2017**, *14*, 29–36. [CrossRef]
5. Hellmann, H.; Krieger, R. Microwave-Activatable Hot-Melt Adhesive. U.S. Patent 4906497A, 4 November 1988.
6. Roth, M.; Padurschel, P.; Niehaus, E.; Hoffmann, G. Hot Melt Adhesive for Microwave Heating. U.S. Patent WO2008071469, 19 June 2008.
7. Srinivasan, D.V.; Idapalapati, S. Review of debonding techniques in adhesively bonded composite structures for sustainability. *Sustain. Mater. Technol.* **2021**, *30*, e00345. [CrossRef]
8. La Rosa, A.D.; Ursan, G.A.; Aradoaei, M.; Ursan, M.; Schreiner, C. Life Cycle Assessment of Microwave Activated Hot-Melt Adhesives. In Proceedings of the 2018 International Conference and Exposition on Electrical and Power Engineering (EPE), Iasi, Romania, 18–19 October 2018. [CrossRef]
9. Sharma, S.; Parne, S.R.; Panda, S.S.S.; Gandi, S. Progress in microwave absorbing materials: A critical review. *Adv. Colloid Interface Sci.* **2024**, *327*, 103143. [CrossRef] [PubMed]

10. Wang, Y.F.; Zhu, L.; Han, L.; Zhou, X.H.; Gao, Y.; Lv, L.H. Recent Progress of One-Dimensional Nanomaterials for Microwave Absorption: A Review. *ACS Appl. Nano Mater.* **2023**, *6*, 7107–7122. [CrossRef]
11. Green, M.; Chen, X. Recent progress of nanomaterials for microwave absorption. *J. Mater.* **2019**, *5*, 503–541. [CrossRef]
12. Zhang, X.J.; Guo, A.P.; Wang, G.S.; Yin, P.G. Recent progress in microwave absorption of nanomaterials: Composition modulation, structural design, and their practical applications. *Iet Nanodielectr.* **2019**, *2*, 2–10. [CrossRef]
13. Wilson, A.J.; Rahman, M.; Kosmas, P.; Thanou, M. Nanomaterials responding to microwaves: An emerging field for imaging and therapy. *Nanoscale Adv.* **2021**, *3*, 3417–3429. [CrossRef] [PubMed]
14. Maleki, S.T.; Babamoradi, M. Microwave absorption theory and recent advances in microwave absorbers by polymer-based nanocomposites (carbons, oxides, sulfides, metals, and alloys). *Inorg. Chem. Commun.* **2023**, *149*, 110407. [CrossRef]
15. Abd El-Gawad, W.M.; Eldesouki, E.M.; Abd El-Ghany, W.A. Development of high performance microwave absorption modified epoxy coatings based on nano-ferrites. *Sci. Rep.* **2024**, *14*, 5190. [CrossRef] [PubMed]
16. Njoku, C.C.; Whittow, W.G.; Vardaxoglou, Y.C. Study on the Variation in Dielectric Properties of Heterogeneous Substrates Composed of Nanomaterials. In Proceedings of the 5th European Conference on Antennas and Propagation, Rome, Italy, 11–15 April 2011; pp. 488–492.
17. Rothwell, W.S. Complex Permittivity of Conductor-Dielectric Mixtures. *IEEE Trans. Microw. Theory Tech-Niques* **1971**, *19*, 413–414. [CrossRef]
18. Ma, S.J.; Zhou, X.W.; Su, X.J.; Mo, W.; Yang, J.L.; Liu, P. A new practical method to determine the microwave energy absorption ability of materials. *Miner. Eng.* **2009**, *22*, 1154–1159. [CrossRef]
19. Zhang, D.; Han, C.; Zhang, H.; Zeng, B.; Zheng, Y.; Shen, J.; Wu, Q.; Zeng, G. The Simulation Design of Microwave Absorption Performance for the Multi-Layered Carbon-Based Nanocomposites Using Intelligent Optimization Algorithm. *Nanomaterials* **2021**, *11*, 1951. [CrossRef] [PubMed]
20. Llave, Y.; Kambayashi, D.; Fukuoka, M.; Sakai, N. Power absorption analysis of two-component materials during microwave thawing and heating: Experimental and computer simulation. *Innov. Food Sci. Emerg. Technol.* **2020**, *66*, 102479. [CrossRef]
21. Li, G.; Zhou, J. COMSOL-Based Simulation of Microwave Heating of Al_2O_3/SiC Composites with Parameter Variations. *Symmetry* **2024**, *16*, 1254. [CrossRef]
22. Silveirinha, M.G.; Fernandes, C.A. Effective permittivity of a medium with stratified dielectric host and metallic inclusions. *IEEE Antennas Propag. Soc. Int. Symp.* **2004**, *4*, 3769–3772. [CrossRef]
23. Vicente, A.N.; Dip, G.M.; Junqueira, C. The step by step development of NRW method. In Proceedings of the SBMO/IEEE MTT-S International Microwave and Optoelectronics Conference (IMOC 2011), Natal, Brazil, 29 October–1 November 2011. [CrossRef]
24. Rothwell, E.J.; Frasch, J.L.; Ellison, S.M.; Chahal, P.; Ouedraogo, R.O. Analysis of the Nicolson-Ross-Weir method for characterizing the electromagnetic properties of engineered materials. *Prog. Electromagn. Res.* **2016**, *157*, 31–47. [CrossRef]
25. Das, S. Matlab Code of Nicholson-Ross-Weir (NRW) Algorithm to Estimate the Permeability and Permittivity of EMI Shielding Materials. 2020. Available online: https://www.researchgate.net/publication/341275777_Matlab_code_of_Nicholson-Ross-Weir_NRW_Algorithm_to_estimate_the_permeability_and_permittivity_of_EMI_shielding_materials#fullTextFileContent (accessed on 28 October 2024).
26. Sahin, S.; Nahar, N.K.; Sertel, K. A simplified Nicolson–Ross–Weir method for material characterization using single-port measurements. *IEEE Trans. Terahertz Sci. Technol.* **2020**, *10*, 404–410. [CrossRef]
27. Dassault Systèmes, CST Studio Suite—Electromagnetic Field Simulation Software. Available online: https://www.3ds.com/products/simulia/cst-studio-suite, (accessed on 28 October 2024).
28. Ciobanu, R.; Schreiner, C.; Damian, R. High frequency electromagnetic energy phenomena in chiral dielectric structures with distributed and localized conductive insertions. *Compos. Part B Eng.* **2019**, *160*, 241–248. [CrossRef]
29. Aradoaei, M.; Schreiner, O.; Lucaci, A.M.; Morosanu, V.; Patrascu, G.; Saridache, A.; Schreiner, C.; Aradoaei, S. Simulating the Interaction of Electromagnetic Radiation with Matter Within Nano/Micro-Conductive Composites. In Proceedings of the International Conference on Electromechanical and Energy Systems (SIELMEN), Craiova, Romania, 11–13 October 2023. [CrossRef]
30. Ciobanu, R.C.; Damian, R.F.; Schreiner, C.M.; Aradoaei, M.; Sover, A.; Raichur, A.M. Simulation of Dielectric Properties of Nanocomposites with Non-Uniform Filler Distribution. *Polymers* **2023**, *15*, 1636. [CrossRef] [PubMed]

applied sciences

MDPI

Article

Analysis of the Matching Media Effects by Microwave Field Distribution Simulations for the Cylindrically Layered Human Arm Model

Tanju Yelkenci

Faculty of Engineering, Turkish-German University, Sahinkaya Cad. No: 106, 34820 Beykoz, Turkey; tanju.yelkenci@tau.edu.tr

Abstract: In this study, a method is presented to determine the matching media parameters that maximize the electromagnetic energy penetrating into the human arm modeled as a radially stratified cylinder. In this context, first, the electromagnetic scattering problem related to the layered cylindrical model in question was solved analytically using cylindrical harmonics. Then, based on this solution, a frequency-dependent functional in terms of the electromagnetic parameters of the matching medium was defined, and the parameters that minimize this functional were determined through the graphs of this functional. In this functional, which depends on the permittivity, conductivity and frequency of the matching medium, one parameter was kept constant at every turn while the other two parameters were optimized. The accuracy of the approach was demonstrated by calculating the electric field amplitudes inside and outside the layers for the parameters determined by the proposed method. The numerical results given in this context demonstrate that if a matching medium is used, the penetrating field increases between 1.3 to 13.96 times compared to the case where the matching medium is absent.

Keywords: matching medium; electromagnetic scattering; multilayered human arm; Debye model; microwave arm imaging; method of moments

1. Introduction

One of the important applications of electromagnetic imaging problems is the detection and imaging of the localized pathological changes in human organs or tissues [1,2]. In recent years, many studies have been carried out on various parts of human body such as the head, breast, arm, etc., which are non-homogeneous and have frequency-dependent electromagnetic parameters [3–6]. When such structures are placed in free space for imaging purposes, due to their constitutive parameters with high dielectric contrast and loss, there will be a problem with the relative poor penetrating of the incident field into the structure and a considerable part of the field will be reflected back. To overcome this difficulty, one of the measures may be taken is the existence of a background space with appropriate parameters. If the contrast difference between the inhomogeneous structure under examination and the medium in which the structure is placed can be reduced enough, as a result of the usage of an appropriate background medium, better transmission and lesser reflection of the incoming field into the structure will be possible, and the field will interact with the structures to be detected and imaged. These types of media are called as "matching media" and a relatively limited number of studies exist in the literature [1–13] since medical microwave imaging is a new developing technique compared to conventional approaches such as MR, X-ray, etc.

Microwave imaging applications have largely concentrated on the head (brain) [4,7,10–12] and female breast [2,3,9,11], but microwave imaging of the arm [1,5,7], limb [1] and bone [6,13] have also drawn interest. The satisfactory imaging quality of the human arm with microwave-based alternative medical technologies requires a background space (matching medium) to directly decrease the reflected field and to transmit the incident field by a sufficient amount at the arm boundary. In recent years, several contributions concerned with microwave imaging of human arm and matching medium have been reported in the literature. For instance, Jofre et al. [7] used a microwave tomographic scanner for biomedical applications using deionized water for matching purposes and presented tomographic slices of an in vivo human forearm. In [1], the authors outlined a process for an appropriate matching fluid and generated images of a human forearm using an experimental microwave tomography system designed for limb imaging. Gilmore et al. [5] presented a pilot study using a microwave tomography system with a matching fluid of deionized water and table salt, and they imaged the forearms of five adult male and female volunteers between the ages of 30 and 48. Amin et al. [6] investigated the application of microwave imaging to monitor the in vivo dielectric properties of human bones and, hence, aid in the monitoring of osteoporosis. Alkhodari et al. [13] conducted a study to evaluate the use of microwave tomography for imaging human bones with varying parameters, where the authors proposed the use of an ultrasound gel instead of the more traditional glycerin/water solution as a matching medium.

Different studies related to the human arm may also be mentioned. Faktorova et al. [14] showed that metamaterial structures can increase the focusation of electromagnetic energy in the human arm to larger deep areas without the undesired radiation of the surrounding area. Azizi et al. [15] studied the behavior of an ultra-wideband antenna and tested the antenna in free space, outside and inside of a numerically modeled human arm.

In the literature, there are studies on layered cylindrical structures that can be analytically solved and human arm models built with these structures. The aim of all these models is to reveal the interaction of electromagnetic waves with the relevant structures, that is, to solve a scattering problem. As an example of the analytical solution methods of the stratified cylinders in the literature, one can mention the work of Yeh and Lindgren [16]. They worked out multilayered cylinders having an arbitrary number of layers, They introduced calculations of the scattering coefficients using successive matrix multiplications. The non-stratified circular cylindrical human arm model implementation can be found in Jensen et al. [17], where they used the recursive Green's function method for the computation of fields scattered by a cylindrical arm model consisting of a circular cylinder representing the muscle and two smaller cylinders representing the bone. The analytical solution presented in this article is quite parallel to that reported by Kai and D'Alessio [18]. Kai and D'Alessio [18] carried out the scattering problem, without matching medium, using a finely stratified cylinder model.

The research reported by Scapaticci et al. [4] is closely similar to the presented study, in which a simple planarly stratified structure was analyzed by the transmission line method for a head phantom having human tissue characteristics, and the matching medium parameters were determined. Catapano et al. [3] investigated a study on microwave tomography of the female breast by similarly exploiting a simple planarly layered breast model. These cited research groups justified the adoption of a such simplified model by reasoning that the scanning wavelength should be small compared to the size of the body, in order to allow the detectability of small tumors and tissue details.

In this study, a new method has been proposed to determine the background medium parameters, which has the main purpose to minimize the energy loss, that is, to maximize the electromagnetic energy passing inside. In this regard, a five-tissue layered cylindrical arm model is illuminated with a line source located outside the arm structure. This classical electromagnetic scattering problem is solved analytically, and the reflected and transmitted wave amplitude coefficients are obtained. A functional is defined in terms of the reflection coefficients of the cylindrical harmonics in the analytical solution depending on the electromagnetic parameters of the background medium, which is also known as matching medium. The constitutive parameters and operating frequency that minimize the functional will be decisive in the choice of the matching medium with optimum characteristics. Through the image plot of the functional, the parameter pairs that minimize this functional are determined. Using these parameters, comparisons of the scattered and transmitted fields are carried out for the cases with and without the matching medium. Testing the accuracy of the approach is based on observing the amplitude of the field in different regions with the calculated parameters. Thus, the numerical results obtained are verified within themselves and their consistency is demonstrated. The numerical results obtained indicate the success of the method.

The remainder of the paper is organized as follows. The modeling of the human arm is presented in Section 2. Green's function formulation of the problem is outlined in Section 3. Section 4 discusses minimizing of the reflection coefficient R_n. Formulation of the five-layered arm model with an internal object is given in Section 5. Section 6 is devoted to numerical results. Finally, concluding remarks are given in Section 7. Throughout the study, a time factor $e^{-i\omega t}$ is assumed and omitted.

2. Modeling of the Human Arm

It is assumed in this research that the human arm is modeled as a radially stratified cylinder of 50 mm radius. The geometry of the arm model consists of five concentric cylindrical layers, of which thicknesses are 1.5 mm (skin), 8.5 mm (fat), 27.5 mm (muscle), 6 mm (bone) and 6.5 mm (bone marrow) [19–21], as shown in Figure 1.

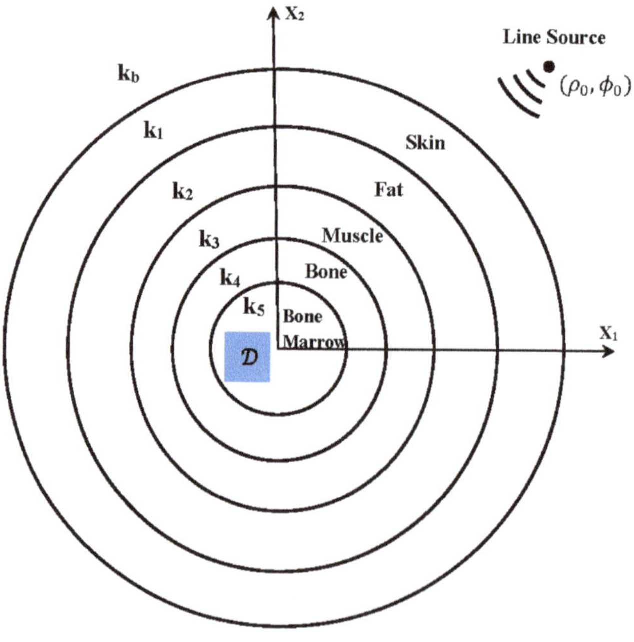

Figure 1. Geometry of the problem with an internal object $\mathcal{D}$.

As it is well known, the electromagnetic properties of human tissues depend generally on frequency and, hence, they are characterized as dispersive media. The constitutive parameters of frequency-dispersive biological tissues are commonly evaluated by a variety of relaxation simulations including Debye and Cole–Cole models. The suitability of the Debye model for fast evaluation of wideband fields was reported in [22] by comparing with the Cole–Cole model. Taking into account the dispersive nature of the tissues in the layers of the arm model, their dielectric properties are determined in this study according to the three-pole Debye model expressed by

$$\epsilon_r(\omega) = \epsilon_\infty + \sum_{k=1}^{3} \frac{\Delta\epsilon_k}{1 - i\,\omega\,\tau_k}, \tag{1}$$

where ω is the angular frequency, ϵ_∞ is the high frequency permittivity, $\Delta\epsilon_k$ is the magnitude of the dispersion and τ_k is the relaxation time constant. These parameters for five tissues considered in this work are extracted for each tissue from the data set reported in [23,24] for the frequency range from 100 MHz to 2 GHz and listed in Table 1.

Table 1. Debye parameters of the arm tissues for the frequency range from 100 MHz to 2 GHz [23,24].

Tissue	ϵ_∞	$\Delta\epsilon_1$	$\Delta\epsilon_2$	$\Delta\epsilon_3$	$\tau_1 \times 10^{-9}$	$\tau_2 \times 10^{-9}$	$\tau_3 \times 10^{-9}$
Marrow	4.162	1.321	0.4162	26.08	0.02899	0.501	10.14
Bone	6.915	5.487	1.653	94.31	0.03378	0.4537	13.19
Muscle	31.45	23.27	6.746	3101	0.02212	0.5729	39.26
Fat	4.113	1.341	0.2853	119.3	0.02728	0.4887	29.28
Dry skin	27.44	12.73	17.1	417.9	0.0366	0.5684	7.581

3. Green's Function Formulation of the Problem

We consider the two-dimensional electromagnetic scattering problem for a human arm model composed of concentric cylindrical layers of skin, fat, muscle, cortical bone and bone marrow. Figure 1 shows the cross-sectional reference geometry of the cylindrical structure, which is located in the transverse direction. The wavenumber of the background medium is k_b. The stratified model is irradiated by a monochromatic electromagnetic wave having the frequency ω excited by an infinitely long line source parallel to the structure. The Green's function $G_b(x, y)$ of the structure satisfies the equation

$$\nabla^2 G_b(x,y) + k_b^2 G_b(x,y) = -\delta(x-y) \tag{2}$$

and has the solution

$$G_b(x,y) = \frac{i}{4} H_0^{(1)}(k_b|x-y|), \tag{3}$$

which is simply the field of the line source located at an arbitrary point $y \in \mathbb{R}^2$ outside the structure. Here, $H_0^{(1)}$ denotes the zeroth order Hankel function of the first kind and can be replaced by using the addition theorem for Hankel functions as follows [25]:

$$H_0^{(1)}(k_b|x-y|) = \frac{i}{4} \begin{cases} \sum\limits_{n=-\infty}^{\infty} H_n^{(1)}(k_b\rho)\, J_n(k_b\rho')\, e^{in(\phi-\phi')}, & \text{if } \rho > \rho' \\ \sum\limits_{n=-\infty}^{\infty} H_n^{(1)}(k_b\rho')\, J_n(k_b\rho)\, e^{in(\phi-\phi')}, & \text{if } 0 < \rho < \rho' \end{cases} \tag{4}$$

where the vectors $\vec{\rho} = (\rho, \phi)$ and $\vec{\rho}_s = (\rho', \phi')$ denote the observation and the source points, respectively. The line source is assumed to be located outside the cylindrical structure. The zeroth order Hankel function appearing in this expression can adequately be replaced by using the addition theorem for Hankel functions as follows:

$$H_0^{(1)}(k_1|x-y|) = \sum_{n=-\infty}^{\infty} H_n^{(1)}(k_1\rho_>)\, J_n(k_1\rho_<)\, e^{in(\phi-\phi')} \tag{5}$$

where $\rho_> = max(\rho, \rho')$ and $\rho_< = min(\rho, \rho')$ [25].

The line source situated at $\rho = \rho_0, \phi = \phi_0$ has the current density

$$J(\rho,\phi) = \frac{I_0}{\rho_0}\, \delta(\rho-\rho_0)\, \delta(\phi-\phi_0), \tag{6}$$

where I_0 denotes the current strength. In the TM case, the incident electric field generated by the line source has only a x_3-component, given by

$$\vec{E}^i(x) = u^i(x)\, \vec{e}_3, \quad \vec{x} = (x_1, x_2) \in \mathbb{R}^2 \tag{7}$$

with

$$u^i(x) = -\frac{k_b\, I_0}{4\,\omega\,\epsilon_0}\, H_0^{(1)}(k_b|\vec{\rho}-\vec{\rho}_s|) \tag{8}$$

where $H_0^{(1)}$ denotes the Hankel function of the first kind with order zero and k_b is the wavenumber of the background medium, and the vectors $\vec{\rho} = (\rho, \phi)$ and $\vec{\rho}_s = (\rho_0, \phi_0)$ denote the observation and the source points, respectively. In what follows is the line source assumed to be located outside the cylindrical structure. The Green's functions in the relevant layers are

$$G(\rho,\varphi,\rho',\phi') = \frac{i}{4} H_0^{(1)}(k_b|\vec{\rho}-\vec{\rho}'|) + \sum_{n=-\infty}^{\infty} R_n H_n^{(1)}(k_b\rho)\, e^{in(\phi-\phi')}, \rho > r_1 \tag{9}$$

and, for respective indices $\ell = 2, 3$ and 4,

$$G(\rho,\phi,\rho',\phi') = \sum_{n=-\infty}^{\infty} \left[T_{\ell\, n}\, H_n^{(1)}(k_\ell\, \rho) + S_{\ell\, n}\, J_n(k_\ell\, \rho)\right] e^{in(\phi-\phi')}, r_\ell < \rho < r_{\ell+1} \tag{10}$$

and

$$G(\rho,\phi,\rho',\phi') = \sum_{n=-\infty}^{\infty} T_{5\,n}\, J_n(k_5\, \rho)\, e^{in(\phi-\phi')}, \rho < r_4 \tag{11}$$

The unknown wave amplitude coefficients $R_n, T_{2n}, S_{2n}, T_{3n}, S_{3n}, T_{4n}, S_{4n}$ and T_{5n} in the above expressions can be obtained by imposing the continuity and the jump conditions

$$G(\rho = \rho' + 0, \phi, \rho', \phi') - G(\rho = \rho' - 0, \rho', \phi') = 0 \tag{12}$$

and

$$\frac{\partial G}{\partial \rho}(\rho = \rho' + 0, \phi, \rho', \phi') - \frac{\partial G}{\partial \rho}(\rho = \rho' - 0, \rho', \phi') = -\frac{1}{\rho'}\, e^{-in\phi'} \tag{13}$$

on each layer surface S, respectively, Ref. [26]. The resulting expressions, when continuity and jump conditions are applied, are as follows:

$$\frac{i}{4}\, J_n(k_b r_1) H_n^{(1)}(k_1\rho') + R_n H_n^{(1)}(k_b r_1) = T_{1n} H_n^{(1)}(k_1 r_1) + S_{1n} J_n(k_1 r_1) \tag{14}$$

$$\frac{i}{4}\, J_n{}'(k_b r_1) H_n^{(1)'}(k_1\rho') + R_n H_n^{(1)'}(k_b r_1) = T_{1n} H_n^{(1)'}(k_1 r_1) + S_{1n} J_n{}'(k_1 r_1) \tag{15}$$

and, for respective indices $\ell = 1, 2$ and 3,

$$T_{\ell\, n} H_n^{(1)}(k_\ell\, r_\ell) + S_{\ell\, n} J_n(k_\ell\, r_\ell) = T_{\ell+1\, n} H_n^{(1)}(k_{\ell+1}\, r_{\ell+1}) + S_{\ell+1\, n} J_n(k_{\ell+1}\, r_{\ell+1}) \tag{16}$$

$$T_{\ell n}H_n^{(1)'}(k_\ell r_\ell) + S_{\ell n}J_n'(k_\ell r_\ell) = T_{\ell+1 n}H_n^{(1)'}(k_{\ell+1} r_{\ell+1}) + S_{\ell+1 n}J_n'(k_{\ell+1} r_{\ell+1}) \quad (17)$$

and

$$T_{4n}H_n^{(1)}(k_4 r_4) + S_{4n}J_n(k_4 r_4) = T_{5n} J_n(k_5 r_5) \quad (18)$$

$$T_{4n}H_n^{(1)'}(k_4 r_4) + S_{4n}J_n'(k_4 r_4) = T_{5n} J_n'(k_5 r_5) \quad (19)$$

where derivatives are considered with respect to the entire argument. The values of the scattered and transmitted wave amplitude coefficients appearing in Equations (14)–(19) will be determined alphanumerically in numerical implementations.

4. Minimizing of the Reflection Coefficient

In the choice of the matching medium, one of the desired issues is the maximization of the incident wave energy that penetrates into the arm structure. This matter assures the interaction of the transmitted field with the arm tissues, which is of major significance for the microwave imaging purposes. A small value of amplitude reflection coefficient will certainly allow more energy to be transmitted into the tissue model [12]. In order to determine the lowest value of the reflection coefficient R_n, one can start from the radar cross section (RCS) definition. For the case of TM polarized incident plane wave, the bistatic cylinder RCS expression with far scattered field approximation is given by

$$\sigma(\phi) = \frac{2\lambda_0}{\pi}\left|\sum_{n=-\infty}^{\infty} R_n(f, \epsilon_{r\,mm}, \sigma_{mm})\, e^{in(\phi-\phi_0)}\right|^2 \quad (20)$$

where ϕ denotes the observation angle [27]. In this expression, R_n is assumed as a function depending on the operating frequency and the constitutive parameters of the matching medium, i.e., relative permittivity and conductivity, respectively. To reduce the total RCS of the cylindrical structure of the arm model, one can define a functional as an integral of the function $\sigma(\phi)$ with respect to the observation angle ϕ :

$$F(f, \epsilon_{r\,mm}, \sigma_{mm}) = \int_0^{2\pi} \sigma(\phi)\, d\phi = 4\lambda_0\left|\sum_{n=-\infty}^{\infty} R_n(f, \epsilon_{r\,mm}, \sigma_{mm})\right|^2 \quad (21)$$

While obtaining this functional, taken into account was the multiplication of the open form of the infinite series in Equation (20) by their conjugate. Regarding one of the three parameters, namely, frequency, relative permittivity and conductivity, as constant and the other two as variables at certain intervals, the image plot of the $F(f, \epsilon_{r\,mm}, \sigma_{mm})$ functional can be easily obtained. The parameter pair that minimizes the functional $F(f, \epsilon_{r\,mm}, \sigma_{mm})$ obtained from the image plot will be the sought-after solution of the problem.

5. Formulation of the 5-Layered Arm Model with an Internal Object

In order to investigate the differences between scattered and transmitted fields in healthy and unhealthy human arms, an unhealthy arm model with an abnormal region (e.g., bleeding, clot or tumor) is created by placing a dielectric structure $\mathcal{D}$ with a rectangular cross-section inside the concentric cylindrical arm model. The configuration is illustrated in Figure 1. Now, assume the unhealthy arm model is illuminated by a line source located outside the inhomogeneities. Let us introduce the field $u_0(x)$, which would be observed in the absence of the body $\mathcal{D}$. When the contribution of the unhealthy tissue $\mathcal{D}$ is denoted by $u_s(x)$, the total field $u(x)$ can be expressed at any point as the following sum:

$$u(x) = u_s(x) + u_0(x). \quad (22)$$

The

$$\nabla^2 u_s(x) + k_b^2\, u_s(x) = -k_b^2\, v(x)\, u(x), \tag{23}$$

where the function $v(x)$ is the so called *object function* and defined by

$$v(x) = \begin{cases} \frac{k^2(x)}{k_b^2} - 1, \text{ if } & x \in \mathcal{D} \\ 0, \text{ if } & x \notin \mathcal{D} \end{cases} \tag{24}$$

In this study, the object $\mathcal{D}$ representing the abnormal tissue is placed in the innermost layer of the cylindrical structure; therefore, in numerical implementations, $k(x)$ is arranged for a medium with the wave number k_5, as depicted in Figure 2. The scattering problem under consideration can be formulated equivalently to the problem of solving the Lippman–Schwinger integral equation

$$u(x) = u_0(x) + k_b^2 \int_{\mathcal{D}} G(x,y)\, v(y)\, u(y)\, dy, \quad x \in \mathcal{D} \tag{25}$$

for u, which is a second-kind Fredholm integral equation. The numerical solution of this equation can be obtained by a method of moments (MoM)-based technique given in [28]. For the numerical evaluation, the dielectric object $\mathcal{D}$ is divided first into M sufficiently small rectangular cells $\mathcal{D}_m$, $m = 1,2,\ldots,M$ assuming the field $u(x)$ constant over each cell. Consequently, the problem can be reduced to the solution of a system of linear equations for the unknown $u(x_n)$:

$$u(x_n) - k_5^2 \sum_{m=1}^{M} u(x_m)\, C_{nm} = u(x_n), \quad n = 1,2,\ldots,M \tag{26}$$

where

$$C_{nm} = v(y_n) \int_{\mathcal{D}_m} G(x_m, y_m) dy \tag{27}$$

is substituted. In the implementation of the MoM-based method, the integration of the singular part $H_0^{(1)}(k_b|x_n - y_m|)$ of the Green's function over the rectangular cells in $\mathcal{D}_m$ can be obtained by the equivalent circular cell approach given in [28] as follows:

$$C_{nm} = \begin{cases} 1 - \frac{i}{2}[\pi k_b a H_0^{(1)}(k_b a) + 2i]\, v(y_m), \; n = m \\ \frac{-i\pi k_b a}{2} J_1(k_b a) H_0^{(1)}(k_b|x_n - y_m|)\, v(y_m), \; n \neq m \end{cases} \tag{28}$$

where a is given by

$$a = \sqrt{\Delta x\, \Delta y / \pi} \tag{29}$$

denoting the radius of the equivalent circular region where Δx and Δy represent the side lengths of a rectangular cell.

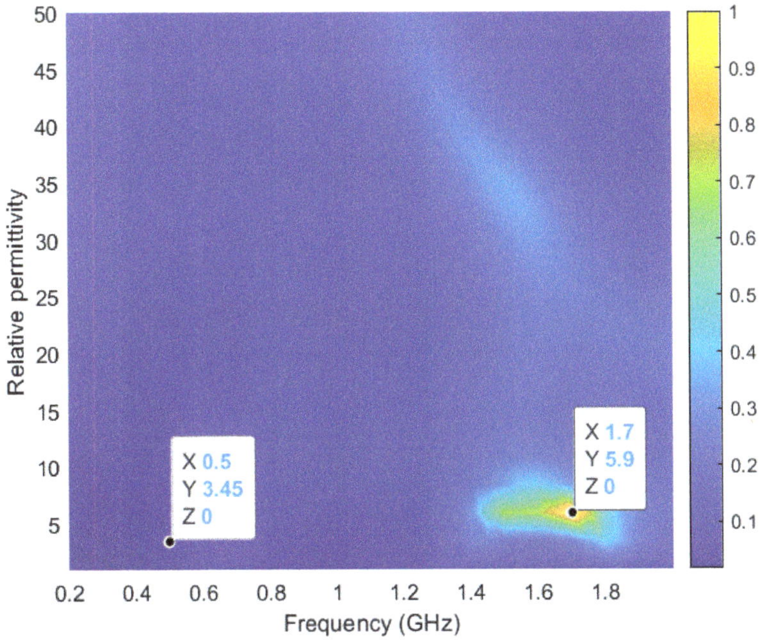

Figure 2. Amplitude variation of the functional $F(f, \epsilon_{r\,mm})$ with $\sigma_{mm} = 0$.

6. Results and Discussion

The numerical results are based on the computations of the analytically formulated scattered and transmitted fields for a radially stratified cylindrical human arm model outlined in the previous sections. The human arm model is illuminated by a line source located at the point $(x_1, x_2) = (0.13, 0)$ m at a single frequency. The scattered field outside the arm model is calculated on a circle of radius 0.3 m and inside the arm model on a circle of radius 0.025 m. An unhealthy arm model having bleeding or a tumor is simulated with an object $\mathcal{D}$ representing the abnormal region placed on the origin in the innermost layer. The object $\mathcal{D}$ is assumed to have a square cross-section and its wave number is taken as 1.5 times the wave number of the innermost layer, i.e., $k_{obj} = 1.5\,k_5$. The added object is chosen of the size 1×1 cm and it is divided into 10×10 cells. The first three examples are devoted to the analysis of the background, which is, in our study, taken as the matching medium, by taking into account its constitutive parameters, namely, its permittivity and conductivity, and the frequency of the source. Here, as a basic approach in our computations, one of the three parameters is supposed to be constant, and the amplitude variation of the functional $F(f, \epsilon_{r\,\mathrm{mm}}, \sigma_{\mathrm{mm}})$ depending on the other two parameters, which are defined on the appropriate ranges, is depicted as an image plot. Points selected from the relevant colored regions in the image plot will indicate the lowest and the highest amplitude values of the functional. To accurately simulate the electrical properties of the human arm tissues at the frequencies from 100 MHz to 2 GHz, the data generated using the three-pole Debye equation provided in Equation (1) are employed in numerical calculations.

6.1. *Analysis for Frequency and Permittivity (Conductivity Is Fixed)*

In the first example, the conductivity of the matching medium is taken as $\sigma_{\mathrm{mm}} = 0$. The image plot depicted in Figure 2 represents the amplitude variation of the functional $F(f, \epsilon_{r\,\mathrm{mm}})$ as a function of the frequency and relative permittivity of the matching medium.

The frequency is varying in the range from 0.2 to 2 GHz and the relative permittivity from 1 to 50, respectively. The points in the yellow-colored regions of the image plot display the frequency and the relative permittivity pairs that correspond to the highest reflected field coefficient values, while the darker blue-colored regions of the image plot comply with $(f, \epsilon_{r\,\mathrm{mm}})$ pairs and appropriate the lowest reflected field values. As labeled on Figure 2, two pairs with values $(f, \epsilon_{r\,\mathrm{mm}}) = (1.7\ \mathrm{GHz}, 5.9)$ from the yellow-colored region and $(f, \epsilon_{r\,\mathrm{mm}}) = (0.5\ \mathrm{GHz}, 3.45)$ from the darker blue-colored region are arbitrarily chosen and they comply with respective highest and lowest amplitude values of the functional $F(f, \epsilon_{r\,\mathrm{mm}})$. In order to illustrate the matching medium effect in terms of the penetrated field, two different background spaces regarding the values from Figure 2 are taken into account. Figure 3 shows graphically the amplitude changes of the total transmitted electric field versus observation angle in the innermost layer of the arm structure along a horizontal line through the origin depending on the respective matching medium parameters.

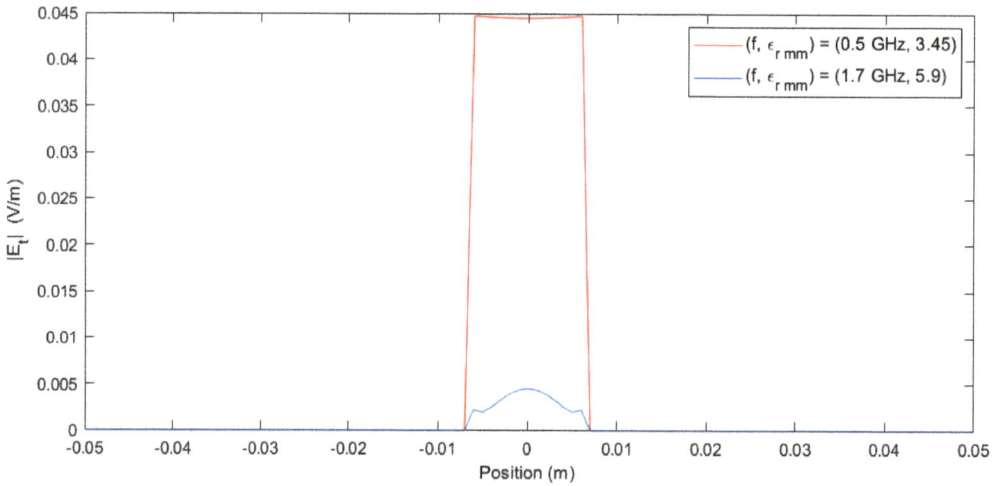

Figure 3. Comparison of the total transmitted field amplitude variations along the line $x_2 = 0$ in the innermost layer of the model for the $(f, \epsilon_{r\,\mathrm{mm}})$ pairs.

It is evident from Figure 3 that the presence of the matching medium owning appropriate parameters from Figure 2 has increased the amplitude of the penetrating electric field in the arm structure. To be able to give a more quantitative measure for the correspondence of two curves, as how close one curve is to the other curve, one of the possibilities is the comparison of the areas under these curves. When the definite integrals of the curves shown in Figure 3 are evaluated via the trapezoid method, the ratio of the areas under the curves gives the value 13.96, which indicates that the considerable part of the energy penetrates into the arm structure. Figure 4 shows the amplitude variation of the scattered field by the internal object placed in the innermost layer of the arm model in the presence and absence of the matching medium.

The scattered electric field $|E_s|$ is calculated outside the arm model on a circle of radius 0.3 m and inside the arm model on a circle of radius 0.025 m. The amplitude changes versus observation angle are plotted in Figure 4 adapting the values of the pair $(f, \epsilon_{r\,\mathrm{mm}}) = (0.5\,\mathrm{GHz}, 3.45)$ in the presence and $(f, \epsilon_{r\,\mathrm{mm}}) = (0.5\,\mathrm{GHz}, 1)$ in the absence of the matching medium, both values from Figure 2. As can be seen from Figure 4, the scattered field changes differ for different background spaces. In this example, when the dielectric permittivity of the matching medium increases, the scattered field amplitude decreases and, hence, the amount of the energy penetrating into the arm structure becomes larger.

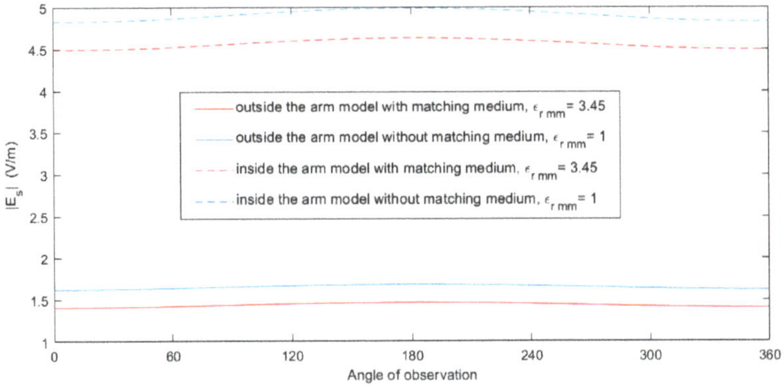

Figure 4. Total scattered field computed outside and inside the arm model having an internal object in presence and absence of the matching medium at $f = 0.5\,\text{GHz}$.

6.2. *Analysis for Frequency and Conductivity (Permittivity Is Fixed)*

In the second example, the relative permittivity of the background medium is chosen as $\epsilon_{r\,\text{mm}} = 40$. Figure 5 represents the image plot of the amplitude variation of the functional $F(f, \sigma_{\text{mm}})$ as a function of the frequency and conductivity of the matching medium, where the frequency is altering in a range from 0.8 to 2 GHz and the conductivity from 0.01 to 0.03 S/m.

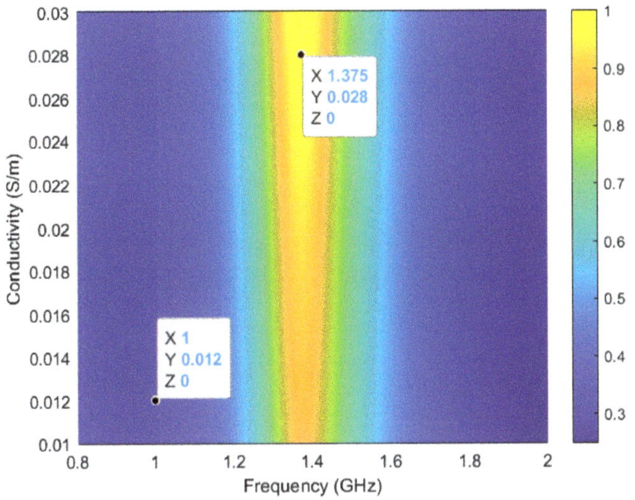

Figure 5. Amplitude variation of the functional $F(f, \sigma_{\text{mm}})$ with $\epsilon_{r\text{mm}} = 40$.

As can be seen from Figure 5, two pairs with values $(f, \sigma_{\text{mm}}) = (1.375\,\text{GHz}, 0.028\,\text{S/m})$ from the yellow-colored region and $(f, \sigma_{\text{mm}}) = (1\,\text{GHz}, 0.012\,\text{S/m})$ from the darker blue-colored region are chosen. These pairs are the corresponding respective highest and lowest amplitude values of the functional $F(f, \sigma_{\text{mm}})$.

Figure 6 represents the amplitude changes of the transmitted electric field in the innermost layer of the arm structure along the line $x_2 = 0$ depending on the respective matching medium parameters taken from Figure 5. The ratio of the areas under the curves plotted in Figure 6 yields in 4.67, from which one can conclude that a noticeable part of the field energy can enter the arm construct. The scattered electric field by the internal object located in the innermost layer of the arm structure is computed on the observation circles of radii 0.3 m and 0.025 m. The amplitude variations versus observation angle are depicted in Figure 7 using the pair $(f, \sigma_{mm}) = (1\,\text{GHz}, 0.012\,\text{S/m})$ from Figure 5, with $\epsilon_{rmm} = 40$ in the presence and $\epsilon_{rmm} = 1$ in the absence of the matching medium, respectively.

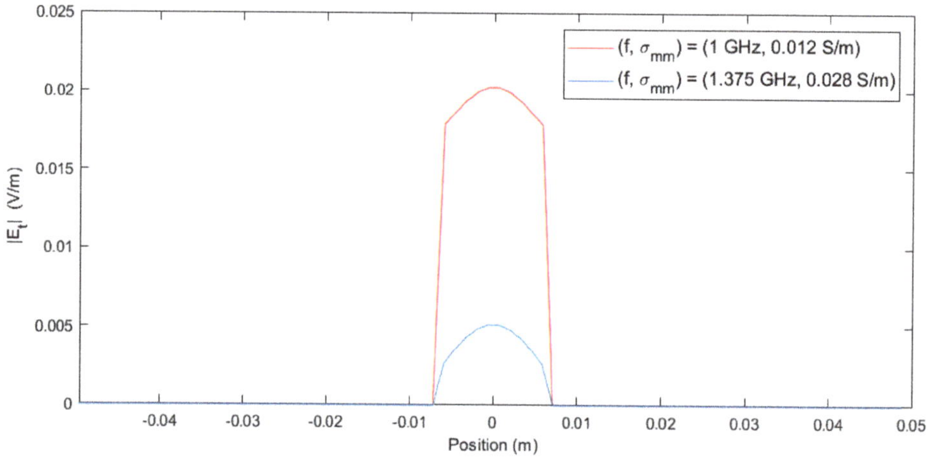

Figure 6. Comparison of the total transmitted field amplitude variations along the line $x_2 = 0$ in the innermost layer of the model for the (f, σ_{mm}) pairs.

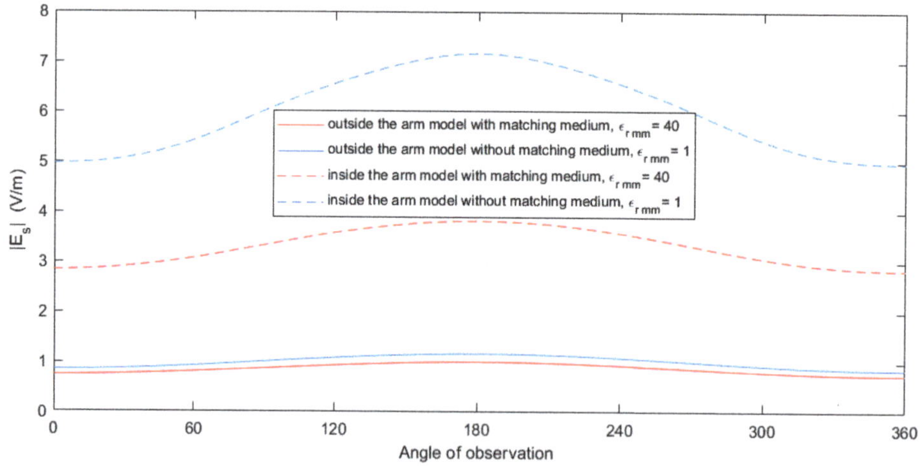

Figure 7. Total scattered field computed outside and inside the arm model having an internal object in presence and absence of the matching medium at $f = 1\,\text{GHz}$.

It is observed from Figure 7 that the scattered field changes both inside and outside the arm model vary for different background spaces.

6.3. Analysis for Permittivity and Conductivity (Frequency Is Fixed)

The final numerical example is devoted to the case where the amplitude variation of the functional $F(\epsilon_{r\,\mathrm{mm}}, \sigma_{\mathrm{mm}})$ depends on the relative permittivity and conductivity parameters of the matching medium, which is illustrated in Figure 8.

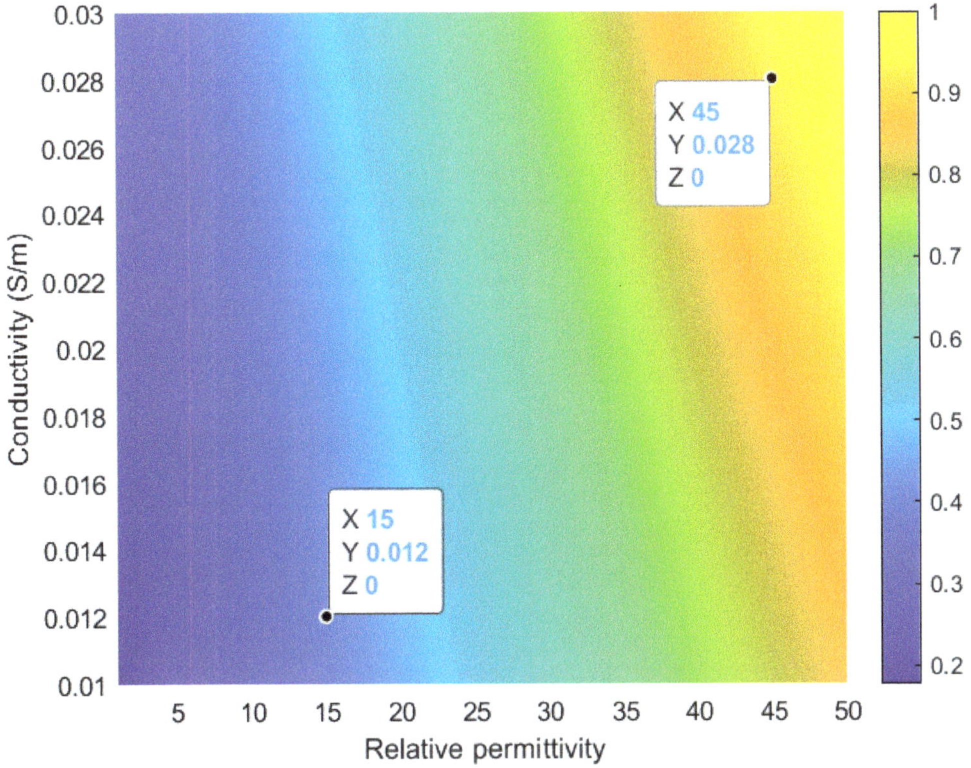

Figure 8. Amplitude variation of the functional $F(\epsilon_{r\,\mathrm{mm}}, \sigma_{\mathrm{mm}})$ with $f = 0.1\,\mathrm{GHz}$.

This image plot is obtained by assuming a constant frequency $f = 0.1\,\mathrm{GHz}$. The relative permittivity is varying in the range from 1 to 50 and the conductivity from $0.01\,\mathrm{S/m}$ to $0.03\,\mathrm{S/m}$, respectively. As labeled in Figure 8, two pairs reading $(\epsilon_{r\,\mathrm{mm}}, \sigma_{\mathrm{mm}}) = (45, 0.028\,\mathrm{S/m})$ from the yellow-colored region and $(\epsilon_{r\,\mathrm{mm}}, \sigma_{\mathrm{mm}}) = (15, 0.012\,\mathrm{S/m})$ from the darker blue-colored region, corresponding to respective highest and lowest amplitude values of the functional $F(\epsilon_{r\,\mathrm{mm}}, \sigma_{\mathrm{mm}})$, are chosen.

Variations of the transmitted electric field in the innermost layer of the arm model along the line $x_2 = 0$ are calculated using two parameter pairs from the image plot of Figure 8 and represented in Figure 9. The ratio of the areas under the transmittance curves presented in Figure 9 is 1.34, which demonstrates that the energy penetration into the arm structure is rather poor. Figure 10 shows the amplitude change of the scattered field by the internal object placed in the innermost layer of the arm model in the presence and absence of the matching medium.

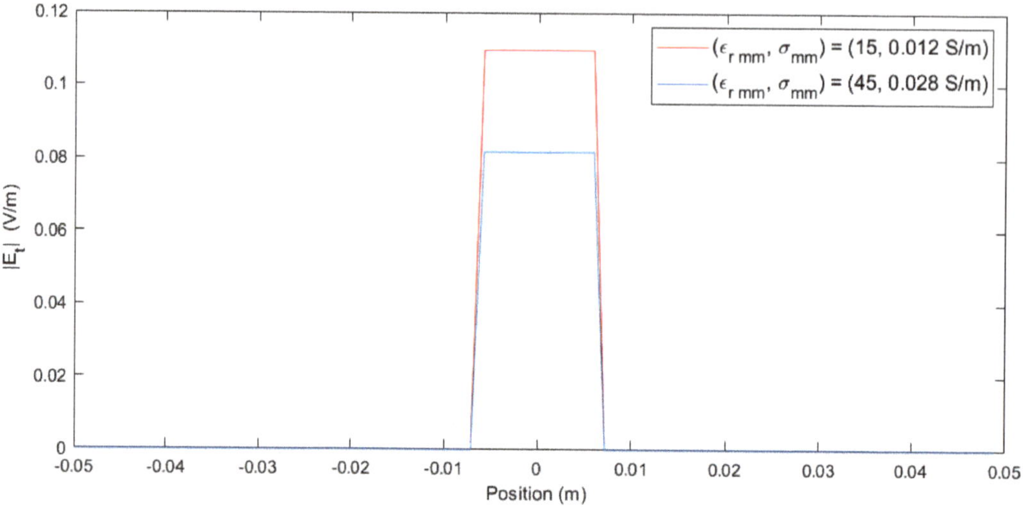

Figure 9. Comparison of the total transmitted field amplitude variations along the line $x_2 = 0$ in the innermost layer of the model for the $(\epsilon_{r\,\mathrm{mm}}, \sigma_{\mathrm{mm}})$ pairs.

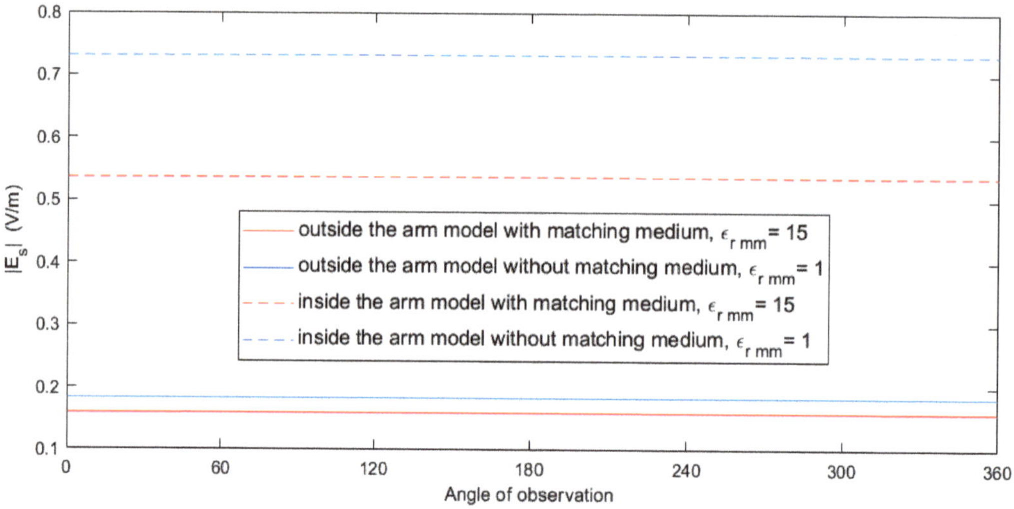

Figure 10. Total scattered field computed outside and inside the arm model having an internal object in presence and absence of the matching medium at $f = 0.1$ GHz.

The scattered electric field $|E_s|$ is calculated on the observation circles of radii 0.3 m and 0.025 m. The amplitude changes versus observation angle are plotted in Figure 10 adapting the values of the pair $(\epsilon_{r\,\mathrm{mm}}, \sigma_{\mathrm{mm}}) = (15, 0.012\,\mathrm{S/m})$ in the presence and $(\epsilon_{r\,\mathrm{mm}}, \sigma_{\mathrm{mm}}) = (1, 0.012\,\mathrm{S/m})$ in the absence of the matching medium, both from Figure 8. The scattered field changes both inside and outside the arm model are shown in Figure 10 and they are dissimilar for different background spaces.

7. Conclusions

In this study, which is essentially theoretically based, by exploiting a simplified human arm model, the factors affecting influence and effectiveness of matching media are analyzed.

The matching medium design was implemented numerically based on the analytical solution for a five-layered cylindrical human arm model. The numerical results have shown that it is inevitable to use a matching medium so that incident electromagnetic waves sufficiently penetrate and interact with the tissues in question. From the results obtained in the preceding analysis, one can draw some useful consequences. For example, the numerical results may contribute to the development of a microwave tomography device by providing the appropriate range of constitutive parameters and operating frequency. The results obtained have demonstrated the accuracy of the method and approach within itself. Undoubtedly, experimental verification of the results and the method of this study is an important problem.

Although this cylindrical model does not perfectly reflect the real human arm geometry, the advantage of this arm model is that it is suitable for analytical calculations. On the other hand, since this model is suitable for analytical solution, which is easy to implement and capable of producing fast solutions from the computational point of view, it can provide advantages over other existing methods [5,19]. It should also be mentioned as a final note that the results obtained by the presented approach can be compared by using commercial softwares, which is left as a future study.

Funding: This research received no external funding.

Institutional Review Board Statement: Not applicable.

Informed Consent Statement: Not applicable.

Data Availability Statement: Data available in a publicly accessible repository.

Conflicts of Interest: The author declares no conflicts of interest.

References

1. Gilmore, C.; Zakaria, A.; LoVerti, J.; Pistorius, S. A Study of Matching Liquid Fluid Loss in a Biomedical Tomography System. *Med. Phys.* **2013**, *40*, 023101. [CrossRef]
2. Lui, H.S.; Fhager, A.; Persson, M. Matching Medium for Biomedical Microwave Imaging. In Proceedings of the Internatinal Symposium on Antennas and Propagation (ISAP), Hobart, Australia, 9–12 November 2015; pp. 569–571, ISBN 978-4-8855-2302-1.
3. Catapano, I.; Donato, L.D.; Crocco, L.; Bucci, O.M.; Morabito, A.F.; Isernia, T.; Massa, R. On quantitative microwave tomography of female breast. *Prog. Electromagn. Res.* **2009**, *97*, 75–93. [CrossRef]
4. Scapaticci, R.; Donato, L.D.; Catapano, I.; Crocco, L. A Feasibility Study on Microwave Imaging for Brain Stroke Monitoring. *Prog. Electromagn. Res.* **2012**, *40*, 305–324. [CrossRef]
5. Gilmore, C.; Zakaria, A.; Pistorius, S.; LoVerti, J. Microwave Imaging of Human Forearms: Pilot Study and Image Enhancement. *Int. J. Biomed. Imaging* **2013**, *2013*, 673027. [CrossRef] [PubMed]
6. Amin, B.; Shahzad, A.; Crocco, L.; Wang, M.; Halloran, M.O.; Gonzalez-Suarez, A.; Elahi, M.A. A Feasibility Study on Microwave Imaging of Bone for Osteoporosis Monitoring. *Med. Biol. Eng. Comput.* **2021**, *59*, 925–936. [CrossRef] [PubMed]
7. Jofre, L.; Hawley, M.S.; Broquetas, A.; Reyes, E.D.; Ferrando, M.; Elias-Fuste, A.R. Medical imaging with a microwave tomographic scanner. *IEEE Trans. Biomed. Eng.* **1990**, *37*, 303–312. [CrossRef]
8. Meaney, P.M.; Pendergrass, S.A.; Fanning, M.W.; Li, D.; Paulsen, K.D. Importance of using a reduced contrast coupling medium in 2D microwave breast imaging. *J. Electromagn. Wave Appl.* **2003**, *17*, 333–355. [CrossRef]
9. Meaney, P.M.; Fox, C.J.; Geimer, S.D.; Paulsen, K.D. Electrical Characterization of Glycerin: Water Mixtures: Implications for Use as a Coupling Medium in Microwave Tomography. *IEEE Trans. Microw. Theory Techn.* **2017**, *65*, 1471–1478. [CrossRef] [PubMed]
10. Koutsoupidou, M.; Groumpas, E.; Karanasiou, I.S.; Christopoulou, M.; Nikita, K.; Uzunoglu, N. The Effect of Using a Dielectric Matching Medium in Focused Microwave Radiometry: An Anatomically Detailed Head Model Study. *Med. Biol. Eng. Comput.* **2017**, *56*, 809–816. [CrossRef] [PubMed]
11. Lui, H.S.; Fhager, A. On the Matching Medium for Microwave-Based Medical Diagnosis. *Biomed. Phys. Eng. Express* **2017**, *4*, 035015. [CrossRef]
12. Lui, H.S.; Fhager, A. On the Matching Medium for Microwave Stroke Diagnosis. *Biomed. Phys. Eng. Express* **2019**, *6*, 045020. [CrossRef]

13. Alkhodari, M.; Zakaria, A.; Qaddoumi, N. Monitoring Bone Density Using Microwave Tomography of Human Legs: A Numerical Feasibility Study. *Sensors* **2021**, *21*, 7078. [CrossRef]
14. Faktorova, D.; Istenikova, K. Increasing of Microwave Energy Focusing Effectiveness by Using Metamaterial Structures. In Proceedings of the The International Conference on Digital Technologies 2013, Zilina, Slovakia, 29–31 May 2013; pp. 81–83. [CrossRef]
15. Azizi, M.S.N.; Salleh, A.; Othman, A.; Mohamad, N.R.; Aris, N.A.; Abu, M.; Hassan, N.; Ramli, M.H.; Ngelayang, T.B. UWB Rectangular Slotted Microstrip Patch Antenna in Proximity of Human Arm Model. *J. Teknol.* **2016**, *78*, 111–115. [CrossRef].
16. Yeh, C.; Lindgren, G. Computing the propagation characteristics of radially stratified fibers: An efficient method. *Appl. Opt.* **1972**, *16*, 483–493. [CrossRef] [PubMed]
17. Jensen, M.A.; Freeze, J.D. A recursive Green's function method for boundary integral analysis of inhomogeneous domains. *IEEE Trans. Antennas Propag.* **1998**, *46*, 1810–1816. [CrossRef]
18. Kai, L.; D'Alessio, A. Finely stratified cylinder model for radially inhomogeneous cylinders normally irradiated by electromagnetic plane waves. *Appl. Opt.* **1995**, *34*, 5520–5530. [CrossRef] [PubMed]
19. Ito, K.; Hotta, Y. Signal Path Loss Simulation of Human Arm for Galvanic Coupling Intra-Body Communication. *J. Adv. Simulat. Sci. Eng.* **2016**, *3*, 29–46. [CrossRef]
20. Song, Y.; Hao, Q.; Zhang, K.; Wang, M.; Chu, Y.; Kang, B. The Simulation Method of the Galvanic Coupling Intrabody Communication with Different Signal Transmission Paths. *IEEE Trans. Instrum. Meas.* **2011**, *60*, 1257–1266. [CrossRef]
21. Wegmueller, M.S.; Kuhn, A.; Froehlich, J.; Oberle, M.; Felber, N.; Kuster, N.; Fichtner, W. An Attempt to Model the Human Body as a Communication Channel. *IEEE Trans. Biomed. Eng.* **2007**, *54*, 1851–1857. [CrossRef] [PubMed]
22. Cruciani, S.; Santis, V.D.; Feliziani, M.; Maradei, F. Cole-Cole vs. Debye Models for the Assessment of Electromagnetic Fields Inside Biological Tissues Produced by Wide Band EMF Sources. In Proceedings of the IEEE Asia-Pacific Symposium on Electromagnetic Compatibility (APEMC), Singapore, 21–24 May 2012; pp. 685–688. [CrossRef]
23. Lumnitzer, R.; Tanner, A.; Elsherbeni, A.Z. Debye Coefficients for Biological Tissues from 100 MHz to 100 GHz. In Proceedings of the IEEE International Applied Computational Electromagnetics Society Symposium (ACES), Monterey, CA, USA, 22–26 March 2020; pp. 1–2. [CrossRef]
24. Lumnitzer, R. Energy Harvesting Near the Human Body Using Finite-Difference Time-Domain Simulations. Ph.D. Thesis, Colorado School of Mines, Golden, CO, USA, 2023; p. 30421532.
25. Harrington, R.F. *Time-Harmonic Electromagnetic Fields*; IEEE Press Series on Electromagnetic Wave Theory; IEEE Press: Piscataway, NJ, USA, 2001; ISBN 0-471-20806-X.
26. Dudley, D.G. *Mathematical Foundations for Electromagnetic Theory*; IEEE Press Series on Electromagnetic Waves; IEEE Press: Piscataway, NJ, USA, 1994; ISBN 0-7803-1022-5.
27. Yelkenci, T. Determination of the Dielectric Coating Parameters to Reduce the Radar Cross-section of the Perfectly Conducting Cylinder. In Proceedings of the 14th International Conference on Electrical and Electronics Engineering (ELECO), Bursa, Turkey, 30 November–2 December 2023. [CrossRef]
28. Richmond, J.H. Scattering by a Dielectric Cylinder of Arbitrary Cross Section Shape. *IEEE Trans. Antennas Propag.* **1965**, *13*, 334–341. [CrossRef]

applied sciences

MDPI

Article

Nonlinear Analysis and Solution for an Overhead Line Magnetic Energy Harvester with an Active Rectifier

Alexander Abramovitz *, Moshe Shvartsas and Alon Kuperman *

School of Electrical and Computer Engineering, Ben-Gurion University of the Negev, P.O. Box 653, Beer-Sheva 84105, Israel; meishe@bgu.ac.il

* Correspondence: alexabr@tauex.tau.ac.il (A.A.); alonk@bgu.ac.il (A.K.)

Abstract: Recently, there has been a significant focus on developing various energy harvesting technologies to power remote electronic sensors, data loggers, and communication devices for smart grid systems. Among these technologies, magnetic energy harvesting stands out as one straightforward method to extract substantial power from current-carrying overhead lines. Due to the relatively small size of the harvester, the high currents in the distribution system quickly saturate its magnetic core. Consequently, the magnetic harvester operates in a highly nonlinear manner. The nonlinear nature of the downstream AC to DC converters further complicates the process, making precise analytical modeling a challenging task. In this paper, a clamped type overhead line magnetic energy harvester with a controlled active rectifier generating significant DC output power is investigated. A piecewise nonlinear analytical model of the magnetic harvester is derived and reported. The modeling approach is based on the application of the Froelich equation. The chosen approximation method allowed for a complete piecewise nonlinear analytical treatise of the harvester's behavior. The main findings of this study include a closed-form solution that accounts for both the core and rectifiers' nonlinearities and provides an accurate quantitative prediction of the harvester's key parameters such as the transfer window width, optimal pulse location, average DC output current, and average output power. To facilitate the study, a nonlinear model of the core was developed in simulation software, based on parameters extracted from core experimental data. Furthermore, theoretical predictions were verified through comparison with a computer simulation and experimental results of a laboratory prototype harvester. Good agreement between the theoretical, simulation, and experimental results was found.

Keywords: magnetic cores; analytical models; saturation magnetization; magnetic flux; magnetic energy harvesting

1. Introduction

Electrical power transmission systems are, perhaps, the world's largest technical installations. Transmission lines span over vast distances, often pass through remote and hardly accessible locations, and are prone to the forces of nature. To alleviate the task of inspection of power systems, monitoring equipment based on smart sensors [1–3], remote cameras [4,5], etc., have recently been developed. To power such electronic devices, several types of magnetic energy harvesters have been proposed [6]. Compared to other energy harvesting methods, a magnetic energy harvester is a mechanically rugged device that has the advantage of higher reliability and greater power density.

An overhead line magnetic energy harvester (OLMEH) is designed to clamp onto a transmission line and convert the magnetic energy generated by the line current into DC power. The OLMEH aims to power electronic equipment or, in most cases, charge a battery. The concept of an OLMEH is similar to that of a current transformer (CT) [7]. Yet, the operational conditions of a CT and an OLMEH are quite different. The CT is operated under a short circuit (or very low impedance) load to keep the core within the linear region, whereas the OLMEH operates with a constant voltage load. To exploit the full range of B-H characteristics, OLMEH's core is usually allowed to run into the saturation regions. The inherent nonlinearity of the magnetic core makes the OLMEH's analysis and design a challenging problem [6–9]. Furthermore, extracting the full available power that the OLMEH can provide requires properly matching the OLMEH parameters to attain the maximum power point conditions [10].

An additional advantage of an OLMEH is that it can harvest the energy from a multi-kV level AC transmission line to power electronic devices operating at a low DC voltage of only a few volts [11–14] while avoiding the application of bulky high-voltage step-down transformers, thus saving hardware and installation costs.

To derive the DC voltage, an OLMEH requires a suitable power electronics interface, which can be either passive or active. Passive OLMEHs utilize uncontrolled diode bridge rectifiers to implement the AC-DC conversion, offering simplicity and higher reliability [10,15]. To maximize the power density, the OLMEH should be matched to the load (typically of a constant voltage type) by appropriate selection of cross-sectional area of the core and number of secondary winding turns. Therefore, such an approach is referred to as "maximum power point matching" (MPPM) and allows for maximizing the harvested OLMEH power for a wide range of primary current magnitudes [16]. Yet, in cases where the CVL is a battery, its voltage is expected to vary depending on its charging state. Thus, only near-optimal matching can be attained.

Active rectification is another viable approach to attain DC power at OLMEH's output. One possible configuration of an OLMEH-based charger with an active rectifier operating under a constant voltage load (CVL) is illustrated in Figure 1. The idea behind this circuit is to short-circuit the OLMEH's secondary winding so to preserve the stored magnetic energy and then release it towards the load under optimal conditions. This helps to increase the power output of the circuit, allows control over the depth of saturation, and, enables the implementation of overcharge and overcurrent protection features. Moreover, the active rectifier allows the application of better maximum power point tracking (MPPT) [17–19], which can help attain a higher power under variable line current conditions.

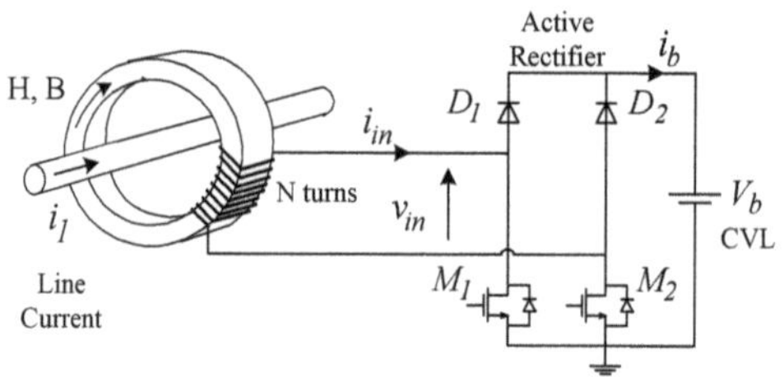

Figure 1. Overhead line magnetic energy harvester with an active rectifier feeding a constant voltage load.

The circuit in Figure 1 was investigated in the recent literature [19–22], which suggests several approaches to deal with the challenging problem of core nonlinearity. For instance, its earlier counterpart [19] applied a cot(*) function to accurately model core nonlinearity. Yet, due to the complex nature of such an approximating function, after a few theoretical steps, the final solutions were obtained only numerically.

This article aims to develop an approximated, yet complete, nonlinear analytical model and solution for an overhead line magnetic energy harvester with an active rectifier (OLMEH-AR), under a constant voltage load, as shown in Figure 1.

The proposed model is derived by applying piecewise nonlinear analysis, relying on the Froelich equation to model the B-H characteristics of the magnetic core. The Froelich approximation is based on a first-order rational function. The advantage of the offered approach is that it lends itself to the analytical treatise. The result is a complete closed-form analytical solution to the problem that is quite accurate, simple, and suited for engineering applications.

The rest of the paper is as follows: simulated waveforms, state analysis, and derivation of equivalent circuits are given in Section 2; also, this section presents the analytical model of the OLMEH-AR and derives the expressions for the power transfer window, output current, and the output power and the conditions for maximum power. Experimental measurements of a laboratory prototype are reported in Section 3 and stand in good agreement with the proposed theoretical predictions.

2. The Applied Methodology

2.1. State Analysis and Equivalent Circuits

Analysis of the OLMEH-AR started with computer simulation (PSIM v. 9.1). Derivation of the computer model of the magnetic core in PSIM v9.1 software is a tedious trial-and-error process; however, the computer simulation model proved to be an indispensable tool in the undertaken study.

Examination of the simulated waveforms of the OLMEH-AR in Figure 2 reveals that its operation period comprises two symmetrical half cycles. The positive half-cycle commences when the secondary current becomes positive, $i_2(t) > 0$ and comprises four states. Note that during the negative half-cycle, a similar order of events occurs (with M_2 interchanging its function with M_1 and D_2 interchanging its function with D_1). Therefore, it is sufficient to analyze only the positive half-cycle.

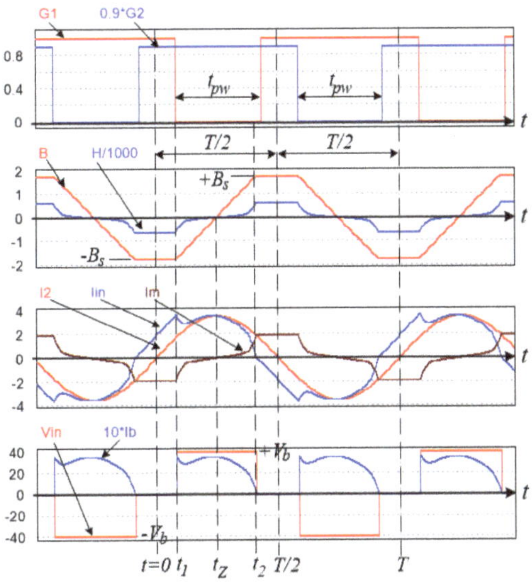

Figure 2. Typical simulated waveforms of an OLMEH-AR under CVL.

State 1 $[0, t_1]$, see Figure 3a, commences at $t = 0$. Here, both active switches conduct and effectively short the secondary winding.

State 2 $[t_1, t_2]$, see Figure 3b, commences at $t = t_1$, when, by the controller's command, the M_1 switch is turned off but M_2 remains on. Here, the rectifier diode D_1 turns on and allows the current to charge the battery. Since a constant battery voltage is applied to the secondary winding, the flux density in the core starts ramping up linearly, while causing the magnetizing current to rise in a nonlinear fashion.

State 3 $[t_2, t_1 + t_{pw}]$, see Figure 3c, commences at $t = t_2$ when the rectifier currents drop to zero, $i_{in} = i_b = 0$. Here, the anti-parallel diode of the M_1 switch starts conducting and shorts the secondary winding.

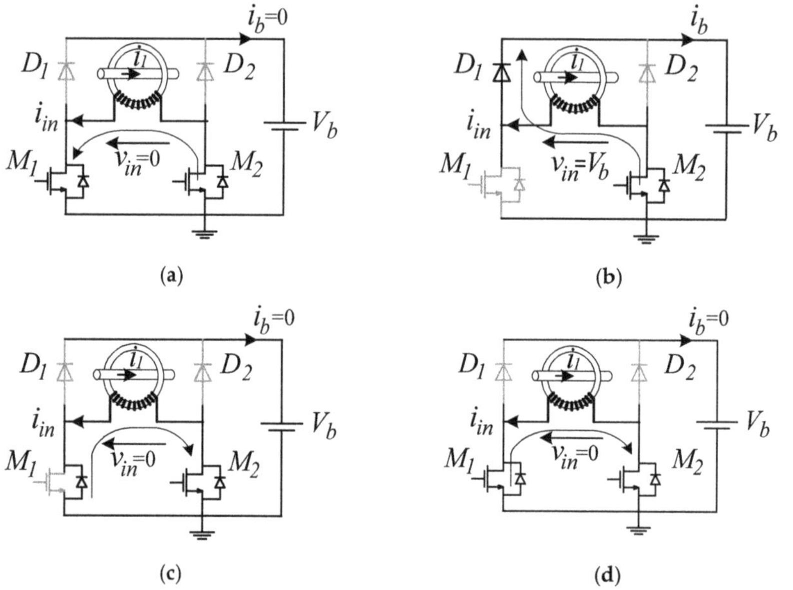

Figure 3. Semiconductor conduction states during State 1 (**a**); State 2 (**b**); State 3 (**c**); and State 4 (**d**).

State 4 $[t_1 + t_{pw}, T/2]$, see Figure 3d, commences when, by the controller's command, switch M_1 is turned on again and helps its anti-parallel diode to keep the winding shorted. Under short circuit conditions, the flux density in the core remains constant until the end of the positive half cycle.

The OLMEH-AR model based on a simplified cantilever model with a saturable magnetizing inductance, L_m, placed on the secondary winding, while neglecting the leakage inductance, is shown in Figure 4a. The ideal transformer $1 : N$ acknowledges the fact that the line conductor acts as a single-turn primary winding, while the secondary one consists of the N turn wound on the OLMEH's core. The simplified equivalent circuit of the short-circuit (SC) state (States 1, 3, and 4) is shown in Figure 4b, and the simplified model of State 2, the only charging state (CS) that allows current to the load, is presented in Figure 4c.

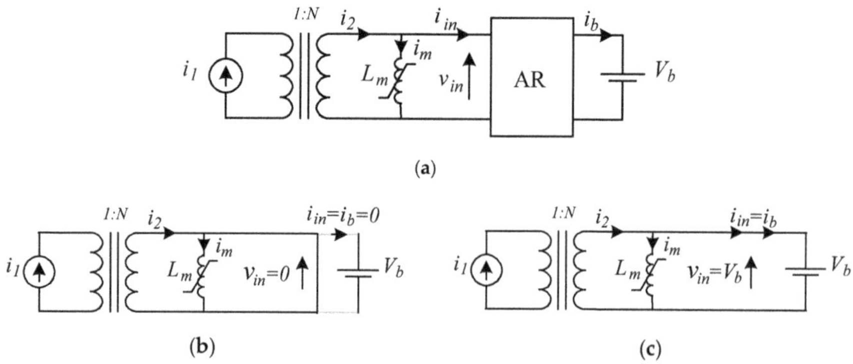

Figure 4. General model of the overhead line magnetic energy harvester with an active rectifier (OLMEH-AR) with a constant voltage load (**a**); its equivalent short circuit States 1, 3, and 4 (**b**) and the equivalent charging State 2 (**c**).

2.2. The Froelich Equation Review

To facilitate the analysis approach, the following simplifying assumptions are adopted.

Firstly, ideal (lossless) rectifiers and ideal switches are assumed with no voltage drop and zero ON resistance. Secondly, the core is assumed to be un-hysteretic (lossless), whose nonlinearity can be described by the Froelich equation:

$$B(H) = \frac{aH}{b+H}, \tag{1}$$

Here, the parameter a is the saturation flux density at infinite field intensity:

$$B_{sat} = \lim_{H\to\infty} \frac{aH}{b+H} = a, \tag{2}$$

whereas b is related to the magnetic permeability of the core at the low field intensity (linear) region

$$\mu(H \ll b) = \frac{a}{b} \tag{3}$$

2.3. Analysis: Power Transfer Window

The following definitions are introduced: N is the number of OLMEH secondary turns, I_{1m} is the line (primary) current amplitude, and $I_{2m} = I_{1m}/N$ is the amplitude of the secondary current. As usual, $f = 1/T = \omega/2\pi$ is the line frequency.

Examining the waveforms in Figure 2 reveals that the magnetic flux density swings from its initial negative saturation value $B(t_1) = -B_s$ towards the positive saturation value $B(t_2) = +B_s$. The transition is linear since it results from the application of the constant voltage V_b to the secondary winding; see Figures 3b and 4c. Therefore

$$B(t_2) = +B_s = \frac{1}{NA_c}\int_{t_1}^{t_2} V_b dt - B_s, \tag{4}$$

whence

$$B_s = \frac{V_b}{2NA_c}(t_2 - t_1), \tag{5}$$

The practical saturation level that the core can reach, B_s, is presumed to be lower than the heavy saturation flux density level, $B_s < B_{sat}$, defined by (2).

The time interval that appears in (5)

$$T_w = (t_2 - t_1), \tag{6}$$

is defined as the power transfer window. To allow the charging current pulse to terminate naturally, the pulse width, t_{pw}, set by the controller, see Figure 2, has to be appropriately matched so that $t_{pw} > T_w$. Switching the MOSFET on, see Figure 3d, decreases the resistance of the conductive path, minimizes the conduction losses during the SC states, and so increases the OLMEH's conversion efficiency.

Combining (5) with the Froelich Equation (1) yields the magnetic field intensity attained by the core at the peak of saturation:

$$H(t_2) = H(B_s) = \frac{bB_s}{a - B_s} = \frac{b\frac{V_b}{2NA_c}(t_2 - t_1)}{a - \frac{V_b}{2NA_c}(t_2 - t_1)} = \frac{N}{l_c} i_m(t_2), \tag{7}$$

The right-hand side of (7) was attained by Ampere's law.

Further examination of the waveforms in Figure 2 suggests that charging State 2 is terminated when the magnetizing current equals the secondary current:

$$i_m(t_2) = i_2(t_2) = I_{2m}\sin(\omega t_2), \tag{8}$$

Hence, substituting (8) into (7) yields

$$\frac{b\frac{V_b}{2NA_c}(t_2 - t_1)}{a - \frac{V_b}{2NA_c}(t_2 - t_1)} = \frac{N}{l_c} I_{2m} \sin(\omega t_2), \quad (9)$$

Since t_1 is set by the controller and assumed to be known, (9) is an equation of a single variable t_2. Yet, (9) is difficult to solve. A straightforward approach to obtain the solution of (9) is to proceed with numerical analysis. An alternative approach offered here is obtaining an analytical solution through approximation of the sine function on the right-hand side of (9) by a rational function, considering that the range of interest lies somewhere in $\frac{2}{3}\pi < \omega t < \frac{5}{6}\pi$. As a compromise between complexity and accuracy, it is suggested to approximate the falling segment of the sine function using

$$\sin(\omega t) \approx \frac{K_1(\pi - \omega t)}{K_2 - \omega t}, \quad (10)$$

With $K_1 = 3.232$ and $K_2 = 6.003$, (10) agrees with the $\sin(\omega t)$ function at 120°, 150°, and 180°, thus providing good accuracy within the range of interest. A comparison plot of the $\sin(\omega t)$ function to the chosen approximation function (10) is shown in Figure 5.

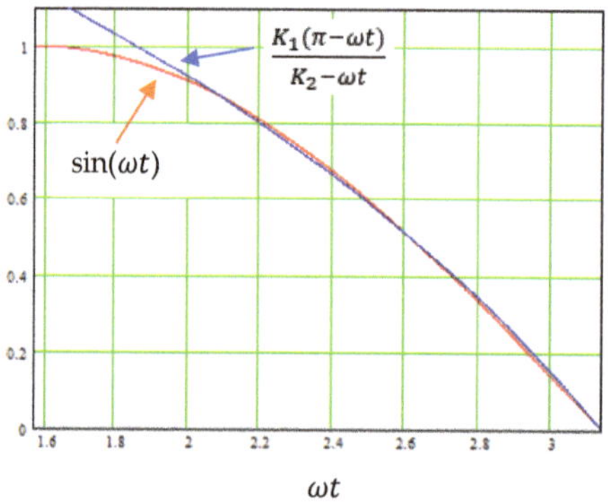

Figure 5. Comparison plot of the $\sin(\omega t)$ function to the chosen approximation function (10) (for K_1 = 3.232 and K_2 = 6.003).

Substitution of (10) into (9) yields

$$\frac{l_c}{N} \frac{b\left[\frac{V_b}{NA_c}(t_2 - t_Z)\right]}{a - \left[\frac{V_b}{NA_c}(t_2 - t_Z)\right]} = I_{2m} \frac{K_1(\pi - \omega t_2)}{K_2 - \omega t_2}, \quad (11)$$

Introducing the definitions of the normalized time

$$t_N = \frac{\omega}{\pi} t, \quad (12)$$

the normalized secondary current

$$I_N = \frac{I_{2m}}{I_{base}} = \frac{N I_{2m}}{b l_c} = \frac{I_{1m}}{b l_c}, \quad (13)$$

the normalized load voltage

$$V_N = \frac{V_b}{V_{base}} = \frac{\pi V_b}{\omega a N A_c}, \quad (14)$$

and applying (12)–(14) to (11) yields the normalized equation:

$$\frac{V_N(t_{2N}-t_{1N})/2}{1-V_N(t_{2N}-t_{1N})/2}=I_N\frac{K_1\pi(1-t_{2N})}{K_2-\pi t_{2N}}, \tag{15}$$

which, after some manipulation, can be written in the form of a standard quadratic equation:

$$\alpha t_{2N}^2+\beta t_{2N}+\gamma=0, \tag{16}$$

where the coefficients are

$$\alpha=\frac{\pi}{2}V_N(1+K_1I_N), \tag{17}$$

$$\beta=-\left[\pi K_1I_N+\frac{\pi}{2}K_1(1+t_{1N})V_NI_N+\frac{1}{2}(\pi t_{1N}+K_2)V_N\right], \tag{18}$$

$$\gamma=\pi K_1I_N+\frac{\pi}{2}K_1t_{1N}V_NI_N+\frac{K_2}{2}t_{1N}V_N, \tag{19}$$

Only the solution for $t_{2N}<1$ (i.e., $t_2<T/2$) has a physical meaning, therefore:

$$t_{2N}=\frac{-\beta-\sqrt{\beta^2-4\alpha\gamma}}{2\alpha}, \tag{20}$$

Translating (20) from the normalized time to the real-time, see (12), gives the termination instant of charging State 2:

$$t_2=\frac{T}{2}t_{2N}, \tag{21}$$

Since the firing instant t_1 is set by the controller and is known and the extinction instant t_2 can be obtained from (21), the power transfer window can be found by (6). This allows for the calculation of the output power of the OLMEH-AR as shown next.

2.4. Analysis: Output Current and Power

The rectifier's input current is the difference between the secondary current, $i_2(t)$, and the magnetizing current, $i_m(t)$; see Figure 4c. Therefore, the charge delivered to the CVL during a half cycle is

$$q_b=\int_{t_1}^{t_2}(i_2(t)-i_m(t))dt, \tag{22}$$

Examining the waveforms in Figure 2 reveals that in the steady-state, the magnetizing current is symmetrical. This means that the charge drawn by the magnetizing current is zero:

$$\int_{t_1}^{t_2}i_m(t)dt=0, \tag{23}$$

This also complies with the principle that for the lossless system in the steady state, the energy captured by an ideal magnetizing inductance is fully recycled.

Combining (22) and (23), the average output current can be found as:

$$I_b=\frac{2}{T}\int_{t_1}^{t_2}I_{2m}\sin(\omega t)dt=\frac{I_{2m}}{\pi}(\cos(\omega t_1)-\cos(\omega t_2)), \tag{24}$$

whence the average output power delivered to the constant voltage load is

$$P_o=V_bI_b=\frac{2\sqrt{2}}{\pi N}V_bI_{1rms}(\cos(\omega t_1)-\cos(\omega t_2)), \tag{25}$$

2.5. Analysis: Maximum Power Conditions

Recall again that the delay time, t_1, in (25) is an independent variable set by the controller's command, whereas t_2, given by (20), (21) is a function of t_1 as well as the

OLMEH's parameters and operation conditions. Thus, the question arises regarding the proper choice of the delay time to optimize/maximize the OLMEH's power output. Symmetry considerations, see Appendix A for details, suggest that the maximum output power can be attained when the power transfer window is aligned symmetrically to the line current peak:

$$t_{2_{mpp}} = \frac{T}{2} - t_{1_{mpp}}, \tag{26}$$

Substituting (26) into (25), the maximum power that can be obtained from the OLMEH-AR under the MPP condition is

$$P_{ompp} = \frac{4\sqrt{2}}{\pi N} V_b I_{1rms} \cos(\omega t_{1mpp}), \tag{27}$$

Whereas the optimum delay time, $t_{1_{mpp}}$, required to implement the MPP condition is found by normalizing the condition (26) and substituting it into (15). Further manipulation of the expression leads to the following quadratic equation:

$$\alpha_{mpp} t_{1Nmpp}^2 + \beta_{mpp} t_{1Nmpp} + \gamma_{mpp} = 0, \tag{28}$$

where

$$\alpha_{mpp} = -2\pi V_N (1 + K_1 I_N), \tag{29}$$

$$\beta_{mpp} = -[(2K_2 - 3\pi)V_N + \pi K_1 (2 - V_N) I_N], \tag{30}$$

$$\gamma_{mpp} = (K_2 - \pi) V_N, \tag{31}$$

Only the positive solution for t_{1Nmpp} has a physical meaning:

$$t_{1Nmpp} = \frac{-\beta_{mpp} - \sqrt{\beta_{mpp}^2 - 4\alpha_{mpp}\gamma_{mpp}}}{2\alpha_{mpp}}, \tag{32}$$

Translating the normalized time (32) to the real-time yields:

$$t_{1mpp} = \frac{T}{2} t_{1Nmpp}, \tag{33}$$

The results of (32), and (33) allow for optimally locating the power transfer window for optimal energy harvesting; see (27).

3. Comparison with the Experimental Results

An experimental laboratory OLMEH prototype, as shown in Figure 1, and its power processing unit were built and tested. The OLMEH's core was constructed by stacking together three pairs of 10H10 C-cores made of silicon steel (EILOR MAGNETIC CORES). The magnetic path length was l_c = 120 mm, and the total core cross-section of A_c = 1800 mm^2. The total number of secondary turns was N = 40.

NTP011N115MC MOSFETs and Schottky rectifiers STP80H100TV were used. The prototype controller was developed and built to allow for adjustment of the delay time, t_1, and the controller's power transfer window width, t_{pw}, independently.

The Froelich parameters of the core a and b were established by curve fitting to the experimental data. Measured B-H characteristics of a silicon steel core sample vs. the fitted BH curve, drawn by using the extracted Froelich parameters, are shown in Figure 6. Here, a = 2.05 [T] and b = 109.4 [At/m] were found, the same as in [10].

Typical waveforms of the experimental OLMEH with an active rectifier at line current I_1 = 100 A rms under constant voltage load, V_b = 45 V, are shown in Figure 7. Here, the waveforms shown are: line current $i_1(t)$, the rectifier input current $i_{in}(t)$, and the gating signal of the M1 switch, $v_g(t)$, generated by the controller. The math channel shows the instantaneous output power $p_o(t)$. Recall that energy transfer to the CVL occurs during the high level of the $v_1(t)$ waveform. Note that the primary voltage, $v_1(t)$, was measured across the line conductor segment of only about 10 cm long. Also, it is worth noting that (during the positive half cycle), the saturation of the core manifests itself by the rapidly falling edge of the $i_{in}(t)$ current. The energy stored by the magnetizing inductor manifests itself by the offset segment seen in the $i_{in}(t)$ waveform, when the $v_1(t)$ is at its zero level. The release of the stored energy is manifested by the down-going step in the $i_{in}(t)$ waveform.

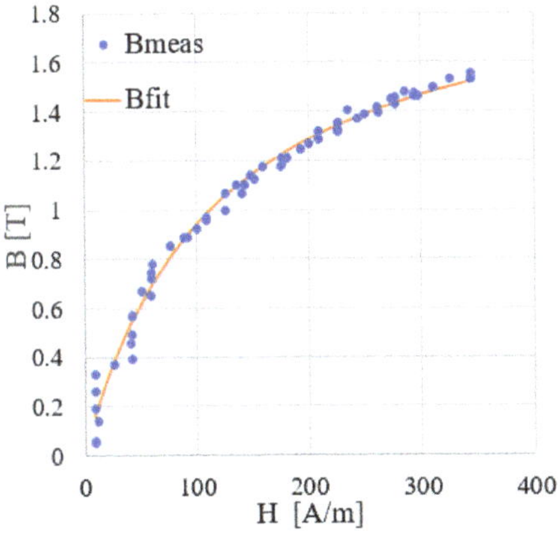

Figure 6. Comparison of the measured BH curve of the silicon steel core sample (EILOR MAGNETIC CORES) to its Froelich approximation [10].

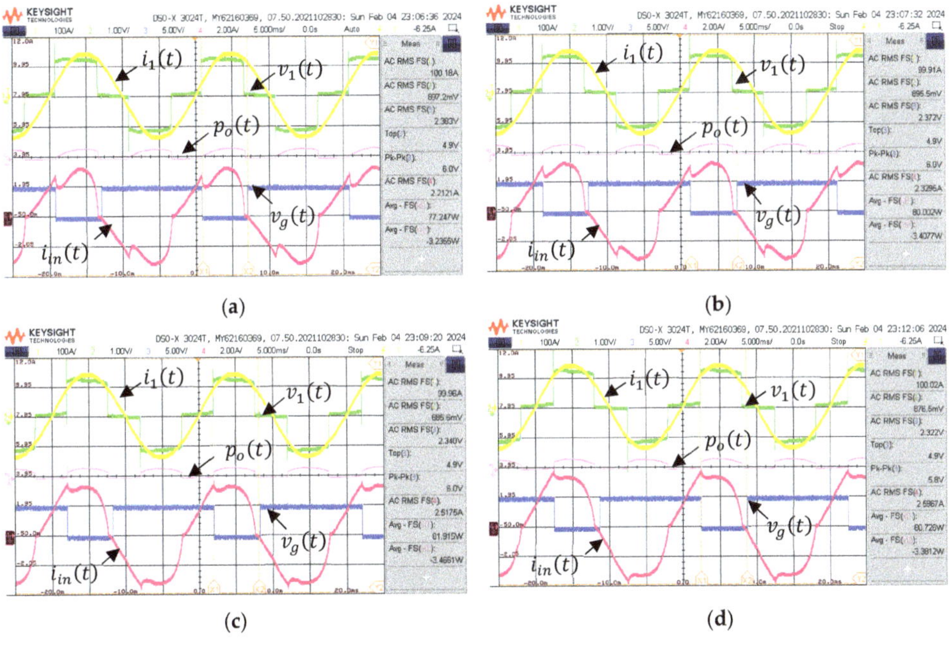

Figure 7. *Cont.*

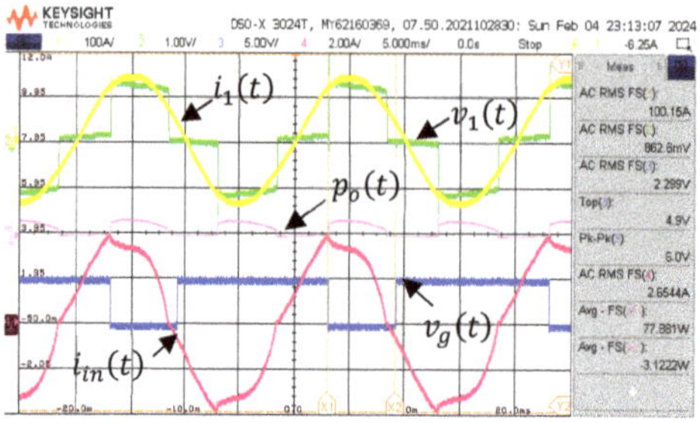

(e)

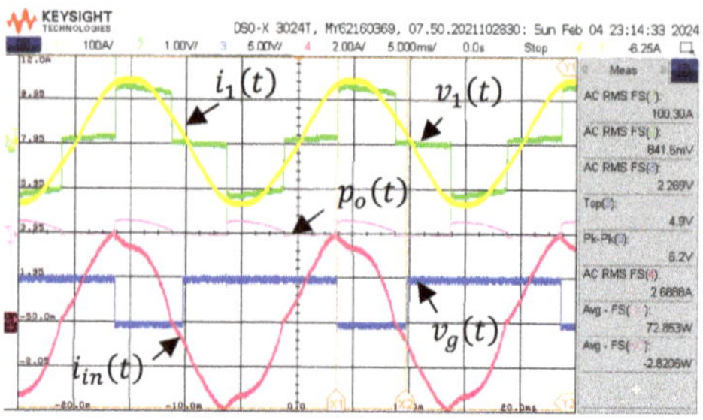

(f)

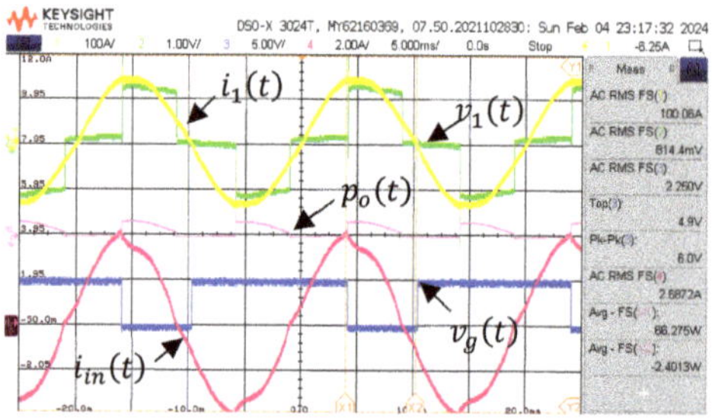

(g)

Figure 7. Typical waveforms of the experimental OLMEH with an active rectifier at line current I_1 = 100 A rms: V_b = 45 V. Delay time: (**a**) t_1 = 1 mS; (**b**) t_1 = 1.5 mS; (**c**) t_1 = 2 mS; (**d**) t_1 = 2.5 mS; (**e**) t_1 = 3 mS; (**f**) t_1 = 3.5 mS; (**g**) t_1 = 4 mS.

The theoretical derivations were first verified by simulation. Figure 8 shows the comparison of the calculated vs. the simulated results for the output power. Here, the assumed line current was I_{line} = 100 A rms and I_{line} = 200 A rms (also N = 40, V_b = 45 V). Good agreement between the results was found.

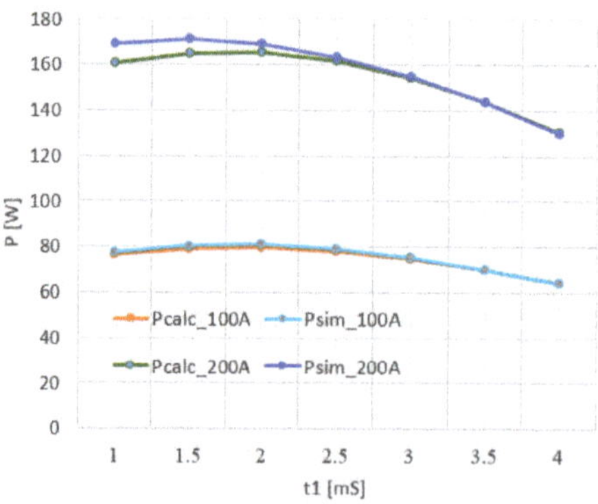

Figure 8. Comparison of the simulated vs. calculated OLMEH's output power as a function of the delay time, t_1, for line current of I_{line} = 100 A rms and I_{line} = 200 A rms, (here, N = 40 and V_b = 45 V).

Comparison of the experimentally measured vs. calculated by (33) and (27) output power as a function of the delay time t_1, for line current I_{line} = 100 A rms and constant load voltage V_b = 45 V, as shown in Figure 9. A good match between the experimental and theoretically predicted power can be observed.

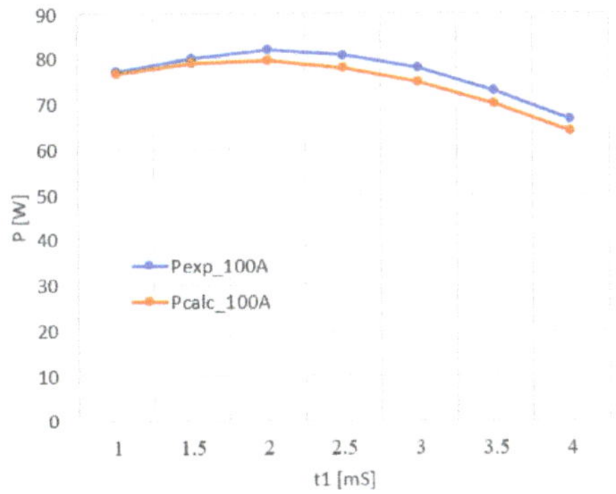

Figure 9. Comparison of the experimentally measured vs. calculated output power as a function of the delay time t_1. (For I_{line} = 100 A rms; V_b = 45 V).

The experimental work revealed that handling the OLMEH prototype brought about variations in its magnetic properties, primarily due to variations in the technological air gap between the two halves of the C-core, which depend on factors like mounting force and surface imperfections. Therefore, some inaccuracies can be seen in the presented comparison plots.

The view of the experimental prototype OLMEH and its test bench are shown in Figures 10a and 10b, respectively, and they are similar to that used in [10].

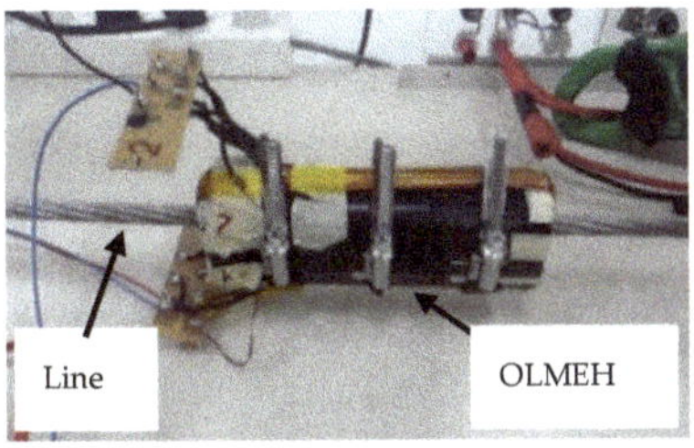

(**a**)

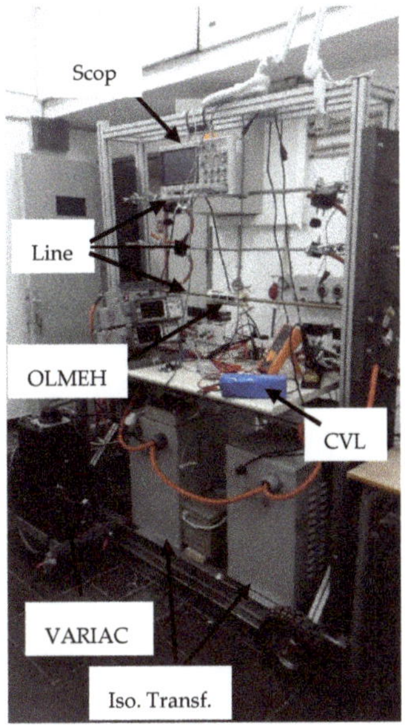

(**b**)

Figure 10. View of the experimental prototype OLMEH (**a**) and its test bench (**b**) [10].

4. Conclusions

This paper aims to offer a closed-form solution for an overhead line magnetic energy harvester with an active rectifier (OLMEH-AR) operating under a constant voltage load.

The article initially examines the Froelich equation, which is subsequently utilized as a fundamental instrument to scrutinize the OLMEH-AR's functionality. The selection of the Froelich equation as the approximation function to characterize the core BH curve is crucial and has been demonstrated to be well suited for facilitating the derivation of a comprehensive analytical model of an OLMEH-AR. The offered model allowed for an extensive analytical study of the OLMEH-AR which yielded approximate analytical solutions for charging and idle periods, the power transfer window, the mean output current, and the mean output power.

An important note drawn from the experimental work is that manipulating the OLMEH-AR prototype affects the magnetic core parameters. This is primarily due to variations in the technological air gap remaining between the two halves of the C-core, which depend on factors such as misalignment, mounting force, and surface imperfections. Therefore, the authors believe that developing a more precise and complex theoretical model, including hysteresis effects, temperature effects, etc., is of limited practical value because of the unavoidable core parameter variations.

As compared to the passive rectifier case, the application of an active rectifier significantly increases the harvester's output power [10]. This comes as a result of two key features, which the passive rectifier does not possess. First is that the controlled rectifier allows for the storage of magnetic energy by short-circuiting the OLMEH's secondary winding, and, secondly, the active rectifier allows aligning the power transfer window to track the optimal power point. The output characteristics of the OLMEH and its proper matching, considering variable line current and load voltage conditions, have also been reported in recent papers [23–25].

While the advantage of the uncontrolled rectifier is its simplicity and robustness, the advantages of the active rectifier are the ability to control the CVL charging current as well as providing overcharge and line overcurrent protection features. However, realizing the active rectifier functions requires a somewhat sophisticated approach, synchronized to the line current control circuit.

Author Contributions: Conceptualization, A.K.; methodology, A.A.; software, A.A.; validation, A.A. and M.S.; formal analysis, A.A.; investigation, A.A.; resources, A.K.; data curation, M.S.; writing—original draft preparation, A.A.; writing—review and editing, A.A.; visualization, A.A. and M.S.; supervision, A.K.; project administration, A.K.; funding acquisition, A.K. All authors have read and agreed to the published version of the manuscript.

Funding: This research received no external funding.

Institutional Review Board Statement: Not applicable.

Informed Consent Statement: Not applicable.

Data Availability Statement: The original contributions presented in the study are included in the articl, further inquiries can be directed to the corresponding author.

Conflicts of Interest: The authors declare no conflicts of interest.

Nomenclature

Throughout this paper, the following definitions of key parameters were adopted.

Time (continuous)	t
State 1 termination instant	t_1
State 2 termination instant	t_2
Flux density zero crossing instant	t_Z
Power line angular frequency	ω
Flux density	$B(t)$
Saturation flux density	B_s
Field intensity	$H(t)$
Saturation field intensity	H_s
Froelich coefficients	a, b
Number of turns	N
Magnetic path length	l_c

Magnetic core area	A_c
Line (primary) current instantaneous	$i_1(t)$
Line (primary) current amplitude	I_{1m}
Secondary current instantaneous	$i_2(t)$
Secondary current amplitude	I_{2m}
Magnetizing current, (instantaneous)	$i_m(t)$
Rectifiers' input (AC side) voltage	$v_{in}(t)$
Rectifiers' input (AC side) current	$i_{in}(t)$
CVL voltage	V_b
CVL current	$i_b(t)$
Average output power	P_o

Appendix A

In our case, the delay angle is $\theta = \omega t_1$ and the extinction angle is $\vartheta + \delta = \omega t_2$. We found the condition on ϑ maximizing the function $\cos(\vartheta) - \cos(\vartheta + \delta)$.

Applying the classical extremum finding procedure:

$$\frac{d}{d\vartheta}(\cos(\vartheta) - \cos(\vartheta + \delta)) = -\sin(\theta) + \sin(\vartheta + \delta) = -\sin\vartheta + \sin\vartheta\cos\delta + \cos\vartheta\sin\delta = 0 \quad \text{(A1)}$$

whence

$$\tan\vartheta = \frac{\sin\delta}{1 - \cos\delta} = \frac{2\sin\frac{\delta}{2}\cos\frac{\delta}{2}}{2\left(\sin\frac{\delta}{2}\right)^2} = \cot\frac{\delta}{2} = \tan\left(\frac{\pi}{2} - \frac{\delta}{2}\right), \quad \text{(A2)}$$

Comparison of the argument on the left hand-side and the right hand-side yields the delay angle:

$$\vartheta = \frac{\pi}{2} - \frac{\delta}{2}, \quad \text{(A3)}$$

whence the extinction angle is

$$\vartheta + \delta = \frac{\pi}{2} + \frac{\delta}{2}. \quad \text{(A4)}$$

Hence, to attain the MPP condition, the pulse edges should be aligned symmetrically relative to $\pi/2$; thus, (26) follows.

References

1. Kang, S.; Yang, S.; Kim, H. Non-intrusive voltage measurement of ac power lines for smart grid system based on electric field energy harvesting. *Electron. Lett.* **2017**, *53*, 181–183. [CrossRef]
2. Moghe, R.; Lambert, F.C.; Divan, D. Smart "stick-on" sensors for the smart grid. *IEEE Trans. Smart Grid* **2012**, *3*, 241–252. [CrossRef]
3. Moghe, R.; Iyer, A.R.; Lambert, F.C.; Divan, D.M. A low-cost wireless voltage sensor for monitoring MV/HV utility assets. *IEEE Trans. Smart Grid* **2014**, *5*, 2002–2009. [CrossRef]
4. Gu, I.Y.H.; Sistiaga, U.; Berlijn, S.M.; Fahlstrom, A. Online detection of snow coverage and swing angles of electrical insulators on power transmission lines using videos. In Proceedings of the 16th IEEE International Conference on Image Processing (ICIP 2009), Cairo, Egypt, 7–10 November 2009; pp. 3249–3252. [CrossRef]
5. Zhong, Y.-P.; Zuo, Q.; Zhou, Y.; Zhang, C. A new image-based algorithm for icing detection and icing thickness estimation for transmission lines. In Proceedings of the IEEE International Conference on Multimedia and Expo Workshops (ICMEW), San Jose, CA, USA, 15–19 July 2013; pp. 1–6. [CrossRef]
6. Riba, J.-R.; Moreno-Eguilaz, M.; Bogarra, S. Energy Harvesting Methods for Transmission Lines: A Comprehensive Review. *Appl. Sci.* **2022**, *12*, 10699. [CrossRef]
7. Moon, J.; Leeb, S.B. Power Electronic Circuits for Magnetic Energy Harvesters. *IEEE Trans. Power Electron.* **2016**, *31*, 270–279. [CrossRef]
8. Enayati, J.; Asef, P. Review and Analysis of Magnetic Energy Harvesters: A Case Study for Vehicular Applications. *IEEE Access* **2022**, *10*, 79444–79457. [CrossRef]
9. Deng, Z.; Dapino, M.J. Review of magnetostrictive vibration energy harvesters. *Smart Mater. Struct.* **2017**, *26*, 103001. [CrossRef]
10. Abramovitz, A.; Shwartsas, M.; Kuperman, A. Nonlinear Analysis and Closed-Form Solution for Overhead Line Magnetic Energy Harvester Behavior. *Appl. Sci.* **2024**, *14*, 9146. [CrossRef]
11. Moon, J.; Leeb, S.B. Wireless sensors for electromechanical system diagnostics. *IEEE Trans. Instr. Meas.* **2018**, *67*, 2235–2246. [CrossRef]

12. Liu, Z.; Li, Y.; Duan, N.; He, Z. An energy management method for magnetic field energy harvesters under wide-range current in railway electrification systems. *IEEE Trans. Ind. Electron.* **2024**, *71*, 5360–5369. [CrossRef]
13. Monagle, D.; Ponce, E.A.; Leeb, S.B. Rule the Joule: An energy management design guide for self-powered sensors. *IEEE Sens. J.* **2024**, *24*, 6–15. [CrossRef]
14. Suntio, T.; Viinamaki, J.; Jokipii, J.; Messo, T.; Kuperman, A. Dynamic characterization of power electronic interfaces. *IEEE J. Emerg. Sel. Top. Power Electron.* **2014**, *2*, 949–961. [CrossRef]
15. Monagle, D.; Ponce, E.; Leeb, S.B. Generalized analysis method for magnetic energy harvesters. *IEEE Trans. Power Electron.* **2022**, *37*, 15764–15773. [CrossRef]
16. Li, Y.; Duan, N.; Liu, Z.; Hu, J.; He, Z. Impedance-matching-based maximum power tracking for magnetic field energy harvesters using active rectifiers. *IEEE Trans. Ind. Electron.* **2023**, *70*, 10730–10739. [CrossRef]
17. Chen, F.; Yang, C.; Guo, Z.; Wang, Y.; Ma, X. A magnetically controlled current transformer for stable energy harvesting. *IEEE Trans. Power Deliv.* **2023**, *38*, 212–221. [CrossRef]
18. Kuperman, A.; Averbukh, M.; Lineykin, S. Maximum power point matching versus maximum power point tracking for solar generators. *Renew. Sustain. Energy Rev.* **2013**, *19*, 11–17. [CrossRef]
19. Moon, J.; Leeb, S.B. Analysis Model for Magnetic Energy Harvesters. *IEEE Trans. Power Electron.* **2015**, *30*, 4302–4311. [CrossRef]
20. Gao, M.; Yi, L.; Moon, J. Mathematical modeling and validation of saturating and clampable cascaded magnetics for magnetic energy harvesting. *IEEE Trans. Power Electron.* **2023**, *38*, 3455–3468. [CrossRef]
21. Vos, M.J. A magnetic core permeance model for inductive power harvesting. *IEEE Trans. Power Electron.* **2020**, *35*, 3627–3635. [CrossRef]
22. Liu, Z.; Li, Y.; Yang, H.; Duan, N.; He, Z. An accurate model of magnetic energy harvester in the saturated region for harvesting maximum power: Analysis, design and experimental verification. *IEEE Trans. Ind. Electron.* **2023**, *70*, 276–285. [CrossRef]
23. Abramovitz, A.; Shvartsas, M.; Orfanoudakis, G.I.; Kuperman, A. Output Characteristics of Passive Magnetic Energy Harvester Feeding a Constant-Voltage-Type Load. *IEEE J. Emerg. Sel. Top. Power Electron.* **2024**, *in press*. [CrossRef]
24. Abramovitz, A.; Shvartsas, M.; Kuperman, A. Enhanced Maximum Power Point Reaching Method for Passive Magnetic Energy Harvesters Operating Under Low Primary Currents. *IEEE Trans. Power Electron.* **2024**, *39*, 6619–6623. [CrossRef]
25. Abramovitz, A.; Shvartsas, M.; Kuperman, A. On the Maximum Power of Passive Magnetic Energy Harvesters Feeding Constant Voltage Loads Under High Primary Currents. *IEEE Trans. Power Electron.* **2024**, *39*, 12076–12080. [CrossRef]

applied sciences

MDPI

Article

Design of a Novel Ultra-Wideband Common-Mode Filter Using a Magnified Coupled Defected Ground Structure

Ding-Bing Lin [1], Mei-Hui Wang [1,*], Aloysius Adya Pramudita [2] and Tjahjo Adiprabowo [2]

[1] Department of Electronic and Computer Engineering, National Taiwan University of Science and Technology, Taipei City 106335, Taiwan; dblin@mail.ntust.edu.tw
[2] School of Electrical Engineering, Telkom University, Bandung 40257, Indonesia; pramuditaadya@telkomuniversity.ac.id (A.A.P.); tjahjo.a@gmail.com (T.A.)
* Correspondence: d10802004@mail.ntust.edu.tw

Abstract: An ultra-wideband common-mode (CM) filter for a gigahertz (GHz) data rate signal is proposed in this paper. The proposed filter was designed only on the printed circuit board (PCB) ground plane; no additional components wererequired. We took advantage of producing second-order transmission zero by an asymmetrical magnified coupled DGS to extend the suppression bandwidth. Full-wave simulations and equivalent circuit models of the DGS resonator were established to predict the suppression performance. The measured differential-mode insertion loss (S_{dd21}) from direct current (DC) to 12.35 GHz was obtained within the −3 dB definitionin the frequency domain. The CM noise was suppressed by more than10 dB in the frequency range from 2.9 GHz to 16.2 GHz. The fractional bandwidth (FBW) reached 139.3%. The proposed filter blocked 62.3% of the CM noise magnitude in the time domain measurement. In addition, the eye diagram measurement proved that good transmission quality was maintained. The proposed filter can be widely implemented to reduce electromagnetic interference (EMI) in radio frequency (RF) andWi-Fi (wireless fidelitystandard) 5 and 6E wireless communication applications.

Keywords: bandstop filter (BSF); common-mode noise; defected ground structure (DGS); electromagnetic interference (EMI); signal integrity (SI)

1. Introduction

At present, electronic products tend to operate at high data rates and complexities. Transmission at a GHz data rate is widely used in modern digital circuits, such as serial advanced technology attachment (SATA) III/6 Gbit/s, universal serial bus (USB) 3.2/10 Gbit/s, and peripheral component interconnect express (PCIe or PCI-e) 4.0/16 GT/s. Differential transmission is the most suited for high-speed data raterequirements. Differential pairs should be symmetrical to maintain good signal integrity (SI) and avoid electromagnetic interference (EMI). However, asymmetric routing is unavoidablein practical situations, even when using smooth bend structures [1], as it results in common-mode (CM) noise because of an imbalanced amplitude or a timing skew on the differential signal paths. CM noise may generate unwanted radiation and cause severe EMI problems, which results in a decreased performance of high-speed circuits [2]. Therefore, we expect the common-mode filter to maintain a good SI and EMI to avoid interfering with the sensor signals in the system.

The conventional method is to use CM choke coils with high-permeability ferrite cores [3] or compact CM filters that were built on multilayer, low-temperature, co-fired ceramics [4]. They suppress the CM noise in the megahertz frequency domain. However, CM chokes do not work well in the frequency domain above 2 GHz. Furthermore, they are not practical for compact circuit designs.

The differential-mode bandpass filter (BPF) and CM noise bandstop filter are better suited for GHz digital circuit systems; several methods are discussed in [5–18]. First, the stopband-extended balanced BPF [5] and balanced dual-band BPF [6] were proposed as design concepts of the resonator for differential signaling. They achieved narrow passbands using adequately designed resonators. Next, wideband or ultra-wideband differential-mode BPFs were proposed in [7–9]. The results showed that differential-mode BPFs were limited in high-speed digital circuit applications because the BPFs did not pass low-frequency terms.

In addition, the authors of [10–18] proposed many CM suppression filter designs using defected ground structures (DGSs). The primary design concept differed from the differential-mode BPF introduced earlier; CM filter designs can pass both passband and low-frequency differential signals. Theoretically, the differential signals were propagated on two coupled microstrip lines in the odd mode. The signal current traveled through one trace and returned through the other trace. In other words, the DGS on the ground plane was negligible in the odd propagation mode because the return current passed through the opposite line of the coupled microstrip line. On the other hand, the CM noises were propagated in the even mode; the return current of the CM noises flowed through the ground plane. Therefore, we designed an inductance (L) and capacitance (C) resonant circuit using the DGS on the ground plane. It acted as an ideal open circuit and reduced the magnitude of CM noise because the DGS blocked the return current.

First, a dumbbell-shaped DGS in [10] obtained a bandstop frequency range from 3.3 GHz to 5.7 GHz using a −20 dB definition. In [11], quarter-wavelength open-circuit stub resonators with 8.5 mm and 6.5 mm lengths achieved a broadband-frequency-range performance from 4.27 GHz to 8.2 GHz using a −20 dB definition.

Next, DGSs in [12–17] obtained a bandstop range using a −15 dB definition. The DGS in [12] used one H- and two C-shaped units to achieve a bandstop frequency range of 3.68 GHz to 8.43 GHz. Similarly, the DGS in [13] obtained a bandstop frequency range of 5.4 GHz to 11.4 GHz. In [14], a CM suppression filter design using two quarter-wavelength resonators in multiple sets of differential pairs achieved a bandstop frequency range of 3.75 GHz to 6.95 GHz. The authors of [15] proposed a reconfigurable filter based on three varactor-loaded compactDGSs, in which the bandstop frequency range was adjustable from 1.8 GHz to 2.9 GHz and from 2.9 GHz to 8.1 GHz. In the design described in [16], the authors symmetrically etched the DGS filter using a UH-shaped patterned ground plane to form three mutually coupled resonators for wideband CM noise suppression. It effectively reduced the CM noise in the frequency range from 3.6 GHz to 9.1 GHz. In [12–16], these designs took advantage of a simple structure and a low fabrication cost, but none could achieve an ultra-wide bandstop frequency range, even with three DGS units. Furthermore, the authors of [17] designed a three-pole DGS filter to reduce the CM noise in the frequency range from 3.2 GHz to 12.4 GHz. However, the physical size of the filter was 10 mm × 10 mm, which is not practical for a compact circuit design. Details of the CM suppression filters are listed in Table 1, including the lower frequency (f_L), upper frequency (f_H), central frequency (f_C), and fractional bandwidth (FBW), where the FBW is defined as the absolute bandwidth divided by the central frequency and is expressed as a percentage of the central frequency. It is given by [19]

$$\text{FBW} = 2 \times \frac{f_1 - f_2}{f_1 + f_2} \tag{1}$$

where f_1 and f_2 are the upper frequency (f_H) and lower frequency (f_L) of the stopband, respectively.

After that, CM suppression filters in [18–20] obtained a bandstop range using a −10 dB definition. In CM rejection filters based on a complementary split-ring resonator design [18], the authors proposed three glide symmetry implementations to achievea maximum FBW of 26.8%. A reconfigurable common-mode (CM) filter with a wide tunable bandwidth and frequency range was proposed in [19]. It obtained three tunable CM suppressions of frequency range from 1.9 to 3 GHz, 2.5 to 5.9 GHz, and 3.3 to 7 GHz, achieving a maximum FBW of 75.5%. In the enhanced coupled DGS design [20], the filter used only two C-shaped DGSs to achieve wideband CM noise suppression. Two C-shaped DGSs were the same size and in opposite directions on the ground plane and were symmetrical to the centerline of the differential microstrip. The simulation results for each half-structure indicated two resonant frequenciesclose to 8 GHz. The enhanced coupling mechanism of the two C-shaped DGSs shifted the first-order resonant frequency to 4.5 GHz and the second-order resonant frequency to 7 GHz. They suppressed the CM noise in the frequency range from 3.7 GHz to 10.8 GHz using the definition of −10 dB, as shown in the simulated result; the stopband FBW reached 97.9%. The physical sizes were 7 mm × 7 mm. However, the simulated results showed that the second-order transmission zero did not bring an extended stopband to suppress the CM noise in the higher-frequency domain due to an insufficient coupling magnitude between the two DGS units. As an extension design of [20], we propose an ultra-wideband CM suppression filter with an asymmetrical magnified coupled DGS in this paper. We take advantage of the second-order transmission zero produced by the asymmetrical magnified coupled DGS, which provides a more pronounced coupling effect than [20].

Table 1. Details of the bandstop filters.

Ref.	f_L (GHz)	f_H (GHz)	f_C (GHz)	FBW (%)
[12]	3.68	8.43	6.1	78.4
[13]	5.4	11.4	8.4	71.4
[14]	3.75	6.95	5.35	59.8
[15]	1.8	2.9	2.35	46.8
	2.9	8.1	5.5	94.5
[16]	3.6	9.1	6.35	86.6
[17]	3.2	12.4	7.8	117.9

Next, the proposed CM suppression filter's design details are presented in Section 2, including full-wave simulations using high-performance electronic design automation (EDA) software ANSYS HFSS 2022 R1 and equivalent circuit modelsusing the circuit simulation tool Agilent ADS. Section 3 presents measurements to validate the simulated results, including frequency and time domain measurements. Section 4 presents discussions. Finally, the conclusion is providedin Section 5.

2. Methods

2.1. Full-Wave Simulation

The ultra-wideband CM suppression filter was simulated on a standard FR4 (εr = 4.4) substrate; the loss tangent was 0.02. The printed circuit board (PCB) thickness was 1.035 mm. We used 1 oz copper on the top layer and layer 2, where layer 2 was ground plane. The computer configurations include Intel i7 CPU, 8 GB RAM, and 500 GB solid-state drive (SSD).

Figure 1 shows the PCB stack-up. Two microstrip lines were on the top layer; their width (W) and spacing (S) were 0.18 mm and 0.36 mm, respectively. We designed W and S to meet a 100 Ω differential pair to maintain good SI in the differential mode. Figure 2 shows the PCB top view. Two DGS units were placed close to each other on the ground plane to form long and narrow bridges on that plane. The detailed geometry parameters of the DGS resonators are also summarized in Table 2.

We first simulated the two half-structure DGS units of the proposed suppression filter using ANSYS HFSS; they are referred to as "Left DGS" and "Right DGS" to simplify the following analysis. Next, we simulated the proposed ultra-wideband CM suppression filter.

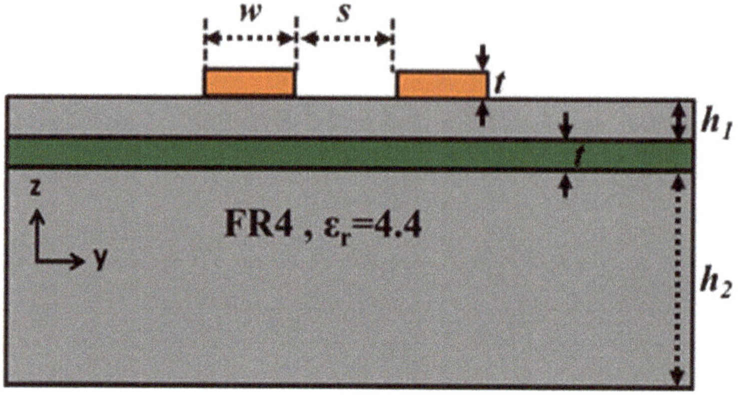

Figure 1. PCB side view.

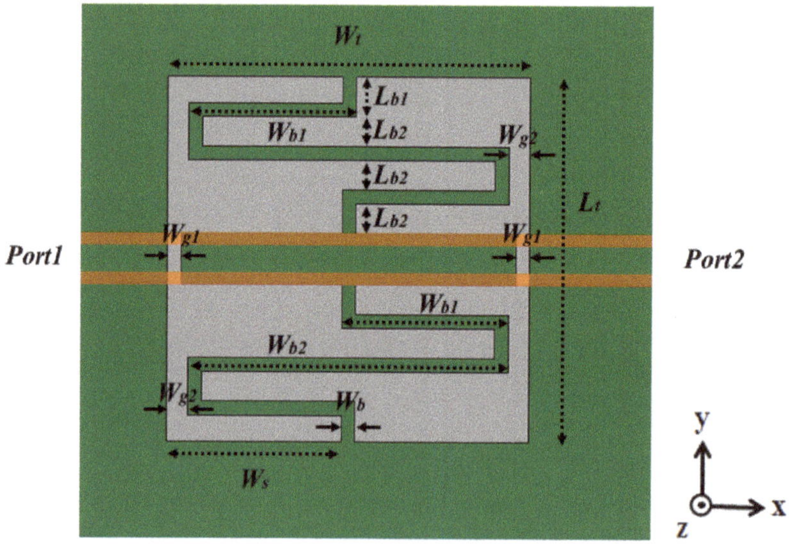

Figure 2. PCB top view.

Table 2. Detailed list of geometry parameters.

Parameters	Value	Parameters	Value	Parameters	Value	Parameters	Value	Parameters	Value
W	0.18	S	0.36	t	0.035	h_1	0.1	h_2	0.865
W_t	5.2	W_s	2.5	W_b	0.2	L_t	5.04	W_{g1}	0.2
W_{b1}	2.4	W_{b2}	4.6	L_{b1}	0.56	L_{b2}	0.4	W_{g2}	0.3

Unit: mm.

Figure 3 demonstrates the full-wave simulations of the differential-mode insertion loss (S_{dd21}) by the blue curve and the common-mode insertion loss (S_{cc21}) by the red curve in the Left DGS. From the S_{cc21} simulated result, we obtained a −3 dB cut-off frequency (f_{CL}) at 2.72 GHz and a resonant frequency (f_{0L}) at 7.3 GHz.

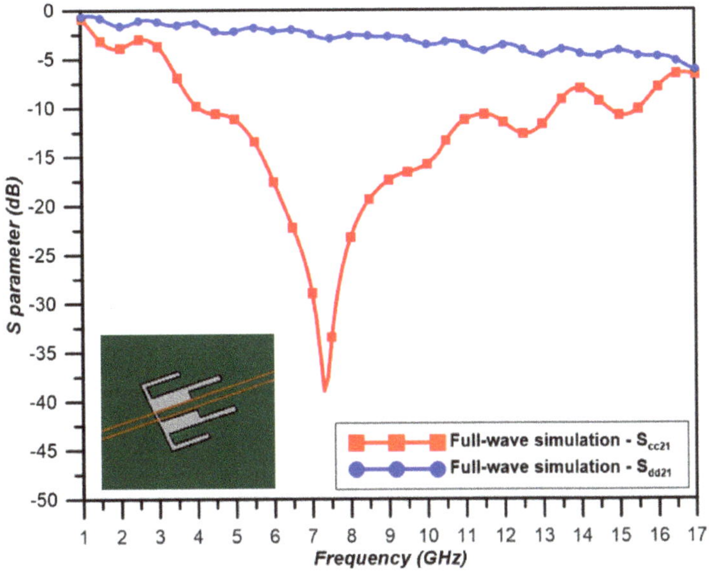

Figure 3. Full-wave simulated insertion loss for the Left DGS.

Similarly, Figure 4 shows the full-wave-simulated S_{dd21} and S_{cc21} in the Right DGS. The simulation results of the −3 dB cut-off frequencies f_{CR1} and f_{CR2} of S_{cc21} were 2.6 GHz and 9 GHz, respectively; the resonant frequencies of the first-order frequency (f_{0R1}) and second-order frequency (f_{0R2}) of 6.6 GHz and 13.7 GHz, respectively, were also obtained.

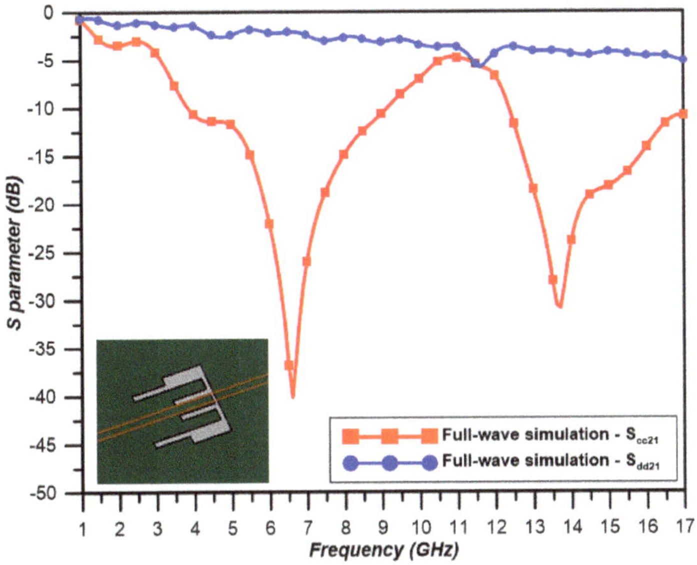

Figure 4. Full-wave simulated insertion loss for the Right DGS.

Finally, Figure 5 shows the S_{dd21} and S_{cc21} of the proposed suppression filter using the magnified coupled DGS. Three resonant frequencies, f_{01}, f_{02}, and f_{03}, were identified, which were 3.2 GHz, 10.3 GHz, and 14.2 GHz, respectively. Since the two DGS units, the

Left DGS and the Right DGS, were placed close to each other to produce a mutual coupling effects, it was easy to determinethe frequency separation and the ultra-wide stopband. All the frequencies of the Left DGS, the Right DGS, and the proposed suppression filter are listed in Table 3.

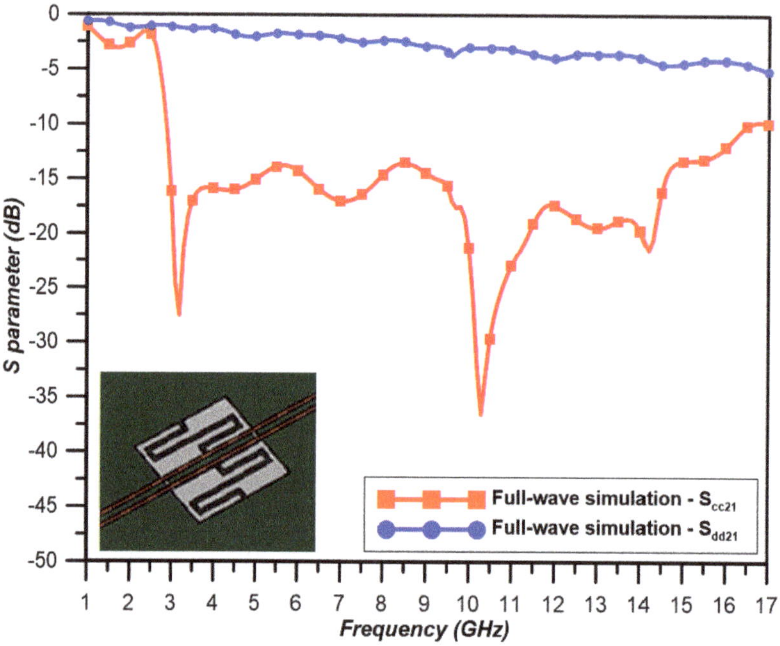

Figure 5. Full-wave-simulated insertion loss for the proposed suppression filter.

Table 3. Cut-off and resonant frequencies of Left DGS, Right DGS, and the proposed suppression filter.

Left DGS	f_{CL}	2.72	f_{0L}	7.3				
Right DGS	f_{CR1}	2.60	f_{CR2}	9.0	f_{0R1}	6.6	f_{0R2}	13.7
The proposedfilter	f_{01}	3.20	f_{02}	10.3	f_{03}	14.2		

Unit: GHz.

2.2. Circuit Simulation

In this section, we report the results of characteristic frequencies to derive the circuit parameters. We built the equivalent circuit model for the Left DGS, the Right DGS, and the proposed CM filter in Agilent ADS, represented to be consistent with the full-wave simulation results.

We considered characteristic impedance (Z_{even}) in an even-mode cascading parallel LC circuit to the ground plane when the CM noise passed through to the DGS. Then, we could extract the equivalent circuit parameters from the simulation results in Section 2.1. First, we modeled a parallel LC-equivalent circuit from a transmission zero in the Left DGS. The parameters of the parallel LC resonator were derived by the resonant frequency (f_{0L}) of 7.3 GHz and the −3 dB cut-off frequency (f_{CL}) of 2.72 GHz. We calculated the parameters of the parallel capacitance (C) and inductance (L) using Formulas (2) and (3), as given by [21],

$$C = \frac{1}{4\pi \times Z_{even}} \left(\frac{f_C}{f_0{}^2 - f_C{}^2} \right) \tag{2}$$

$$L = \frac{1}{4\pi^2 f_0{}^2 C} \tag{3}$$

since

$$f_0 = \frac{1}{2\pi\sqrt{L \times C}} \tag{4}$$

The equivalent circuit model of the Left DGS is shown in Figure 6. The even-mode impedance Z_{even} equaled 49.46 Ω. The equivalent circuit model of the Left DGS resulted in the following parallel capacitance and inductance parameters: C_{L1} and L_{L1} were 0.095 pF and 4.98 nH, respectively. Figure 7 shows the comparison of the full-wave simulation parameter S_{cc21} for the red curve and the equivalent circuit model for the black curve.

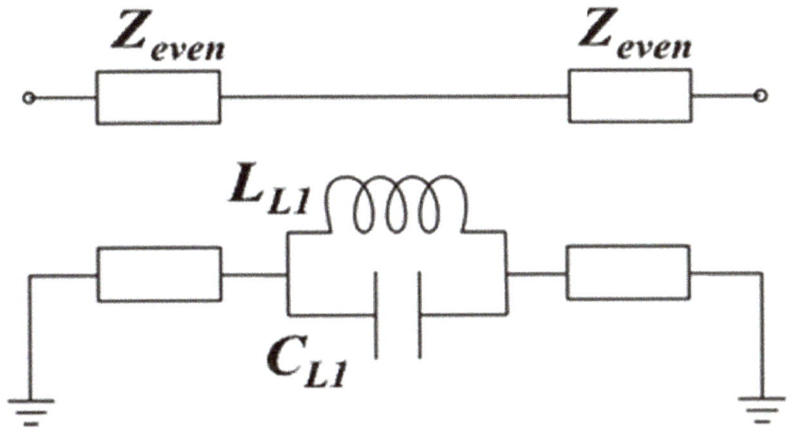

Figure 6. The parallel LC-equivalent circuit model of the Left DGS.

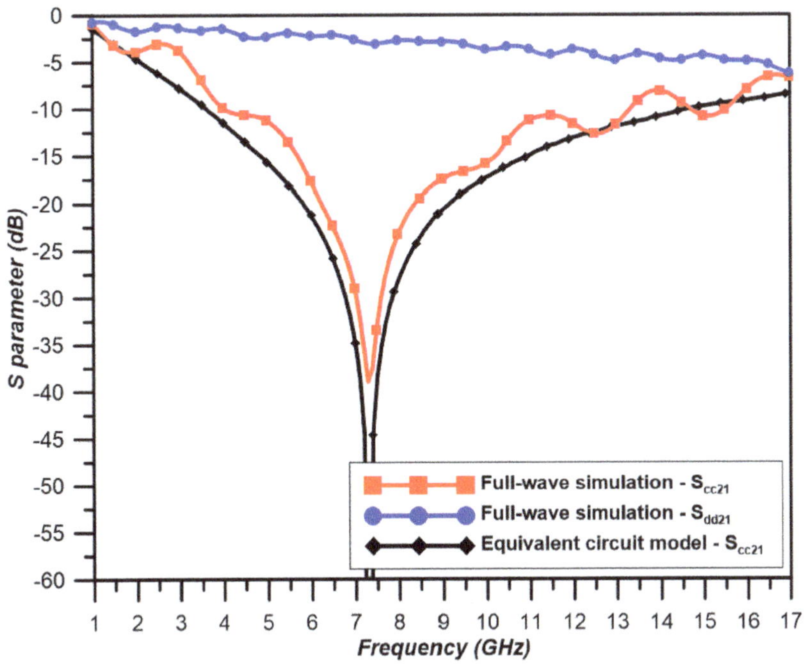

Figure 7. S-parameter of equivalent circuit model and full-wave simulation for the Left DGS.

Similarly, the equivalent circuit model of the Right DGS is shown in Figure 8. It consists of two LC elements in parallel, since two transmission zeros were obtained from the Right DGS simulation. We obtained the circuit parameters C_{R1}, C_{R2}, L_{R1}, and L_{R2}, which were 0.114 pF, 0.136 pF, 5.12 nH, and 0.99 nH, respectively. Figure 9 shows the comparison of the insertion loss of the full-wave simulation and the equivalent circuit model.

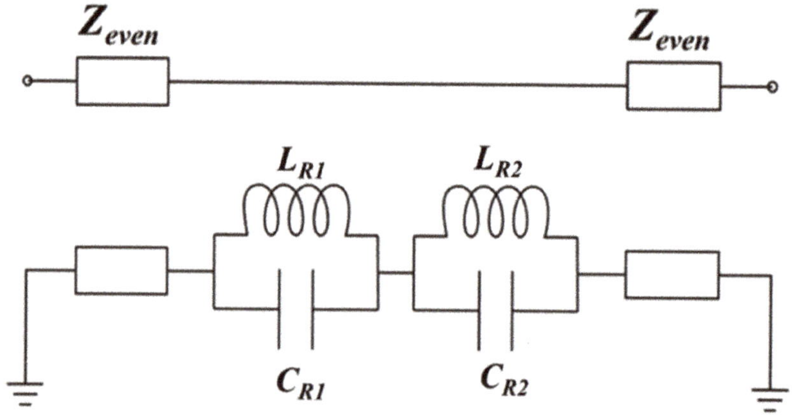

Figure 8. The parallel LC-equivalent circuit model of the Right DGS.

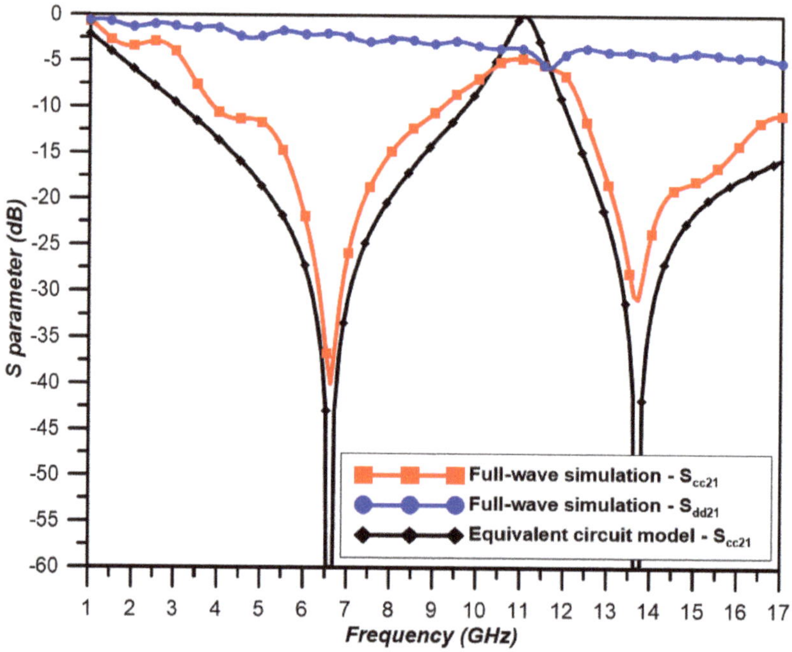

Figure 9. S-parameter of equivalent circuit model and full-wave simulation for the Right DGS.

Following this, we used the general formulation (5) for extracting the coupling coefficient k, regardless of whether the coupling was electric, magnetic, or mixed; k_x is the mixed coupling coefficient and is given as Formula (6), where k_m and k_e are the magnetic and electric coupling coefficients, respectively. The coupling coefficient k is given by [22].

$$k = \pm\frac{1}{2}\left(\frac{f_{02}}{f_{01}} + \frac{f_{01}}{f_{02}}\right)\sqrt{\left(\frac{{f_{p2}}^2 - {f_{p1}}^2}{{f_{p2}}^2 + {f_{p1}}^2}\right)^2 - \left(\frac{{f_{02}}^2 - {f_{01}}^2}{{f_{02}}^2 + {f_{01}}^2}\right)^2} \tag{5}$$

$$k_x = k_e - k_m \tag{6}$$

We analyzed the coupling characteristics of the DGS resonator using a single-ended microstrip line model [21], as shown in Figure 10. The peak frequencies corresponding to the resonant characteristics are obtained by S21 simulation in Figure 11; three peak frequencies, f_{p1}, f_{p2}, and f_{p3}, were at 4.9 GHz, 7.7 GHz, and 13.1 GHz, respectively. In addition, we obtained the resonant frequencies of the Left DGS and the Right DGS—6.6 GHz, 7.3 GHz, and 13.7 GHz for f_{0R1}, f_{0L}, and f_{0R2}, respectively—as reported in Section 2.1.

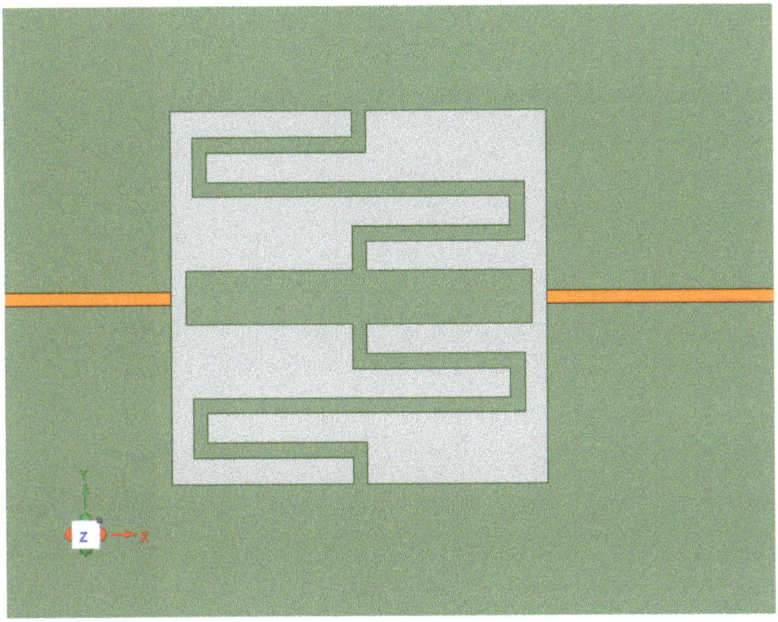

Figure 10. The single-ended microstrip line model.

The coupling coefficients k_{x1} and k_{x2} of the asynchronously tuned coupled resonators were obtained as 0.413534 and 0.612251, respectively, for which k_x is a crucial design parameter. Using k_x and Z_{even}, we built the equivalent circuit model of the proposed suppression filter. Figure 12 shows the equivalent circuit model of the proposed suppression filter. Figure 13 compares the insertion loss of the equivalent circuit model and the full-wave simulation. Finally, we summarize the detailed parameters of the equivalent circuit model in Table 4. From the results, this equivalent circuit model still effectively predicted the ultra-wide stopband produced by the proposed suppression filter.

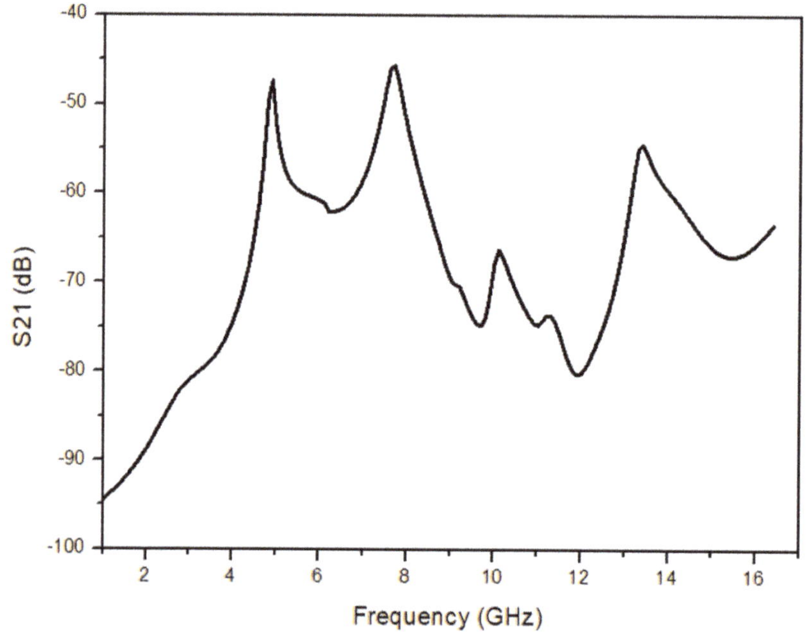

Figure 11. The S21 simulation of the single-ended microstrip line model.

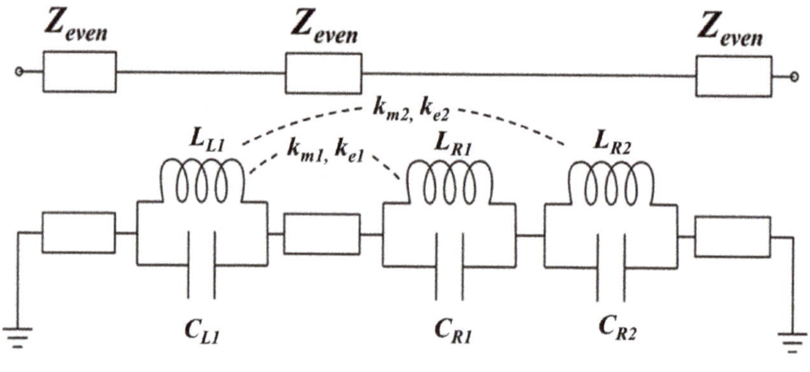

Figure 12. The equivalent circuit model of the proposed suppression filter.

Table 4. Parameters of the equivalent circuit model.

Z_{even}	49.46 Ω	C_{L1}	0.095 pF	L_{L1}	4.98 nH		
C_{R1}	0.114 pF	L_{R1}	5.12 nH	C_{R2}	0.136 pF	L_{R2}	0.99 nH
f_{p1}	4.9 GHz	f_{p2}	7.7 GHz	f_{p3}	13.1 GHz		
f_{0R1}	6.6 GHz	f_{0L}	7.3 GHz	f_{0R2}	13.7 GHz		
k_{x1}	0.413534	k_{e1}	0.202218	k_{m1}	0.211316		
k_{x2}	0.612251	k_{e2}	0.261125	k_{m2}	0.351126		

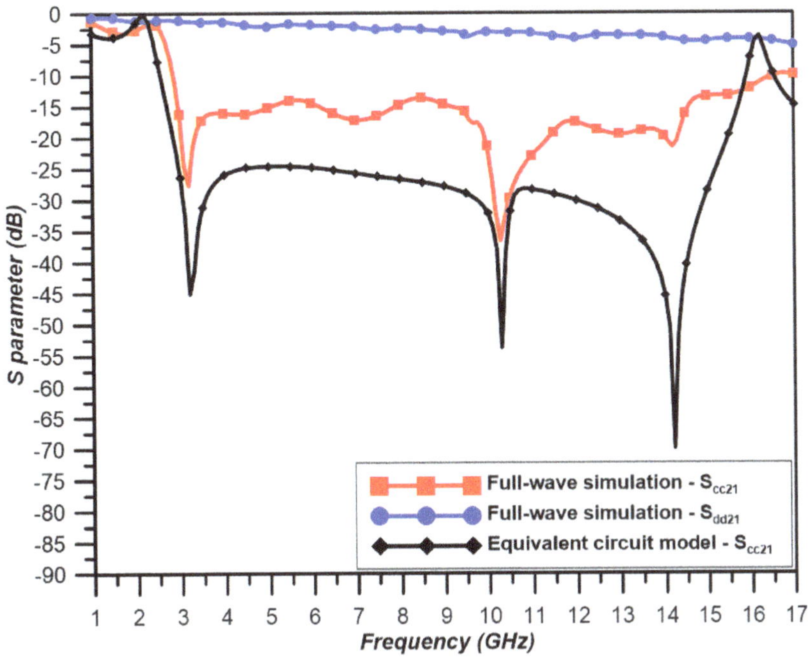

Figure 13. S-parameter of equivalent circuit model and full-wave simulation for the proposed filter.

3. Results

We fabricated four PCB test samples using a chemical etching process to verify our proposed design. The PCB size was 60 mm × 40 mm. First, two PCBs are shown in Figure 14a,b: a reference board with a solid ground plane and a proposed filter board. As shown in Figure 14b, the proposed filter was located in the center of the PCB. Later, we used the Keysight E5071C ENA Vector Network Analyzer to measure the S-parameters in the frequency domain. We also measured and observed the differences in the eye diagrams in the time domain. Additionally, we measured and observed the noise magnitude for time-domain performance verification using another two PCBs with a 5 mm delay line, as shown in Figure 15a,b.

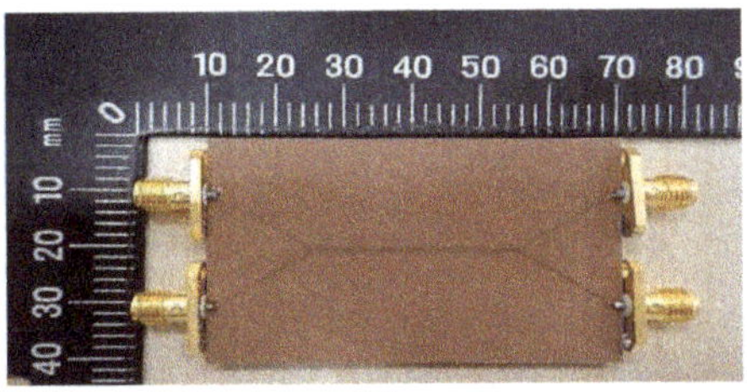

(a)

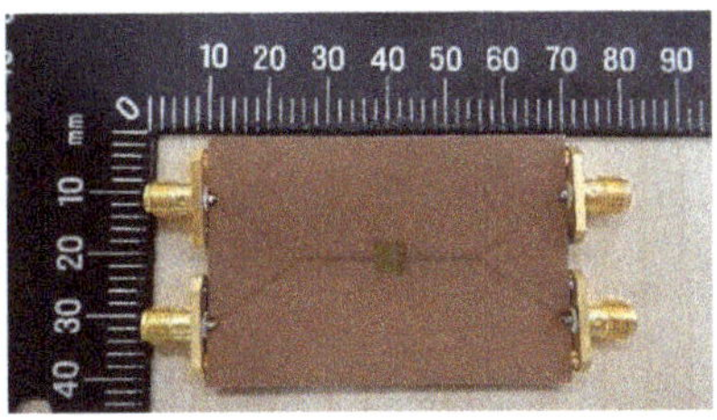

(b)

Figure 14. Photographs of the fabricated prototype: (**a**) the reference board and (**b**) the proposed filter board.

Figure 15. Photographs of the fabricated prototype with a 5 mm delay line: (**a**) the reference board and (**b**) the proposed filter board.

3.1. Frequency-Domain Performance

As shown in Figure 16, we obtained the S_{dd21} measurement from DC to 12.35 GHz within the −3 dB definition. In addition, the CM noise stopband is shown as the S_{cc21} measurement. The frequency range was from 2.9 GHz to 16.2 GHz using a −10 dB definition.

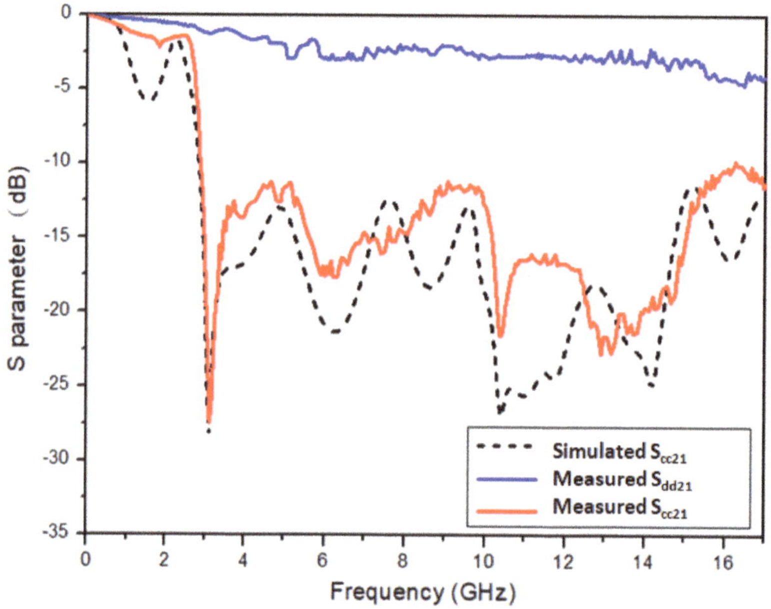

Figure 16. S-parameter of measurements and full-wave simulation.

3.2. Time-Domain Performance

In the following, we also used an eye diagram and a CM noise magnitude (V_{common}) measurement to verify the capability of the proposed filter in the time domain. We measured the eye diagram to verify the proposed suppression filter that maintained good SI for differential signal transmission. First, we excited an 8 Gbps signal to ports 1 and 2 with a 2^7-1 pseudo-random binary sequence (PRBS), where the amplitude was ±0.5 V and the rise time was 50 ps. The eye diagram measurement is shown in Figure 17 from ports 3 and 4.

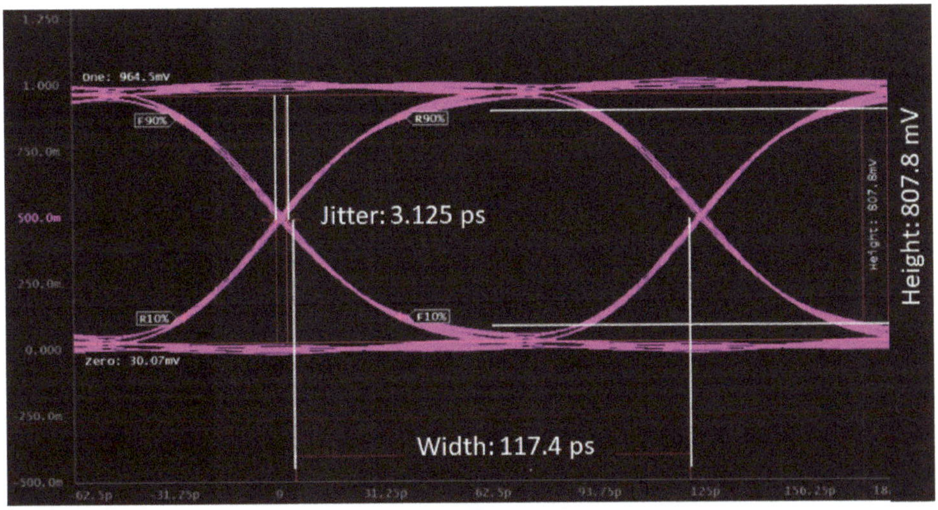

(**a**)

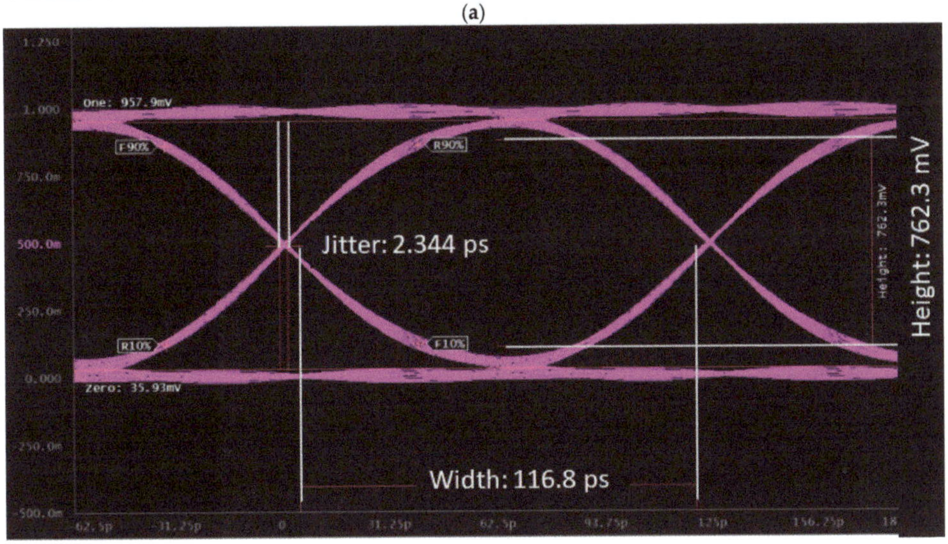

(**b**)

Figure 17. Measurement of the eye diagrams: (**a**) the reference board and (**b**) the proposed suppression filter board.

Table 5 summarizes the detailed parameters of the eye diagram, including the eye height, width, and timing jitter. As shown by the results, this system still maintained a good transmission quality, although there was a small amount of degradation on the filter board measurement.

Table 5. Detailed parameters of the eye diagram.

	Eye Height	Eye Width	Timing Jitter
Reference Board	807.8 mV	117.4 ps	3.125 ps
Filter Board	762.3 mV	116.8 ps	2.344 ps

Furthermore, we excited CM noise to one of the differential lines with an implemented 5 mm delay line using a two-step function to measure the CM noise magnitude. Figure 18 shows the CM noise magnitude measurements for the reference and the proposed suppression filter boards with a 5 mm delay line. The results showed that the peak voltage was 138 mV and 52 mV, respectively. The measurement results of the CM noise showed that the proposed filter suppressed 62.3% of the magnitude.

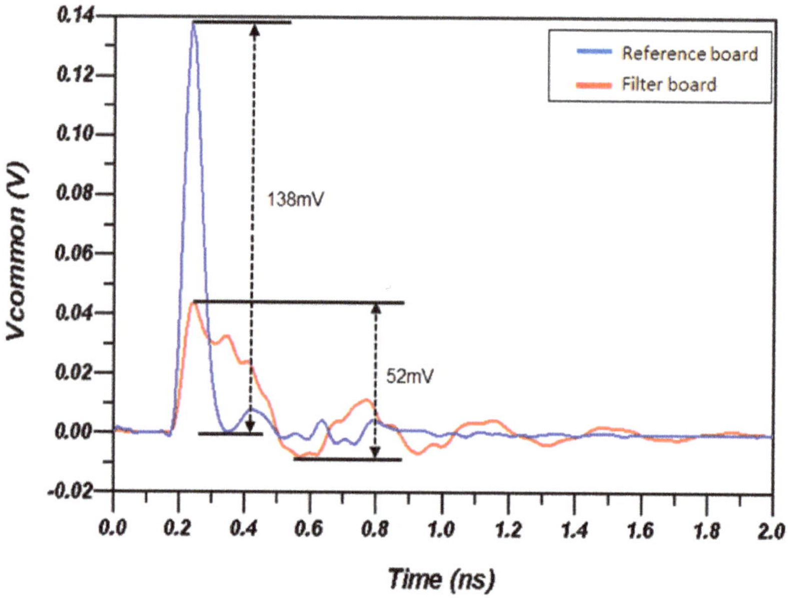

Figure 18. Measurements of CM noise magnitude.

4. Discussion

A parallel resonant circuit stores circuit energy in the inductor's magnetic field and the capacitor's electric field. Currents through inductors and capacitors are always equal and opposite. We can adjust the geometry parameters of the capacitor (W_{g1}) and inductor (W_b, W_{b1}, and W_{b2}) in Figure 2 to obtain the resonant frequency. The smaller W_{g1} has larger capacitance. Thinner W_b or longer W_{b1} and W_{b2} produce larger inductance. In addition, larger capacitance or inductance leads to lower frequency from Formula (3) in Section 2.2. As seen in Figures 7, 9 and 13, the S_{cc21} of the equivalent circuit model agrees very well with the full-wave simulation result. It shows differences in the high-frequency domain since the full-wave simulation is realistic and different from the equivalent circuit model; it should be considered in additional parasitic capacitances and inductances, such as PCB vias. In addition, the equivalent circuit model does not include some distribution effects, such as complex boundary conditions.

In Section 2.2, we quickly predict the first- and second-order transmission zero using the circuit simulation tool Agilent ADS. The Left and Right DGS are considered asynchronously tuned coupled resonator circuits [22] because two DGSs obtain three different resonant frequencies. The coupling coefficient changes and affects the frequency characteristics because of the stored electromagnetic field distributions. In other words, the bandwidth is extended due to the asynchronously tuned coupling.

According to the S_{cc21} measurement in Section 3.1, from 2.9 GHz to 16.2 GHz, the fractional bandwidth (FBW) of the proposed design reaches 139.3%, where the stopband's central frequency is 9.55 GHz. The filter's physical size is 5.04 mm × 5.2 mm. The filter's electrical size is only 0.292 $\lambda_g \times$ 0.302 λ_g, where λ_g is the guided wavelength of the stopband central frequency. The guided wavelength is given by

$$\lambda_g = \frac{\lambda_0}{\sqrt{\varepsilon_{re}}}, \quad (7)$$

where λ_0 is the free space wavelength at operation frequency. The effective dielectric constant (ε_{re}) is obtained by the dielectric constant (ε_r), PCB thickness (h), and trace width (W). This work fallsunder the case of $W/h \geqq 1$; the ε_{re} is given by [23],

$$\varepsilon_{re} = \frac{\varepsilon_r + 1}{2} + \frac{\varepsilon_r - 1}{2} \frac{1}{\sqrt{1 + 12\frac{h}{W}}}. \quad (8)$$

The measured value has ripples and attenuations in Figure 16, possibly due to impedance mismatch between the PCB and connector; additionally, there are tolerances in PCB manufacturing, and the FR4 substrates have limited performance at high frequencies above 10 GHz. However, S_{cc21} still shows a good agreement between the measurements and simulations. This demonstrates that the CM noise is suppressed in an ultra-wide stopband.

We asymmetrically etched the DGS filter in the proposed filter to form three mutually coupled resonators, resulting in a pronounced ultra-wideband suppression effect. Table 6 lists the detailed comparison with [17–20]. The proposed work reveals the smallest physical size; compared with [19], the physical size is reduced by 85%. The area at the stopband central frequency is the most compact. In addition, the CM noise suppression band is the most extensive range, and a high frequency reaches 16.2 GHz. Compared with previous research [20], the FBW is increased by 41.3%.

Table 6. Detailed comparison with related DGSs.

	[17]	[18]	[19]	[20]	This Work
Physical size (mm × mm)	10 × 10	—	9.2 × 19	7 × 7	5.04 × 5.2
Electrical size ($\lambda_g \times \lambda_g$)	—	—	0.22 × 0.46	0.305 × 0.305	0.292 × 0.301
Area	100 mm^2/ —	—/ —	174.8 mm^2/ 0.1012 λ_g^2	49 mm^2/ 0.0929 λ_g^2	26.21 mm^2/ 0.0881 λ_g^2
$S_{cc21} < -10$ dB (GHz)	3–13	2.4–3.15	1.9–7.3	3.7–10.8	2.9–16.2
FBW	125%	Max. 26.8%	Max. 75.5%	98%	139.3%

Table 7 compares the noise magnitude with previous research. Compared with [20], the suppression ratio increases by 20.3%. This demonstrates that the proposed suppression filter possesses a more excellent CM noise-reduction capability.

Table 7. Comparison of the noise magnitude.

	[20]	This Work
Reference Board	170 mV	138 mV
Filter Board	98.6 mV	52 mV
Suppression Ratio	42%	62.3%

5. Conclusions

This paper proposes an ultra-wideband CM noise suppression filter using a magnified coupled DGS. We appropriately designed the resonant frequency using an asynchronously tuned coupled resonant circuit. In the frequency domain, the measurements demonstrated that the S_{dd21} from DC to 12.35 GHz was measured within the −3 dB definition and that the CM noise could be reduced by more than −10 dB from 2.9 to 16.2 GHz. In addition, the FBW of the stopband reached 139.3%. In the time domain, the measurement of the CM noise magnitude was reduced by 62.3%.

Although most high-speed interfaces use stripline routing, we could still place the proposed filter near the connector. The proposed filter is characterized by having two simple geometry structures, no additional components, the most compact size on the PCB, and a considerable FBW. As far as we know, this is the first filter to achieve CM noise suppression exceeding 130% FBW in ultra-wideband. Finally, the proposed filter can be widely implemented to reduce the electromagnetic interference in RF and wireless communication applications with Wi-Fi 5 and 6E. In addition, it can be used in satellite and microwave telecommunication applications with the C and X bands, as part of the Ku band.

Author Contributions: Conceptualization, D.-B.L.; methodology, M.-H.W.; software, M.-H.W.; validation, T.A.; formal analysis, A.A.P.; investigation, T.A.; resources, D.-B.L.; data curation, A.A.P.; writing—original draft preparation, M.-H.W.; writing—review and editing, D.-B.L., T.A. and A.A.P.; visualization, M.-H.W.; supervision, D.-B.L.; project administration, D.-B.L.; funding acquisition, D.-B.L. All authors have read and agreed to the published version of the manuscript.

Funding: This research was funded by the Ministry of Science and Technology (Taiwan), grant number (MOST 110-2221-E-011-052).

Institutional Review Board Statement: Not applicable.

Informed Consent Statement: Not applicable.

Data Availability Statement: No new data were created or analyzed in this study. Data sharing is not applicable to this article.

Acknowledgments: The authors would like to thank Yi-Hsien Lee, Department of Electronic Engineering, National Taipei University of Technology, Taiwan, R.O.C., for providing the idea of defected ground structure in this research.

Conflicts of Interest: The authors declare no conflict of interest.

References

1. Lin, D.-B.; Wang, M.-H.; Huang, Y.-L. Smooth Bend Structures Using Hybrid Compensation for Common-Mode Noise Reduction. *Appl. Sci.* **2022**, *12*, 6479. [CrossRef]
2. Fornberg, P.E.; Kanda, M.; Lasek, C.; Piket-May, M.; Hall, S.H. The impact of a nonideal return path on differential signal integrity. *IEEE Trans. Electromagn. Compat.* **2002**, *44*, 11–15. [CrossRef]
3. Gavenda, J.D. Measured effectiveness of a toroid choke in reducing common-mode current. In Proceedings of the National Symposium on Electromagnetic Compatibility, Denver, CO, USA, 23–25 May 1989; pp. 208–210.
4. Tseng, B.C.; Wu, L.K. Design of Miniaturized Common-Mode Filter by Multilayer Low-Temperature Co-Fired Ceramic. *IEEE Trans. Electromagn. Compat.* **2004**, *46*, 571–579. [CrossRef]
5. Wu, C.-H.; Wang, C.-H.; Chen, C.H. Stopband-Extended Balanced Bandpass Filter Using Coupled Stepped-Impedance Resonators. *IEEE Microw. Wirel. Compon. Lett.* **2007**, *17*, 507–509. [CrossRef]
6. Jin, S.; Quan, X. Novel Balanced Dual-Band Bandpass Filter Using Coupled Stepped-Impedance Resonators. *IEEE Microw. Wirel. Compon. Lett.* **2010**, *20*, 19–21. [CrossRef]
7. Wang, X.-H.; Xue, Q.; Choi, W.-W. A Novel Ultra-Wideband Differential Filter Based on Double-Sided Parallel-Strip Line. *IEEE Microw. Wirel. Compon. Lett.* **2010**, *20*, 471–473. [CrossRef]
8. Lin, W.-J.; Houng, M.-P.; Lin, D.-B.; Tang, I.-T. Investigation in open circuited metal lines embedded in defected ground structure and its applications to UWB filters. In Proceedings of the 2010 IEEE 11th Annual Wireless and Microwave Technology Conference (WAMICON), Melbourne, FL, USA, 26 January 2010; pp. 1–4.
9. Feng, W.; Che, W. Novel Wideband Differential Bandpass Filters Based on T-Shaped Structure. *IEEE Trans. Microw. Theory Tech.* **2012**, *60*, 1560–1568. [CrossRef]
10. Liu, W.-T.; Tsai, C.-H.; Han, T.-W.; Wu, T.-L. An Embedded Common-Mode Suppression Filter for GHz Differential Signals Using Periodic Defected Ground Plane. *IEEE Microw. Wirel. Compon. Lett.* **2008**, *18*, 248–250. [CrossRef]
11. Lin, D.-B.; Chen, Y.-X.; Lee, Y.-H.; Zheng, L.-Z. Transmission line model for designing common-mode suppression filter on multi differential signal pairs. In Proceedings of the 2017 Asia-Pacific International Symposium on Electromagnetic Compatibility (APEMC), Seoul, Republic of Korea, 20 June 2017; pp. 201–203.

12. Pang, Y.; Feng, Z. A compact common-mode filter for GHz differential signals using defected ground structure and shorted microstrip stubs. In Proceedings of the 2012 International Conference on Microwave and Millimeter Wave Technology (ICMMT), Shenzhen, China, 5–8 May 2012; pp. 1–4.
13. Gao, X.-K.; Lee, H.M.; Gao, S.; Liu, E.-X.; Png, C.E. A compact common-mode noise suppression filter for high speed differential signals using defected ground structure. In Proceedings of the 2015 Asia-Pacific Symposium on Electromagnetic Compatibility (APEMC), Taipei, Taiwan, 26–29 May 2015; pp. 685–688.
14. Lin, D.-B.; Zhong, G.-J.; Chen, Y.-X. Wideband common-mode suppression filter design for multi differential signal pairs. In Proceedings of the 2015 Asia-Pacific Symposium on Electromagnetic Compatibility (APEMC), Taipei, Taiwan, 26–29 May 2015; pp. 53–55.
15. Zeng, Z.; Chen, S.J.; Fumeaux, C. A Reconfigurable Filter Using Defected Ground Structure for Wideband Common-Mode Suppression. *IEEE Access* **2019**, *7*, 36980–36990. [CrossRef]
16. Wu, S.-J.; Tsai, C.-H.; Wu, T.-L.; Itoh, T. A Novel Wideband Common-Mode Suppression Filter for Gigahertz Differential Signals Using Coupled Patterned Ground Structure. *IEEE Trans. Microw. Theory Tech.* **2009**, *57*, 848–855. [CrossRef]
17. Zeng, Z.; Yao, Y.; Zhuang, Y. A Wideband Common-Mode Suppression Filter with Compact-Defected Ground Structure Pattern. *IEEE Trans. Electromagn. Compat.* **2015**, *57*, 1277–1280. [CrossRef]
18. Fernández-Prieto, A.; Santos, V.d.; Martel, J.; Río, J.L.M.d.; Mesa, F.; Quevedo-Teruel, O.; Boix, R.R.; Medina, F. Glide Symmetry Applied to the Design of Common-Mode Rejection Filters Based on Complementary Split-Ring Resonators. *IEEE Trans. Circuits Syst. II Express Briefs* **2023**, *70*, 1911–1915. [CrossRef]
19. Zeng, Z.; Bai, L. A Compact Bandwidth and Frequency-Range Reconfigurable CM Filter Using Varactor-Loaded Dual-Slot-Based Defected Ground Structure. *IEEE Trans. Electromagn. Compat.* **2023**, *10*, 5639. [CrossRef]
20. Lin, D.-B.; Lee, Y.-H. A wideband common-mode suppression filter using enhanced coupled defected ground structure. In Proceedings of the 2016 IEEE International Symposium on Electromagnetic Compatibility (EMC), Ottawa, ON, Canada, 25–29 July 2016; pp. 791–796.
21. Ahn, D.; Park, J.-S.; Kim, C.-S.; Kim, J.; Qian, Y.; Itoh, T. A design of the low-pass filter using the novel microstrip defected ground structure. *IEEE Trans. Microw. Theory Tech.* **2001**, *49*, 86–93. [CrossRef]
22. Hong, J.-S. *Microstrip Filters for RF/Microwave Applications*, 2nd ed.; Wiley: New York, NY, USA, 2011.
23. Hammerstad, E.O. Equations for Microstrip Circuit Design. In Proceedings of the 1975 5th European Microwave Conference, Hamburg, Germany, 1–4 September 1975; pp. 268–272.

applied sciences

MDPI

Communication

A Filtering Switch Made by an Improved Coupled Microstrip Line

Xiangsuo Fan [1,2], Xiaokang Chen [1,*], Wenhao Xu [1], Lingping Feng [3], Ling Yu [1] and Haohao Yuan [1]

1 School of Automation, Guangxi University of Science and Technology, Liuzhou 545006, China; 100002085@gxust.edu.cn (X.F.); 221068407@stdmail.gxust.edu.cn (W.X.); 100001839@gxust.edu.cn (L.Y.); 100000917@gxust.edu.cn (H.Y.)

2 Guangxi Collaborative Innovation Centre for Earthmoving Machinery, Guangxi University of Science and Technology, Liuzhou 545006, China

3 School of Microelectronics, Xi'an Jiaotong University, Xi'an 710049, China; fenglinping@xjtu.edu.cn

* Correspondence: 221068342@stdmail.gxust.edu.cn

Abstract: In this paper, we propose a new filtering switch with excellent working performance made using an optimized coupled microstrip line. Upon analyzing the RF (radio frequency) front-end's system structure, the switching device was simplified to a diode, which was connected to the microstrip circuit we designed to become a filter switch with both filtering and shutdown functions. First, we obtained an equivalent schematic of this filtering switch based on the relevant microstrip line theory. This switch consists of two coupled microstrip circuits, parallel-coupled feed lines and coupled-line stub-load resonators (CLSs), and a PIN diode. Second, the operating principle is described by the switching of the operating states, with ideal shutdown performance in the off state and considerable selectivity and excellent out-of-band rejection performance in the filtered state. Finally, a prototype filtering switch with a center frequency of 0.8 GHz was designed and tested. After subsequent optimization and improvement, the simulation and test performance results were noticeably consistent, consequently verifying the performance requirements of this filtering switch in two operating states in the center frequency band.

Keywords: filtering switch; coupled microstrip line; working performance

1. Introduction

As enhancements in communication technology become increasingly mature, the working performance and integration necessities of RF (radio frequency) systems' front-end devices are also increasing [1–3]. In modern communication systems, an RF (radio frequency) system's front end is usually composed of filters and switches designed separately and then cascaded. This cascading method will cause matching distortion, which undoubtedly increases the production cost and design difficulty of the RF (radio frequency) system's front-end circuit structure. The solution to this problem is to design an integrated circuit with the required performance, i.e., a filtering switch with both filtering and shutdown functions [4].

Figure 1 depicts our idea for the RF (radio frequency) front-end system, namely, integrating the filter and switch modules in the RF (radio frequency) front-end system to form a filtering switch. The filtering switch is a microwave-integrated circuit, which can change the circuit's working state by changing the external diode's on/off state. This design method aims to improve the circuit's integration, avoid the matching distortion of the filter and switch cascade in the RF (radio frequency) front end, and reduce the integrated circuit's design cost. There are different types of filtering switches, such as those designed with external PIN diodes for harmonic filter switching [5], those designed with cavity-coupled bandpass filters for switching between filter and switch functions [6], and those with an integrated divider and filter as one circuit [7,8]. There are filter switches that are integrated after designing a triplexer using the aggregate component theory [1] and integrated filter switches based on the fractal stub-loaded resonator (F-SLR) design [9]. In our paper, we combine a theoretical analysis of a coupled microstrip line and PCB fabrication process to design and produce a filtering switch with two functions of being switched between filter and off.

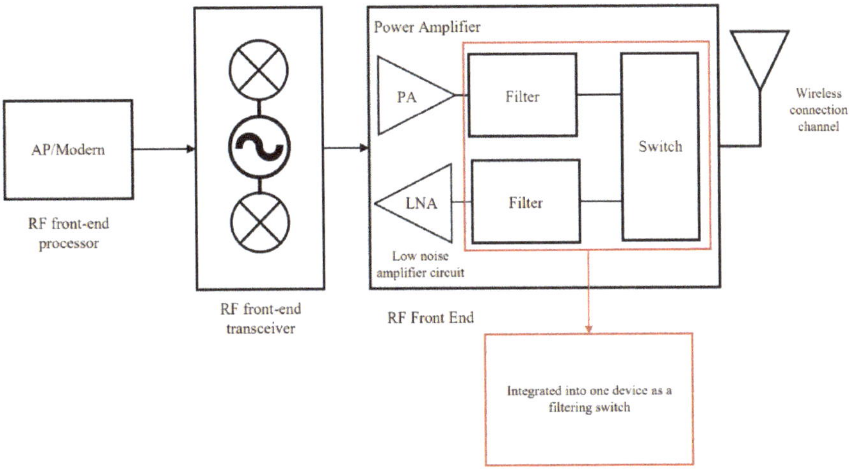

Figure 1. RF (radio frequency) front-end system diagram.

Previously, many researchers have labored on the design and commissioning of filtering switches. The filtering switch proposed in [10] has a compact size and clear passband selectivity; however, it has a relatively high insertion loss (IL) in pass-state operation and less than optimal filtering operating performance. Although the view that the transmission zeros (TZs) generated by the filtering switch enhance the filtering switch's through-state selectivity to some extent was introduced in [11], the selectivity impact exhibited by using the filter switch in the filtered state is still relatively unsatisfactory. Although the filtering switch mentioned in [2] exhibits excellent shutdown rejection, it is not the best for use in the filter function, and while the filtering switch designed in [12] has excellent turn-off rejection performance and excessive integration, its relative bandwidth (FBW) is only 8.96%, with a connection insertion loss of 1.72 dB, which indicates an unsatisfactory band pass filtering effect.

To avoid the above filtering switch problems and to fill knowledge gaps in the related fields, in this paper, we combine an improved filter response theory of coupled microstrip lines [13] with the basic design techniques of filtering switches [14–16]. We also studied related research in similar fields, such as MEMS switch-based filters [17], ferrite resonant filters [18], and enhanced gain in waveguide antennas [19]. Based on these theoretical approaches, we propose a filtering switch designed according to the filtering effect of coupled microstrip lines.

Our design idea is to choose parallel-coupled feeder lines and coupled-line stub-load resonators (CLSs). At a wider operating bandwidth, the parallel-coupled feeder lines can reduce the insertion loss (IL) and return loss (RL), and the coupled-line stub-load resonators (CLSs) can produce symmetrically controllable transmission zeros (TZs).

Compared to the design idea of using aggregate elements in [1], the design using these two coupled microstrip lines is easier to implement. There is no need to consider the specific length setting of the short route, and the design is less difficult. Compared to the design using a fractal stub-loaded resonator (F-SLR) proposed in [9], coupled-line stub-loaded resonators (CLSs) are able to achieve good bandwidth more easily while providing transmission zeros. Moreover, the results achieved with the help of simulation and the actual performance of the work performed will match a bit better and be easier to optimize and adjust.

Combining the benefits of the two coupled microstrip lines for the design ensures the reduction in insertion loss under the filtering effect, as well as the filtered state's selectivity through the symmetrical transmission zeros (TZs). When the diode is in on-state bias, the filtering switch is in the filter state with excessive frequency selectivity, extensive rejection band, and extraordinarily low IL. When the diode is in off-state bias, the filtering switch is in the off-state mode with true off-state suppression.

2. Materials and Methods

We first perform a theoretical analysis of the coupled microstrip line given in Figure 2.

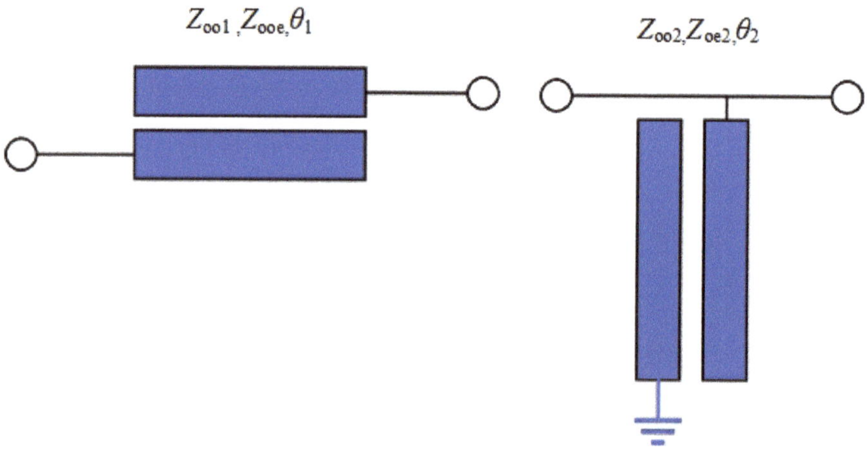

Figure 2. Parallel-coupled feed lines (**left**); coupled-line stub-loaded resonators (**right**).

Figure 2 shows the equivalent schematic diagram of the coupled microstrip lines required to compose the filtering switch, where the parallel-coupled feed lines exhibit bandpass filtering effects, and where the coupled-line stub-loaded resonators mainly produce transmission zeros to enhance the passband selectivity in the filtered state.

First, the Z-parameter matrix of the parallel-coupled feed line is obtained primarily based on the coupled microstrip line's parity mode analysis principle:

$$Z = \begin{bmatrix} Z_{11} & Z_{12} \\ Z_{21} & Z_{22} \end{bmatrix} = \tag{1}$$

$$\begin{bmatrix} -\frac{j}{2}(Z_{oe1} + Z_{oo1})cot\theta_1 & -\frac{j}{2}(Z_{oe1} - Z_{oo1})csc\theta_1 \\ -\frac{j}{2}(Z_{oe1} - Z_{oo1})csc\theta_1 & -\frac{j}{2}(Z_{oe1} + Z_{oo1})cot\theta_1 \end{bmatrix} \tag{2}$$

It is easy to see that the enter impedance Z_1 of this coupled microstrip line is:

$$Z_1 = \sqrt{Z_{11}^2 - \frac{Z_{11}Z_{21}^2}{Z_{22}}} = \tag{3}$$

$$\frac{\sqrt{(Z_{oe1} - Z_{oo1})^2 - (Z_{oe1} + Z_{oo1})^2 cos\theta_1^2}}{2sin\theta_1} \tag{4}$$

The $ABCD$ matrix M_1 of the parallel-coupled feed lines can be obtained from the Z-parameter matrix:

$$M_1 = \begin{bmatrix} A_1 & B_1 \\ C_1 & D_1 \end{bmatrix} = \tag{5}$$

$$\begin{bmatrix} \frac{Z_{11}}{Z_{21}} & \frac{Z_{11}Z_{22} - Z_{12}Z_{21}}{Z_{21}} \\ 1/Z_{21} & Z_{22}/Z_{21} \end{bmatrix} \tag{6}$$

The $ABCD$ matrix M_2 of the coupled-line stub-load resonators (CLSs) is:

$$M_2 = \begin{bmatrix} A_2 & B_2 \\ C_2 & D_2 \end{bmatrix} = \tag{7}$$

$$\begin{bmatrix} \frac{Z_{oe2} + Z_{oo2}}{Z_{oe2} - Z_{oo2}} cos^2\theta_2 & \frac{j}{2} \frac{(Z_{oe2} + Z_{oo2})^2}{Z_{oe2} - Z_{oo2}} sin^2\theta_2 - \frac{j2Z_{oe2}Z_{oo2}}{(Z_{oe2} - Z_{oo2})sin\theta_2} \\ \frac{j2}{Z_{oe2} - Z_{oo2}} & \frac{Z_{oe2} + Z_{oo2}}{Z_{oe2} - Z_{oo2}} cos^2\theta_2 \end{bmatrix} \tag{8}$$

The equal impedance Z_2 of the coupled-line stub-load resonators can be acquired from the M_2 matrix:

$$Z_2 = \frac{B_2}{D_2} = \tag{9}$$

$$\frac{j}{2}(Z_{oe2} + Z_{oo2})tan\theta_2 - \frac{j2Z_{oe2}Z_{oo2}}{(Z_{oe2} + Z_{oo2})sin\theta_2 cos\theta_2} \tag{10}$$

According to [13], the transmission zeros (f_a, f_b) of the bandpass filter in the skip state are acquired from the equal impedance at $Z_2 = 0$.

$$f_a = \frac{2}{\pi} arcsin\sqrt{1 - K_2^2} < f_0 \tag{11}$$

$$f_b = 2f_0 - f_a > f_0 \tag{12}$$

where f_0 is the center frequency at which the device operates, and K_2 is the coupling coefficient of the coupled-line stub-load resonators. Compared with the traditional quarter-wavelength stub, the coupled-line stub-load resonators (CLSs) can produce two symmetric transmission zeros to improve the passband selectivity. Compared to stub-loaded resonators, coupled-line stub-load resonators (CLSs) can exhibit better passband effects at the same operating frequency.

Figure 3 shows the equivalent schematic of the filter switch based on the above analysis of coupled microstrip lines. The physical structure of the proposed filtering switch consists of two coupled microstrip lines, a parallel-coupled feed line, and coupled-line stub-load resonators (CLSs), where the coupled microstrip line TL_1 is connected to a PIN diode to achieve a good on/off control effect. The equivalent schematic diagram that makes up this filtering switch is given in Figure 3.

Figure 4 depicts the equal circuit used for the filtering switch's PIN diode and the bias's specific states. The filtering switch behaves as a bandpass filter characteristic when the switch is in on-state bias, and as a shutdown characteristic when it is in off-state bias.

First, based completely on the impedance evaluation and response of the coupled microstrip lines cited above, we conclude that the return loss (RL) of this filtering switch underneath on-state bias is the main decision, with the aid of the parallel-coupled feed lines TL_1 and TL_2; this is proven in Figure 2, while the transmission zero and out-of-band rejection are decided with the aid of the coupled-line brief-reduce loading resonators TL_3 and TL_4.

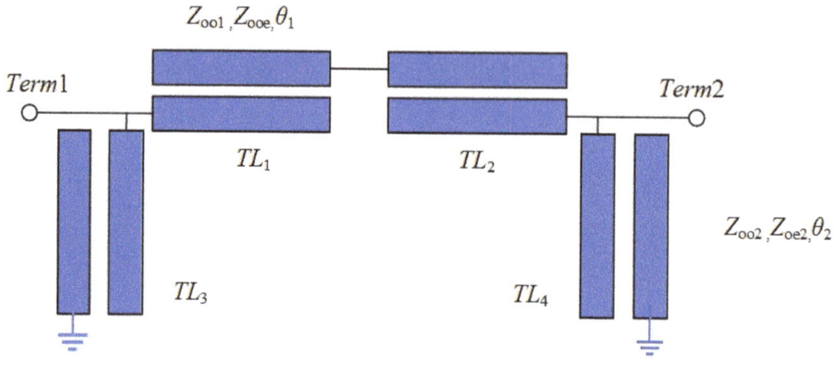

Figure 3. Equivalent circuit schematic.

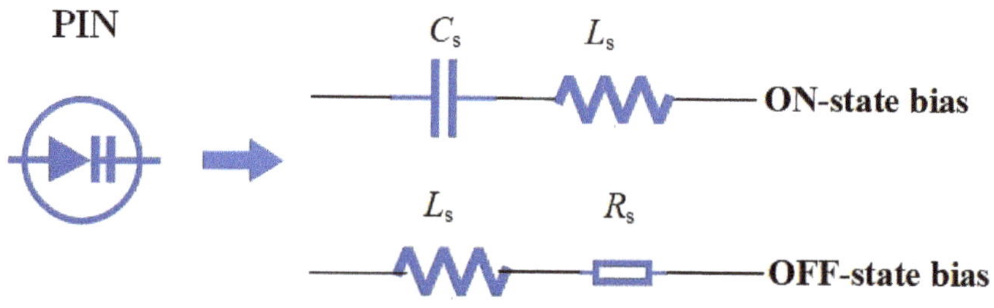

Figure 4. Equivalent circuit of Skyworks SMP1345-079LF PIN diode.

Second, according to our analysis of the individual coupled microstrip lines [13], we derive the final system matrix M_d.

$$M_d = \begin{bmatrix} A_d & B_d \\ C_d & D_d \end{bmatrix} = \tag{13}$$

$$\begin{bmatrix} 1 & 0 \\ 1/Z_2 & 1 \end{bmatrix} \begin{bmatrix} A_1 & B_1 \\ C_1 & D_1 \end{bmatrix} \begin{bmatrix} D_1 & B_1 \\ C_1 & A_1 \end{bmatrix} \begin{bmatrix} 1 & 0 \\ 1/Z_2 & 1 \end{bmatrix} \tag{14}$$

Finally, the filtering switch's S-parameters are calculated from the system matrix M_d.

$$S_{11} = \frac{A_d + B_d/Z_0 - C_d Z_0 - D_d}{A_d + B_d/Z_0 + C_d Z_0 + D_d} \tag{15}$$

$$S_{21} = \frac{2}{A_d + B_d/Z_0 + C_d Z_0 + D_d} \tag{16}$$

3. Results

3.1. Design and Measurement

Figure 5 shows the physical layout of our proposed prototype filter switch based on the performance analysis of the selected microstrip line (Figure 2) and the construction of the schematic diagram (Figure 3). Based on the physical layout of Figure 5, the filter switch's physical prototype is fabricated (Figure 6).

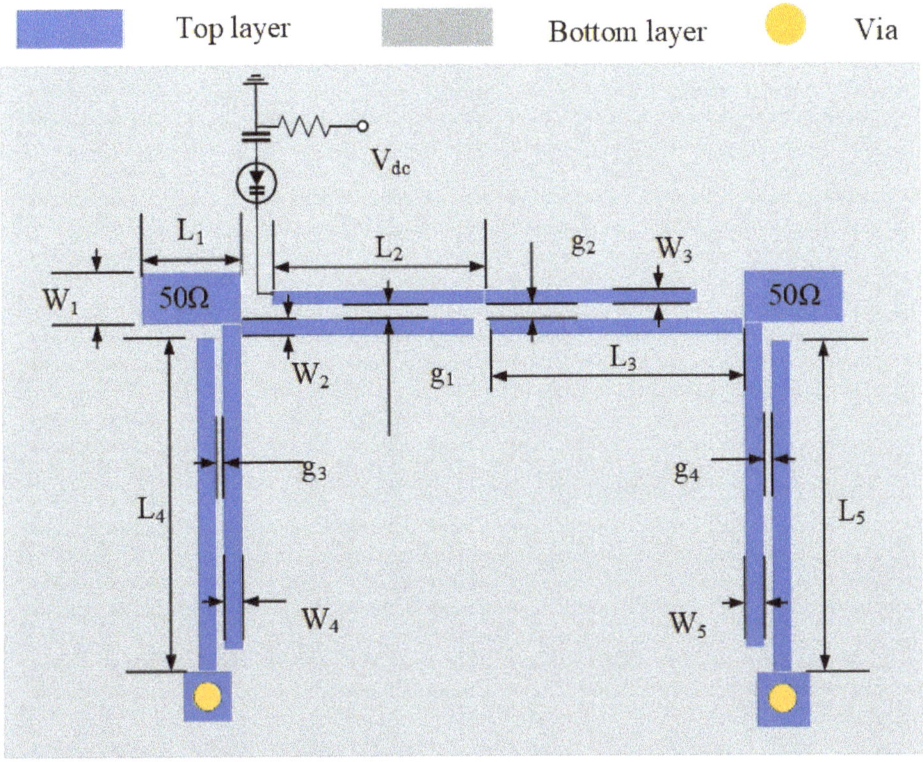

Figure 5. Filtering switch's physical layout.

Figure 6. Photograph of the fabricated filtering switch.

To effectively control the implementation of the on- and off-state function of this filtering switch in practical engineering, PIN diodes are connected to the parallel-coupled feed line. Figure 5 depicts the typical design of the filtering switch, with the complete circuit used being fabricated on a Rogers _RT_Duroid5880 (ε_{re} = 2.2, h = 0.8 mm, tan δ = 0.0009) substrate. Murata capacitors C_b = 470 pF are used for DC blocking off and RF sign pass devices. In this paper, a PIN diode (SMP1345-079LF) from Skyworks is used, and the equal circuit [10] is given in Figure 4, i.e., L_s = 0.7 nH, R_s = 2 Ω, C_s = 0.15 pF. An external DC voltage equal to 16 V is applied to the diode through a 1 kΩ resistor, which adjusts the on-state or off-state of the diode, thus achieving the filtering switch's on-and-off effect.

Under the simulation and debugging of the ADS 2019 software , the impedances of the coupled microstrip lines we chose to analyze in even and odd modes are Z_{oe1} = 205 Ω, Z_{oo1} = 83.7 Ω, Z_{oe2} = 186.1 Ω, and Z_{oo2} = 76 Ω. According to the actual situation, the external equivalent switching device will produce an excitation effect on the filter circuit using the ADS software to optimize the overall circuit, and the physical dimensions obtained after the system processing are as follows: W_1 = 2.38, L_1 = 17.8, W_2 = 0.25, L_2 = 73.1, g_1 = 0.193, W_3 = 0.30, L_3 = 68.828, g_2 = 0.17, W_4 = 0.33, L_4 = 71.45, g_3 = 0.18, W_5 = 0.4, L_5 = 70.98, and g_4 = 0.14 (unit: mm).

3.2. Simulated and Actual Measured Data

The simulated and measured parameters of the filtered changes are proven in Figure 7. Figure 7a indicates the simulated and measured transmission and reflection coefficients in the filtered state. Figure 7b indicates its corresponding off state. In this paper, we measured the frequency response of the fabricated filtering switch by using a Keysight E5071C. The outcomes are in suitable agreement with the ADS simulation results.

The measured results, inclusive of the impact of the I/O SMA connector, show that the in-band return loss is higher than 15 dB for a core frequency of 0.8 GHz, while the IL = 0.65 dB has a −3 dB FBW of 44.5%, and the measured turn-off rejection is higher than 27 dB in the passband of the filter state. Out-of-band rejection indicates that the filter change has state-of-the-art frequency selectivity and overall performance in the on state.

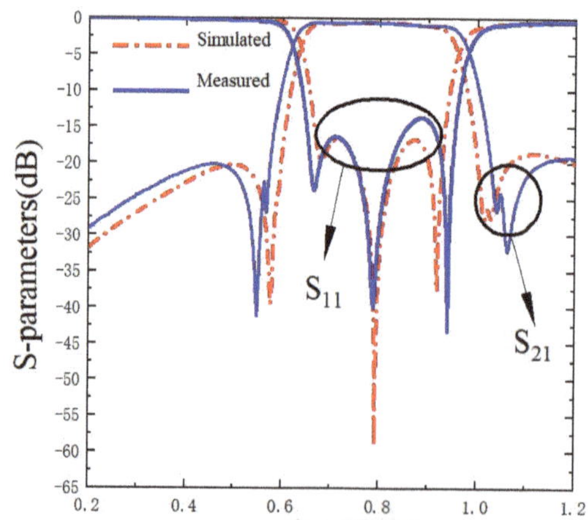

(a)

Figure 7. *Cont.*

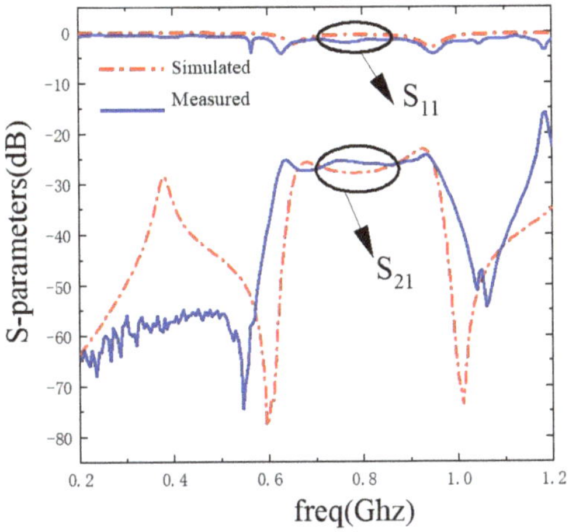

(b)

Figure 7. Simulated and actual measured filtering switch S-parameters' (**a**) on state (**b**) off state.

4. Discussion

On the one hand, the coupling of the microstrip lines is adjusted by the bandpass filter response of the parallel-coupled microstrip lines to gain an accurate bandpass filter effect. The g_1 and g_2 values in Figure 5 decide the bandwidth of the filtering effect. However, the coupling of the brief reduce-load resonator can be adjusted to generate the preferred passband transmission zeros (TZs) to enhance the filter function's passband selectivity. The microstrip line coupling decided via the g_3 and g_4 values in Figure 5 can similarly manipulate the vicinity of the transmission zeros. So, when the filtering switch is operating in the on-state filter state, it has good passband selectivity and out-of-band rejection. When the filtering switch is operating in the off state, the input and output ports are approximately fully reflected, and good off-band rejection is easily obtained by changing the PIN diode's bias.

Because we chose to design a lower center frequency, the physical size of the prototype shown in Table 1 is inevitably too large; however, this is only at the prototype stage, and we will conduct more research to further reduce the size and production cost while ensuring the operating performance. Due to our limited experimental resources, unavoidable tolerances occurred during the fabrication of the prototype. The prototype in Figure 7b showed differences between the out-of-band simulation and the actual measurement in the off state, but it did not affect the operating condition in the operating bandwidth. In addition to this, the filtering switch prototype operates at a 0.8 Ghz center frequency and 44.5% FBW in the on state. In this case, part of the passband range is within the low frequency band of 700 MHz, and the relative bandwidth of 44.5% can cover most of the band; therefore, the prototype is expected to be applied to the 700 MHz band in 5G communication.

Table 1. Comparison with other reported works. −: not given.

Ref.	CF (GHz)	IL (dB)	FBW (%)	RL (dB)	OSS (dB)	SIZE (λ_g^2)	Feasibility of 5G Band Operation
[2]	0.903	1.58	10.6	20	37.6	0.0355	Yes
[2]	0.914	1.36	11.4	−	>40	0.0355	Not
[10]	1.246	1.9	15.3	<15	32.8	0.0201	Not
[10]	0.901	1.8	14.3	<15	27.8	0.0191	Not
[11]	0.991	1.08	14.73	20	48.2	0.029874	Not
[12]	0.994	1.72	8.96	−	>40.6	0.033115	Not
This work	0.8	0.65	44.5	16.3	27.3	0.1178	Yes

For the uncontrollable losses that appear above, we will refer to the work on LQG control of linear lossless positive-real systems [20] in our subsequent work to optimize and improve our prototype design, so that it can achieve better operating performance in the future.

5. Conclusions

The performance of the filtering switch made with the PCB fabrication technique was compared with that of the filtering switch discussed earlier, and the information is summarized in Table 1. In comparison with the performance of the filtering switches reported in [2,10–12], the suggested filtering switch in this study exhibits strong selectivity in the on state, as well as a high degree of in-band insertion loss. In the off state, it possesses good out-of-band rejection capabilities. Furthermore, this filtering switch has a low in-band insertion loss. In this paper, we proposed a filtering switch based mainly on coupled microstrip lines, and we experimentally verified its theoretical overall performance. The proposed filtering switch has good frequency selectivity, low insertion loss in the on state, and good overall rejection performance in the off state. Its operating frequency also satisfies the specific 5G frequency band, which means that it has good application prospects for future 5G communication systems.

Author Contributions: Conceptualization, X.F. and X.C.; methodology, X.F., X.C., W.X. and L.F.; software, X.C., W.X. and L.F.; validation, X.F. and X.C.; formal analysis, X.F., X.C., W.X., L.F., L.Y. and H.Y.; investigation, X.F., X.C., W.X. and L.F.; writing—original draft preparation, X.F., X.C., W.X. and L.F.; writing—review and editing, X.F., X.C., W.X., L.F., L.Y. and H.Y. All authors have read and agreed to the published version of the manuscript.

Funding: This work was supported in part by the National Natural Science Foundation of China under Grant 62001170, China; Guangxi Science and Technology Planning Project (AD22080042); the Guangdong Basic and Applied Basic Research Funding Regional Joint Fund for Youth Project under Grant 2019A1515110417; the China Postdoctoral Science Foundation under Grant 2022M712513; and the Joint Foundation of Key Laboratory of Shanghai Jiao Tong University Xidian University, Ministry of Education and Shaanxi Key Laboratory of Deep Space Exploration Intelligent Information Technology under Grant No. 2021SYS04; as well as the Innovation Project of Guangxi Graduate Education (YCSW2023492, YCSW2023477).

Conflicts of Interest: The authors declare no conflict of interest.

Abbreviations

The following abbreviations are used in this manuscript:

RF	radio frequency
TZ	transmission zero
CLS	coupled-line stub-load resonator
CF	center frequency
RL	return loss
IL	insertion loss

References

1. Xu, J.; Chen,Z.-Y.; Wan, H. Lowpass–bandpass triplexer integrated switch design using common lumped-element triple-resonance resonator technique. *IEEE Trans. Ind. Electron.* **2019**, *67*, 471–479. [CrossRef]
2. Xu, J.; Cai, Q.-H.; Feng, Z.-Y. BPF-integrated spdt switches with improved performance using frequency selective star-junction matching circuit and switched magnetic coupling technique. *IEEE Trans. Microw. Theory Tech.* **2019**, *68*, 1452–1461. [CrossRef]
3. Xu, J.; Cai, Q.-H.; Chen, Z.-Y.; Du, Y.-Q. Quasi-lumped-element filter-integrated single-pole double-throw switch. *IEEE Trans. Microw. Theory Tech.* **2017**, *65*, 4564–4571. [CrossRef]
4. Xu, J.-X.; Xue, Y.-M.; Gao, L.; Zhang, X.Y. Switchable filtering circuit with single-and multi-channel operations. *IEEE Trans. Circuits Syst. II Express Briefs* **2020**, *67*, 2958–2962. [CrossRef]
5. Jing, Y.; Cui, H. A design of tunable filter integrated spdt switch based on pin diode. In *Journal of Physics: Conference Series*; IOP Publishing: Bristol, UK, 2022; Volume 2290, p. 012007.

6. Li, Y.C.; Fang, X.; Xue, Q.; Chen, S. Dual-mode filtering switches based on hybrid microstrip-cavity structures. *IEEE Trans. Microw. Theory Tech.* **2021**, *69*, 3853–3860. [CrossRef]
7. Li, H.-Y.; Xu, J.-X.; Yang, Y.; Zhang, X.Y. Novel switchable filter- ing circuit with function reconfigurability between spqt filtering switch and four-way filtering power divider. *IEEE Trans. Microw. Theory Tech.* **2019**, *68*, 867–876. [CrossRef]
8. Shen, G.; Zhu, H.; Zeng, D.; Xue, Q.; Che, W. A ka-band high-power switchable filtering power combiner mmic in 100-nm gan-on-si. *IEEE Trans. Ind. Electron.* **2021**, *69*, 10467–10477. [CrossRef]
9. Xu, J.; Liu, F.; Feng, Z.-Y.; Guo, Y.-F. Diplexer-Integrated SPDT Switches With Multiple Operating Modes Using Common Fractal Stub-Loaded Resonator. *IEEE Trans. Microw. Theory Tech.* **2021**, *69*, 1464–1473. [CrossRef]
10. Xu, J. Compact switchable bandpass filter and its application to switch-able diplexer design. *IEEE Microw. Wirel. Compon. Lett.* **2015**, *26*, 13–15. [CrossRef]
11. Xu, J.; Wan, H.; Chen, Z.-Y. Sharp skirt bandpass filter-integrated single-pole double-throw switch with absorptive off-state. *IEEE Trans. Microw. Theory Tech.* **2018**, *67*, 704–771. [CrossRef]
12. Xu, J.; Liu, F.; Feng, Z.-Y. Single-/dual-band bandpass filter- integrated single-pole double-throw switch using distributed coupling tri- mode resonators. *IEEE Trans. Microw. Theory Tech.* **2019**, *68*, 741–749. [CrossRef]
13. Chu, Q.-X.; Qiu, L.-L. Wideband balanced filters with high selectivity and common-mode suppression. *IEEE Trans. Microw. Theory Tech.* **2015**, *63*, 3462–3468. [CrossRef]
14. Deng, P.-H.; Chen, Y.-T.; Liu, R.-C.; Chao, S.-F.; Dai, L.-C. Design of a switchable selectivity bandpass filter based on diode-loaded resonators. In Proceedings of the 2013 European Microwave Conference, Nuremberg, Germany, 6–10 October 2013; pp. 1199–1202.
15. Makimoto, M.; Yamashita, S. *Microwave Resonators and Filters for Wireless Communication: Theory, Design and Application*; Springer Series in Advanced Microelectronics; Springer Science & Business Media: Berlin, Germany, 2001; Volume 4.
16. Xu, J.; Zhu, Y. Tunable bandpass filter using a switched tunable diplexer technique. *IEEE Trans. Ind. Electron.* **2016**, *64*, 3118–3126. [CrossRef]
17. Donelli, M.; Manekiya, M.; Tagliapietra, G.; Iannacci, J. A reconfig- urable pseudohairpin filter based on mems switches. *Sensors* **2022**, *22*, 9644. [CrossRef] [PubMed]
18. Shirkolaei, M.M.; Aslinezhad, M. Substrate integrated waveguide filter based on the magnetized ferrite with tunable capability. *Microw. Opt. Technol. Lett.* **2021**, *63*, 1120–1125. [CrossRef]
19. Mohammadi Shirkolaei, M.; Ghalibafan, J. Magnetically scannable slotted waveguide antenna based on the ferrite with gain enhancement. *Waves Random Complex Media* **2021**, 1–11. [CrossRef]
20. Bucolo, M.; Buscarino, A.; Fortuna, L.; Frasca, M. Lqg control of linear lossless positive-real systems: The continuous-time and discrete- time cases. *Int. J. Dyn. Control* **2022**, *10*, 1075–1083. [CrossRef]

applied sciences

MDPI

Article

An S–K Band 6-Bit Digital Step Attenuator with Ultra Low Insertion Loss and RMS Amplitude Error in 0.25 μm GaAs p-HEMT Technology

Quanzhen Liang [1,2,3], Kuisong Wang [1,3], Xiao Wang [1,3], Yuepeng Yan [1,3] and Xiaoxin Liang [1,3,*]

[1] Communication Center, Institute of Microelectronics of the Chinese Academy of Sciences, Beijing 100029, China; liangquanzhen@ime.ac.cn (Q.L.); wangkuisong@ime.ac.cn (K.W.); wangxiao@ime.ac.cn (X.W.); yanyuepeng@ime.ac.cn (Y.Y.)

[2] University of Chinese Academy of Sciences, Beijing 101408, China

[3] Beijing Key Laboratory of New Generation Communication RF Technology, Beijing 100029, China

* Correspondence: liangxiaoxin@ime.ac.cn; Tel.: +86-198-0031-9190

Featured Application: The DSA proposed in this paper is well suited for wideband phasedarray systems.

Abstract: This paper presents an ultra-wideband, low insertion loss, and high accuracy 6-bit digital step attenuator (DSA). To improve the accuracy of amplitude and phase shift of the attenuator, two innovative compensation structures are proposed in this paper: a series inductive compensation structure (SICS) designed to compensate for high frequency attenuation values and a small bit compensation structure (SBCS) intended for large attenuation bits. Additionally, we propose insertion loss reduction techniques (ILRTs) to reduce insertion loss. The fabricated 6-bit DSA core area is only 0.51 mm^2, and it exhibits an attenuation range of 31.5 dB in 0.5 dB steps. Measurements reveal that the root-mean-square (RMS) attenuation and phase errors for the 64 attenuation states are within 0.18 dB and 7°, respectively. The insertion loss is better than 2.54 dB; the return loss is better than −17 dB; and the input 1 dB compression point (IP1 dB) is 29 dBm at IF 12 GHz. To the best of our knowledge, this chip presents the highest attenuation accuracy, the lowest insertion loss, the best IP1dB, and a good matching performance in the range of 2–22 GHz using the 0.25 μm GaAs p-HEMT process.

Keywords: digital step attenuator; GaAs; insertion loss reduction technique; series inductive compensation structure; small bit compensation structure

1. Introduction

Attenuators serve as a crucial component in RF communication, radar, and measurement systems. Their primary function is to provide amplitude control, including linearity adjustment and damage protection [1,2]. The phased array system, with its high precision beam pointing, fast beam synthesis, and scanning capability, is widely used in modern RF-integrated systems such as radar remote sensing and low orbit broadband satellite communications. In phased-array systems, attenuators are required for high accurate amplitude control to obtain lower side lobe levels, better null points, and higher beam sharpness. Digital step attenuators (DSAs) show superior switching speed, reduced power consumption, enhanced linearity, minimal current variation, precise amplitude control, and diminished amplitude/phase change characteristics when compared to variable gain amplifiers. Furthermore, digital attenuators demonstrate higher attenuation accuracy, an improved voltage standing wave ratio, and a broader attenuation dynamic range when compared to continuous variable attenuators [3–6].

In passive DSAs, three primary topologies have been investigated in the literature: distributed, switched path, and switched T-/π-type attenuators [7–13], where, TL is a λ/4 microstrip line, Ms is a series transistor, Mp is a parallel transistor, Rs is a series branch resistor, Rp is a parallel branch resistor, and Vc and $\overline{Vc}$ are a pair of complementary voltages for controlling the on and off of the transistor.

Switched path attenuators use single-pole-double-throw (SPDT) switches to steer the signal path between a thru line and a resistive network, as shown in Figure 1a. This topology provides low phase variation over attenuation states, but it shows high insertion losses at reference states due to the cumulative losses of all SPDT switches for a multi-bit design, and it occupies a large chip area [7,8].

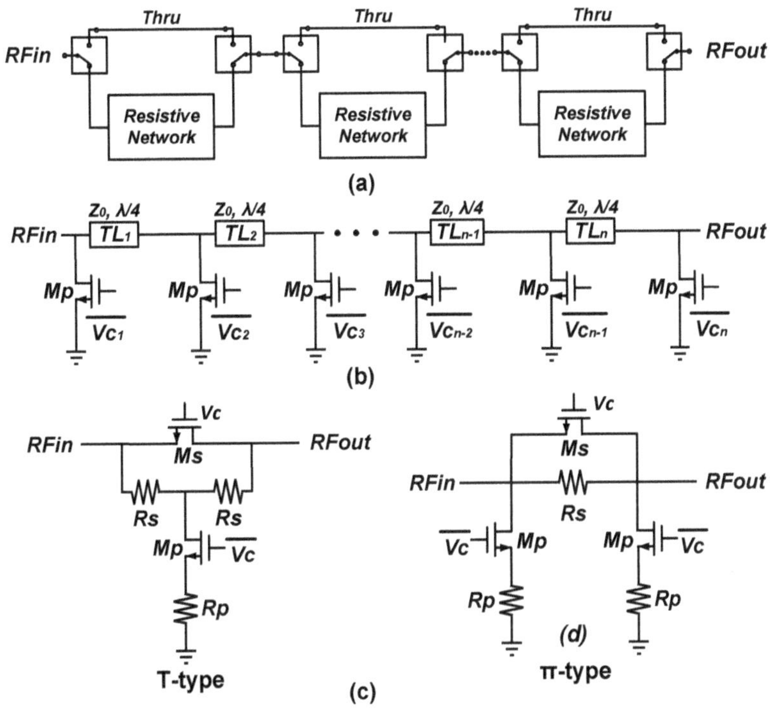

Figure 1. Topologies of DSAs. (**a**) Switched path attenuator. (**b**) Distributed attenuator. (**c**) Switched T/π attenuators.

Distributed attenuators do not have series switches in the signal path and therefore have the advantage of low insertion loss, as shown in Figure 1b. However, they show lower maximum attenuation values and larger chip sizes [9,10].

Switched T-/π-type attenuators show low IL, high maximum attenuation, and compact size, as shown in Figure 1c. However, they still show relatively large attenuation variation and phase variation at higher frequencies, which can limit fine amplitude control in broadband applications [11–13].

The primary source of errors in the amplitude and phase of switched T-/π attenuators is attributed to the parasitic capacitance (Coff) of switching transistors [14]. In order to minimize the Coff-induced errors and improve the accuracy of the attenuator amplitude and phase, several compensation structures have been proposed recently. References [15,16] introduced a parallel capacitive compensation structure that transforms the attenuator into a two-pole, two-zero system, significantly reducing phase error. However, the insertion loss of this structure rapidly deteriorates with frequency increase. At 6 GHz, the insertion loss reaches 2.3 dB and 4 dB, respectively. Consequently, this structure is unsuitable for ultra-wideband attenuators. Adding a tail capacitor in the shunt branch was reported in [17], but this series capacitor limits the bandwidth, counteracting the benefits of the switched-type topology. The inductive compensation structure shows a lower insertion loss compared to the capacitive compensation structure. A parallel inductive compensation structure was introduced by reference [18]. However, the inductor is parallel to the resistance network, resulting in a smaller attenuation value. To address this issue, additional series resistance needs to be incorporated, thereby increasing circuit complexity. Furthermore, as frequency increases, the impact of the phase lag in the parallel inductive compensation structure gradually diminishes. Therefore, this structure is only suitable for low-frequency applications. Consequently, the challenge arises with increasing bandwidth to maintain high linearity, low insertion loss, compact area, and minimal amplitude and phase error concurrently.

In this paper, we propose two innovative compensation structures: a series inductive compensation structure (SICS) designed to compensate for high frequency attenuation values and a small bit compensation structure (SBCS) for large attenuation bits. Additionally, this study adopts a simplified T-type structure and low-pass compensation structure, which not only reduces the insertion loss but also greatly improves the matching performance. Building upon the aforementioned innovations, the proposed 6-bit DSA shows ultra-low insertion loss, excellent attenuation accuracy, high power capacity, and commendable matching performance.

This paper is structured as follows:

In Section 2, we give an in-depth analysis of the conventional switch T-/π structure, deriving optimal resistance values for each attenuation topology to avoid undesired impedance mismatch.

In Section 3, we explain the limitations of conventional digital attenuator design.

In Section 4, we propose a series inductive compensation structure and a small bit compensation structure to effectively extend the operating bandwidth of the attenuator and reduce the additional phase shift.

In Section 5, we propose two methods to reduce the insertion loss, which reduce the insertion loss by 50%.

In Section 6, we cascade the six attenuation bits to maximize performance over the entire bandwidth (from 2 GHz to 22 GHz).

In Section 7, we present the measurement results of the chip and compare with other recently published broadband digital attenuators. The measurement results show that the DSA proposed in this paper can achieve ultra-wideband operation from 2 GHz to 22 GHz with high amplitude tuning accuracy (i.e., <0.18 dB RMS amplitude error) and low phase error (i.e., <7° RMS phase error). Therefore, the proposed DSA can meet the requirements of multi-band phased array systems.

2. Design Method of Conventional Switched Type Attenuators

Conventional attenuators of the switched type typically show three prevalent topologies: T-type [19–25], simplified T-type [4–6,26,27], and π-type [24–31] structures. Each topology is capable of functioning in two distinct states: reference state and attenuation state. The disparity in loss between these two states constitutes the relative attenuation. Given that these three topologies display varying attenuation characteristics, their selection necessitates careful consideration to design specific attenuation bits. Although the resistances of the three topologies can be found in [6,32], calculations ignoring transistor losses are inaccurate for switching attenuators. In this section, we derive the optimal attenuation resistance values considering transistor losses.

2.1. Conventional T-Type Attenuator

Figure 2 shows the conventional T-type attenuation structure and its equivalent circuits.

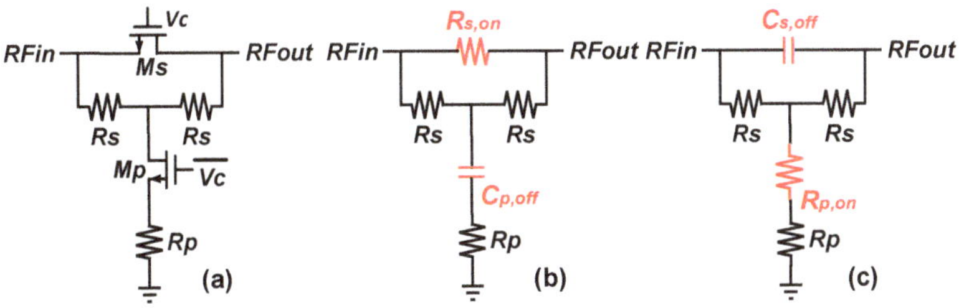

Figure 2. (**a**) Schematic diagram of the conventional T-type attenuation structure and its equivalent circuits. (**b**) Reference state. (**c**) Attenuation state.

In the reference state, transistor *Ms* is on, and transistor *Mp* is off. The insertion loss is determined by the on-resistance of transistor *Ms* and the series resistance *Rs*. Consequently, the corresponding transmission S-parameter can be expressed as follows:

$$S_{21,REF-T} = 1 - \frac{R_s R_{s,on}}{(2R_s + R_{s,on})Z_0 + R_s R_{s,on}} \tag{1}$$

where *Rs,on* represents the on-resistance of the transistor *Ms*, and *Z0* is the characteristic impedance.

In the attenuation state, transistor *Ms* is off, and transistor *Mp* is on; the insertion loss is determined by the T-type attenuation network and the on-resistance of transistor *Mp*. Consequently, the corresponding transmission S-parameter can be expressed as follows:

$$S_{21,ATT-T} = \frac{2Z_0(R_p + R_{p,on})}{2(Z_0 + R_s)(R_p + R_{p,on}) + 2Z_0R_s + {Z_0}^2 + {R_s}^2} \tag{2}$$

where *Rp,on* represents the on-resistance of the transistor *Mp*.

The relative attenuation $\Delta S21$ can be expressed as follows:

$$\Delta S_{21} = \frac{S_{21,REF-T}}{S_{21,ATT-T}} \tag{3}$$

Furthermore, it is imperative that the *S*11 of both the reference state and attenuation state be as minimal as possible to ensure impedance matching. Notably, the transistor *Ms*, typically larger in size, is designed to minimize insertion loss. Consequently, this results in a superior impedance matching of the reference state. The *S*11 of attenuation state can be expressed as follows:

$$S_{11,ATT-T} = \frac{2R_s(R_p + R_{p,on}) + {R_s}^2 - {Z_0}^2}{2Z_0(R_p + R_{p,on}) + 2Z_0R_s + 2R_s(R_p + R_{p,on}) + {R_s}^2 + {Z_0}^2} \tag{4}$$

When the value of $S_{11,ATT-T} = 0$, it allows for the computation of the following:

$$R_s = \frac{Z_0(1 - S_{21,ATT-T})}{(1 + S_{21,ATT-T})} \tag{5}$$

$$R_p = \frac{2Z_0S_{21,ATT-T}}{(1 - S^2_{21,ATT-T})} - R_{p,on} \tag{6}$$

Based on (1), (3), (5), and (6), we can calculate *Rs* and *Rp* under the optimal impedance matching condition.

2.2. Simplified T—Type Attenuator

When the value of attenuation is small, the resistance of the series resistor *Rs* in the T-type structure is also small. Taking the 0.5 dB attenuation bit as an example, we find that the resistance of the series resistor *Rs* in the T-type structure is only 1.34 Ω, which means that the microstrip line can be used to replace the series resistor *Rs*.

Figure 3 shows the simplified T-type structure and its equivalent circuits. In this configuration, *Cp,off* represents the equivalent capacitance of the parallel transistor in its off state, while *Rp,on* denotes the on resistance during the parallel transistor's on state. The simplified T-type design eliminates both the series resistor *Rs* and the series transistor *Ms* found in conventional T-type structure. This allows for targeted attenuation by solely adjusting *Rp* and *Mp* [26,27]. Notably, this architecture boasts reduced insertion loss and a compact layout area.

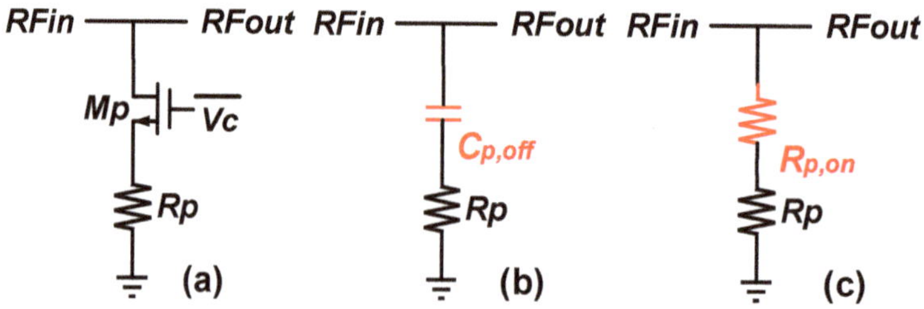

Figure 3. (**a**) Schematic diagram of the simplified T-type attenuation structure and its equivalent circuits. (**b**) Reference state. (**c**) Attenuation state.

In the reference state, the transistor *Mp* is turned off, and the signal to ground shows a high resistance state; at this time, the insertion loss tends to be close to zero. Conversely, in the attenuation state, the transistor *Mp* is on, and the signal leaks to the ground to achieve the attenuation. The resistance value of Rp can still be obtained by (6).

2.3. Conventional π-Type Attenuator

As can be seen from (4), as attenuation increases, *Rp* diminishes, and the return loss of the T-type structure deteriorates. Notably, when the attenuation reaches 8 dB, there is a marked deterioration in the return loss, indicating that the T-type structure is not suitable for large attenuation bits. In contrast, the π-type structure shows greater suitability for large attenuation bits. Figure 4 shows the conventional π-type structure and its equivalent circuits.

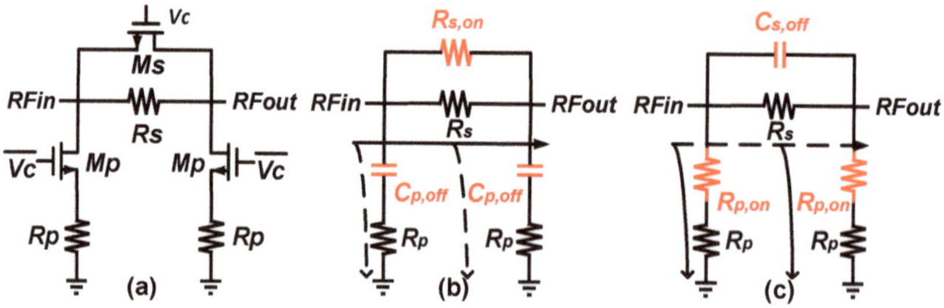

Figure 4. (**a**) Schematic diagram of the π-type attenuation structure and its equivalent circuits. (**b**) Reference state. (**c**) Attenuation state.

In the reference state, transistor *Ms* is on, and transistor *Mp* is off. The insertion loss is determined by the on-resistance of transistor *Ms* and the series resistance *Rs*. Consequently, the corresponding transmission *S*-parameter can be expressed as follows:

$$S_{21,REF-\pi} = 1 - \frac{R_s R_{s,on}}{2(R_s + R_{s,on})Z_0 + R_s R_{s,on}} \tag{7}$$

In the attenuation state, transistor *Ms* is off, and transistor *Mp* is on; the insertion loss is determined by the T-type attenuation network and the on-resistance of transistor *Mp*. Consequently, the corresponding transmission *S*-parameter can be expressed as follows:

$$S_{21,ATT-\pi} = \frac{2Z_0(R_p + R_{p,on})^2}{(2Z_0 + R_s)(R_p + R_{p,on})^2 + (2R_p + 2R_{p,on} + R_s){Z_0}^2} \tag{8}$$

The relative attenuation Δ*S*21 can be expressed as follows:

$$\Delta S_{21} = \frac{S_{21,REF-\pi}}{S_{21,ATT-\pi}} \tag{9}$$

The *S*11 of attenuation state can be expressed as follows:

$$S_{11,ATT-\pi} = \frac{R_s(R_p + R_{p,on})^2 - {Z_0}^2(2R_p + 2R_{p,on} + R_s)}{(R_s + 2Z_0)(R_p + R_{p,on})^2 + 2Z_0(R_s + Z_0)(R_p + R_{p,on}) + R_s {Z_0}^2} \tag{10}$$

Similarly, when the value of *S*11,*ATT*-*π* = 0, it allows for the computation of the following.

The relative attenuation Δ*S*21 can be expressed as follows:

$$R_s = \frac{Z_0(1 - S^2_{21,ATT-\pi})}{2S_{21,ATT-\pi}} \tag{11}$$

$$R_p = \frac{Z_0(1 + S_{21,ATT-\pi})}{(1 - S_{21,ATT-\pi})} - R_{p,on} \tag{12}$$

Based on (7), (8), (11), and (12), we can calculate *Rs* and *Rp* under the optimal impedance matching condition.

3. Limitations of Conventional Structures

In general, attenuation bits with large attenuation values contribute large amplitude and phase variations and limit the bandwidth. In this section, the limitations of the conventional structure are analyzed using a conventional *π*-type attenuator as an example.

Figure 4 shows the equivalent circuit of the conventional *π*-type attenuator. Considering the parasitic capacitance of the transistor, the equivalent circuit of the reference state can be considered as a low-pass filter with phase lag, and the parasitic capacitance *Cp,off* of the shunt transistor will cause the leakage of the high-frequency signal to ground. Conversely, the equivalent circuit of the attenuation state can be considered as a high-pass filter with phase advance, and the high-frequency signal can be directly output through the parasitic capacitance *Cs,off* of the series transistor. Consequently, the presence of *Cs,off* and *Cp,off* leads to a relative attenuation value that is less than the ideal one, accompanied by an additional phase shift exceeding 0°.

The transmission matrix for a conventional *π*-type structure in the reference state is given by the following (13):

$$\begin{pmatrix} A & B \\ C & D \end{pmatrix} = \begin{pmatrix} 1 + \frac{\omega^2 {C_{p,off}}^2 R_p R_{s,on} R_S}{R_s + R_{s,on}} + \frac{j\omega C_{p,off} R_{s,on} R_s}{R_s + R_{s,on}} & \frac{R_s R_{s,on}}{R_s + R_{s,on}} \\ 2\omega^2 {C_{p,off}}^2 (R_p - \frac{R_s R_{s,on}}{R_s + R_{s,on}}) + j2\omega C_{p,off} & 1 + \frac{\omega^2 {C_{p,off}}^2 R_p R_{s,on} R_s}{R_s + R_{s,on}} + \frac{j\omega C_{p,off} R_{s,on} R_s}{R_s + R_{s,on}} \end{pmatrix} \tag{13}$$

The corresponding transmission S-parameter is given by the following (14):

$$S_{21} = \left(\frac{2}{A + \frac{B}{Z_0} + Z_0 C + D}\right) = \frac{1}{1 + \frac{R_{s,on} R_s}{R_{s,on} + R_s}\left(\omega^2 {C_{p,off}}^2 R_P + j\omega C_{p,off} + \frac{1}{2Z_0}\right) + \omega Z_0 C_{p,off}\left(\omega C_{p,off} R_P - \frac{\omega C_{p,off} R_{s,on} R_s}{R_{s,on} + R_s} + j\right)} \quad (14)$$

where *Cp,off* is the parasitic capacitance in the off state of the parallel transistor.

The transmission phase in the reference state can be expressed as follows:

$$\varphi_R \approx -\tan^{-1}\frac{2\omega Z_0^2 C_{p,off}(R_{s,on} + R_s)^2 (R_{s,on} R_s + R_{s,on} Z_0^2 + R_s Z_0^2)}{R_{s,on} R_s} \quad (15)$$

Similarly, the transmission phase in the attenuation state can be expressed as follows:

$$\varphi_A \approx -\tan^{-1}\frac{\omega C_{s,off}(Z_0^2 + R_{p,on} R_p)}{2R_{p,on} Z_0^2 + 2R_p Z_0^2} \quad (16)$$

where $\omega^2 {C_{s,off}}^2 {R_s}^2$ and $\omega^2 {C_{p,off}}^2 {R_p}^2$ are omitted to simplify the calculations.

In order for the transmission phase difference to be zero, the following equation must be satisfied:

$$\Delta\varphi = \varphi_A - \varphi_R = 0 \quad (17)$$

The transmission phase φ_R is less than or equal to 0, while φ_A is greater than or equal to 0°. Therefore, to achieve zero transmission phase difference, the following relationship should be satisfied:

$$\varphi_A = \varphi_R = 0 \quad (18)$$

However, the parasitic capacitance of the transistor cannot be zero, so the phase of the attenuation state of the conventional π-type topology is always ahead of the phase of the reference state. Additionally, as the frequency increases, the reactance of the capacitor decreases, which can exacerbate the amplitude and phase errors between the two states.

4. Design of the Proposed Structures

In this section, we propose two innovative compensation structures: a series inductive compensation structure (SICS) designed to compensate for high frequency attenuation values and a small bit compensation structure (SBCS) for large attenuation bits. The proposed compensation structures effectively broaden the bandwidth of the attenuator and reduce the high-frequency additional phase shift. Furthermore, the compensation structures are equally applicable to T-type attenuators.

4.1. Series Inductive Compensation Structure (SICS)

Figure 5 shows the modified π-type attenuator with a series inductive compensation structure and its equivalent circuits for the reference and attenuation states.

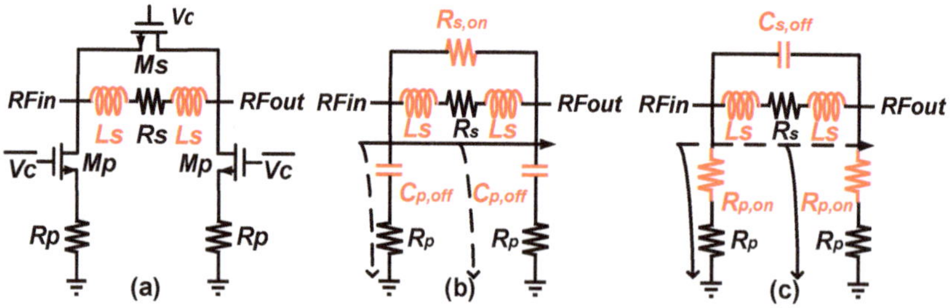

Figure 5. (**a**) Schematic diagram of the modified $\pi-$type attenuator and its equivalent circuits. (**b**) Reference state. (**c**) Attenuation state.

For the amplitude, when the resistance *Rs* is connected in series with two inductors *Ls*, the resultant total impedance becomes $Rs + 2j\omega Ls$. Notably, the total impedance value escalates in correlation with the frequency, thereby compensating for the attenuation values at high frequencies.

For the phase, the transmission phases φ_R and φ_A in the reference and attenuation states of SICS can be written as (1) and (2), respectively.

$$\varphi_R = -\tan^{-1}\left(\frac{2((R_{s,on}{}^2 + 4\omega^2 L_s C_{p,off} Z_0(Z_0 + R_{s,on}))L_s + (Z_0 R_{s,on}{}^2 + R_s R_{s,on}(R_{s,on} + 4Z_0) + R_s{}^2(Z_0 + R_{s,on}))Z_0 C_{p,off})\omega}{(2Z_0 + R_s)R_{s,on}{}^2 + 4R_{s,on}R_s{}^2 + 2(R_s{}^2 + 2R_s R_{s,on})Z_0 - 2\omega^2 L_s(2C_{p,off} R_{s,on}{}^2 Z_0 - 2L_s R_{s,on} - 4L_s Z_0)}\right) \quad (19)$$

$$\varphi_A = \tan^{-1}\left(\frac{(C_{s,off} R_s{}^2 + 12\omega^2 L_s{}^2 C_{s,off} - 2L_s)(R_p + Z_0)\omega}{(R_s R_p + Z_0 R_s + 4Z_0 R_p)(1 - 4\omega^2 L_s C_{s,off})}\right) \quad (20)$$

The inductor required to realize (17) is derived as follows:

$$L_S \cong \frac{1 - \sqrt{1 - 16\omega^2 R_s{}^2 C_{s,off}{}^2}}{2\omega^2 C_{s,off}} \quad (21)$$

Figure 6 shows the simulated relative attenuation and additional phase shift versus frequency for different *Ls* values of the 8 dB modified π-type attenuator. As the value of compensation inductance *Ls* increases, both the relative attenuation value and additional phase shift progressively approach ideal values. Selecting the appropriate *Ls*, the attenuation bandwidth can be broadened, and the additional phase shift can be optimized. This paper uses microstrip lines instead of series inductors to achieve compensation functions.

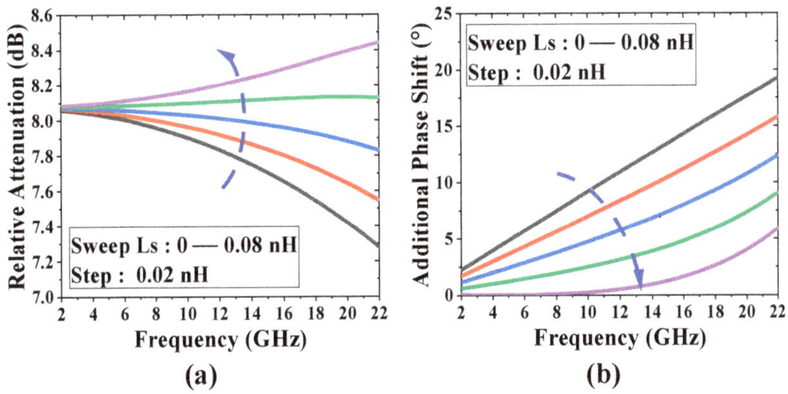

Figure 6. Simulated (**a**) relative attenuation and (**b**) additional phase shift using SICS for different *Ls* values.

It is worth noting that *Ls* has little effect on the low-frequency performance, which is critical for designing ultra-wideband attenuators.

4.2. Small Bit Compensation Structure (SBCS)

The conventional method to achieve a 16 dB attenuator is to cascade two 8 dB conventional π-type attenuators. However, there are many parasitic parameters in the π-type structure, and the parasitic effect will cause errors in the high frequency attenuation and additional phase shift. Cascading two π-type attenuators will accumulate these errors and deteriorate matching performance. The matching performance of the conventional T-type structure deteriorates with the increase of attenuation, which will lead to the deterioration of the performance of the whole 6-bit DSA.

Figure 3a shows the simplified T-type structure, which has the dual capability of compensating attenuation and minimizing insertion loss. Notably, the attenuation values of both the T-type and π-type structures diminish with increasing frequency. In contrast, the attenuation value of the simplified T-type structure escalates with frequency, enabling a small bit compensation through its utilization. Specifically, a modified π-type and T-type attenuator is employed to achieve most of the attenuation firstly, and a simplified T-type attenuator is then inserted between them to compensate for the high-frequency attenuation and further improve the matching performance, as shown in Figure 7.

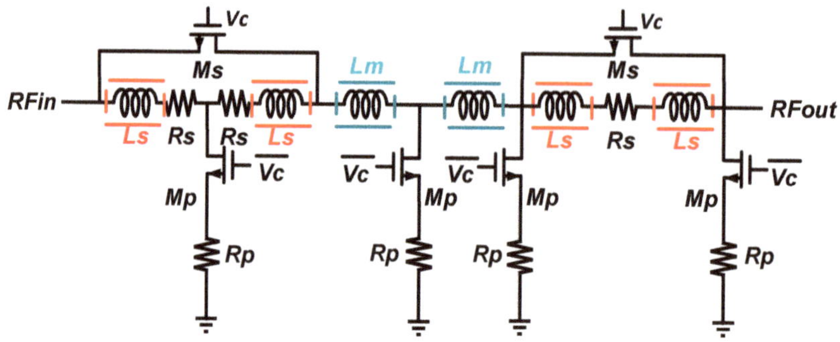

Figure 7. Schematic of 16 dB attenuator with small bit compensation structure.

Figure 8 shows the comparison of simulation results for these two structures, where, the dashed line represents the simulation results of the conventional structure, and the solid line represents the simulation results of the small bit compensation structure proposed in this paper. The simulation results show that the implementation of SBCS enhances the high-frequency attenuation value of a 16 dB attenuator by 1.2 dB, significantly broadening the operating bandwidth of the attenuator. Furthermore, it substantially improves the return loss.

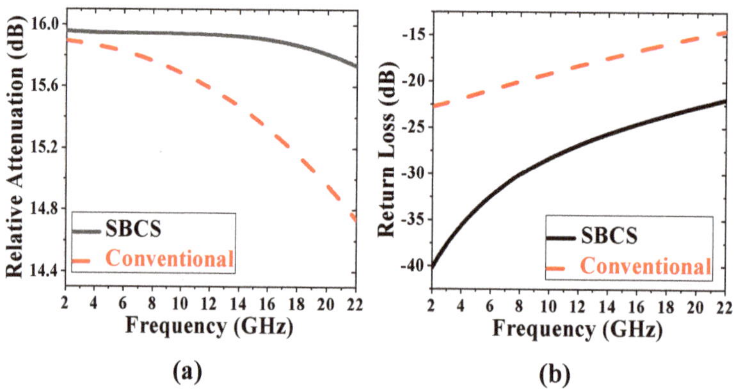

Figure 8. Simulation results of conventional structure and SBCS. (**a**) Relative attenuation and (**b**) return loss.

The SBCS offers an effective design strategy for the creation of attenuators with large attenuation value.

5. Insertion Loss Reduction Techniques (ILRTs)

The primary sources of DSA insertion loss are twofold: the transistor on-resistance and reflection between multiple attenuation bits. Consequently, this paper proposes ILRTs to minimize circuit insertion loss and enhance matching performance.

5.1. Simplified T-Structure

The insertion loss of a conventional T-type attenuator is related to the on-resistance of the series transistors. While the on-resistance of series switches is unavoidable in a switching T-type topology, in some cases, the series transistors can be removed without loss of functionality.

Figures 2a and 3a show the T-type and simplified T-type structure, respectively. The simplified T-type structure removes the series transistor Ms and series resistors *Rs*, which greatly reduces the insertion loss.

Figure 9 shows the insertion loss of the simplified T-type and the T-type attenuators, where, the dashed line shows the simulation results for the conventional T-structure and the solid line shows the simulation results for the simplified T-structure. The result shows that the insertion loss of the simplified T-type structure is reduced by 1.15 dB. In this paper, both 0.5 dB and 1 dB attenuation bits use the simplified T-type structure, which reduces the insertion loss of the 6-bit DSA by 2.3 dB.

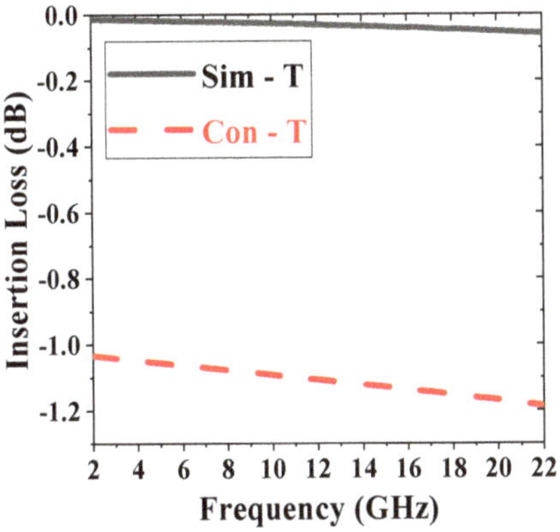

Figure 9. Comparison of insertion loss between T-type and simplified T-type attenuators.

5.2. *Low-Pass Inductive Compensation Structure*

When cascading attenuation bits, the parasitic capacitance of transistors causes mismatch, which results in reflected signal power and causes unnecessary loss. Since the transistor parasitic capacitance forms a high-pass structure, we aim to construct a low-pass filter to compensate for its tendency to change; thus, we connect inductors in series at the input and output of the π-type structure to realize the low-pass filtering function, and the π-type topology of the structure employing low-pass inductive compensation is shown in Figure 10a.

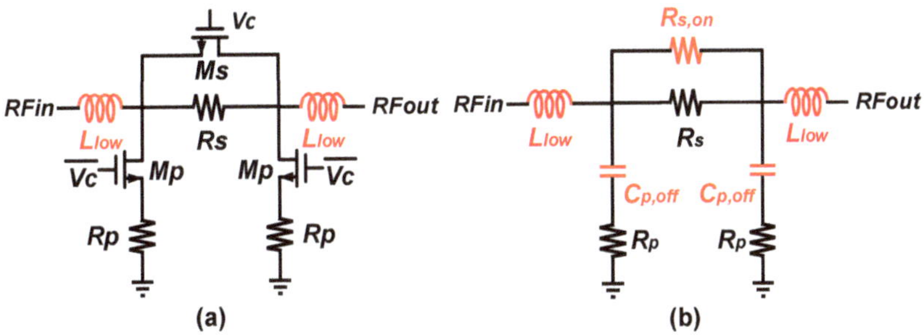

Figure 10. (**a**) π-structure with low-pass inductive compensation and its equivalent circuit. (**b**) Reference state.

Figure 10b shows the equivalent circuit of Figure 10a in its reference state. *Llow*, *Cp,off*, *Rs,on* and *Rs* form a low-pass filter.

Figure 11 shows the insertion loss before and after using low-pass inductive compensation structure to the conventional 8 dB π-type attenuator. Low-pass inductive compensation structure reduces the insertion loss from −1.36 dB to −0.92 dB, effectively compensating for the circuit's capacitive losses. This paper uses microstrip lines instead of inductors to achieve compensation functions.

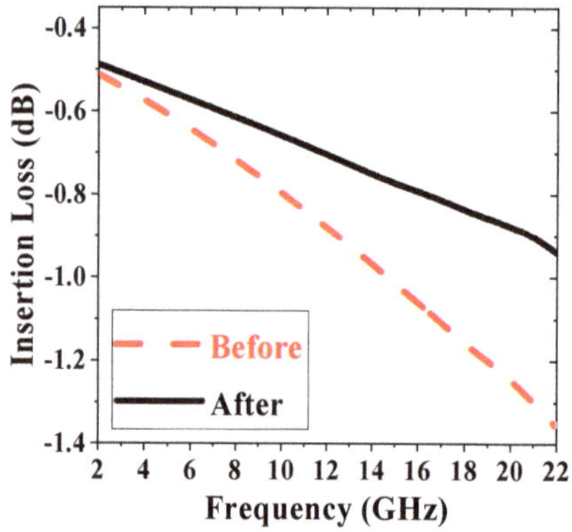

Figure 11. Comparison of insertion loss between the π-type attenuator before and after using low-pass inductive compensation structure.

6. Implementation of The Six-Bit DSA

Figure 12 shows the complete schematic of the 6-bit DSA, which has an attenuation range of up to 31.5 dB in 0.5 dB steps. The 0.5 dB and 1 dB attenuation bits use a simplified T-type structure to minimize insertion loss; the 2 dB and 4 dB attenuation bits use a modified T-type structure based on SICS; the 8 dB attenuation bit uses a modified π-type structure based on SICS; and the 16 dB attenuation bits use an innovative multi-type structure based on SBCS to minimize amplitude errors and phase fluctuations. Since the parasitic capacitance associated with the transistors and the interconnection structures presents a negative imaginary impedance, this shifts S_{11} and S_{22} down along the 50 Ω circle on the Smith chart. Inserting series inductors *Lm* between bits improves the matching characteristics, thereby maximizing operational bandwidth. In this paper, transmission lines are used instead of inductors. In addition, the transistors in each attenuation bit are optimized to achieve a balance between insertion loss and phase error.

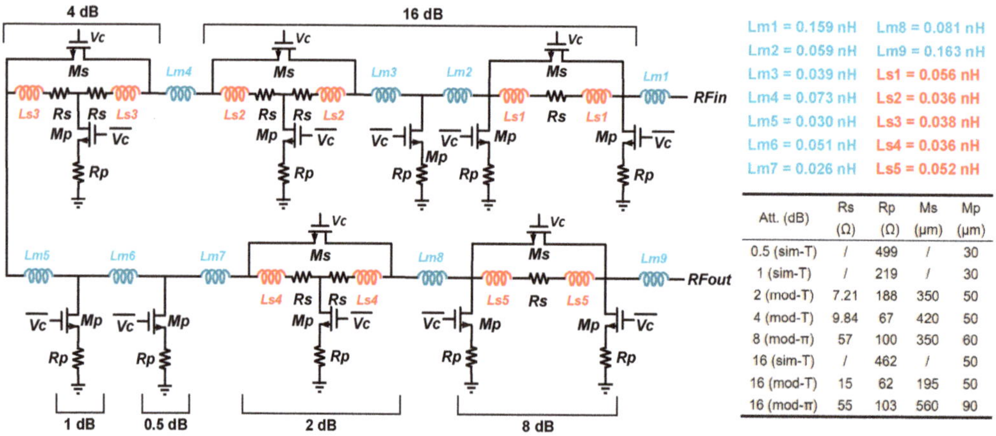

Att. (dB)	Rs (Ω)	Rp (Ω)	Ms (μm)	Mp (μm)
0.5 (sim-T)	/	499	/	30
1 (sim-T)	/	219	/	30
2 (mod-T)	7.21	188	350	50
4 (mod-T)	9.84	67	420	50
8 (mod-π)	57	100	350	60
16 (sim-T)	/	462	/	50
16 (mod-T)	15	62	195	50
16 (mod-π)	55	103	560	90

Figure 12. Schematic of the 6-bit DSA.

7. Measurement Results

Figure 13 shows the chip micrograph of the proposed 6-bit DSA implemented by a 0.25 μm GaAs process, with a core area of 0.51 mm^2. The chip integrates a positive voltage controller and realizes compatibility with conventional CMOS TTL control voltage.

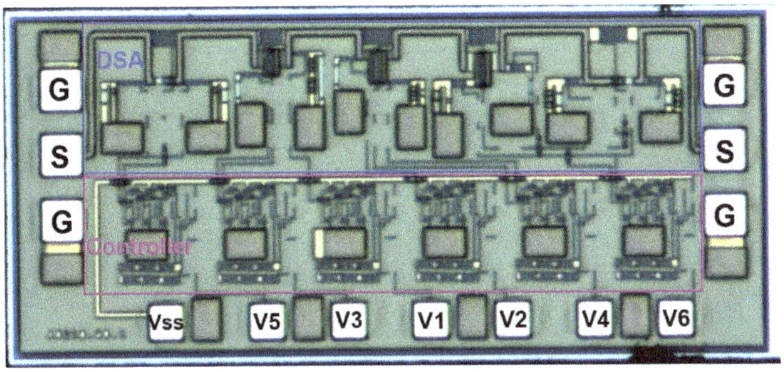

Figure 13. Micrograph of the proposed 6-bit DSA.

To evaluate the performance of the DSA, the root mean square (RMS) amplitude/phase error is defined as follows:

$$RMS\_Amplitude\_Error = \sqrt{\frac{1}{2^n - 1}\sum_{i=1}^{i=2^n-1}(RA_{i\_measured} - RA_{i\_ideal})^2} \quad (22)$$

$$RMS\_Phase\_Error = \sqrt{\frac{1}{2^n - 1}\sum_{i=1}^{i=2^n-1}(\varphi_{i\_measured} - \varphi_{ref})^2} \quad (23)$$

where *RAi_measured* and *RAi_iedal* are the relative attenuation values measured at state i and the ideal relative attenuation value of state *i*, respectively. Similarly, *φi_measured* and *φref* are the additional phase shift measured at state *i* and the reference state, respectively.

The electrical properties of the chip were measured on-chip using a Cascade probe station, an ACP-GSGI50 microwave probe, an Agilent 5227B vector network analyzer, a digital multi-channel power supply, and a DC bias power supply. The DC bias voltage applied to the chip, along with the truth table for the digitally controlled power supply, is detailed in Table 1, where 0 represents 0 V, and 1 represents 5 V. This configuration enables the chip to attain 64 states by changing the control voltages.

Table 1. Bias Voltage and Truth Table.

Bias Voltage/V	Digitally Controlled Voltage						Status
	Vc1	Vc2	Vc3	Vc4	Vc5	Vc6	
−5	0	0	0	0	0	0	0 dB
−5	0	0	0	0	0	1	0.5 dB
−5	0	0	0	0	1	0	1 dB
−5	0	0	0	1	0	0	2 dB
−5	0	0	1	0	0	0	4 dB
−5	0	1	0	0	0	0	8 dB
−5	1	0	0	0	0	0	16 dB
−5	1	1	1	1	1	1	31.5 dB

Figure 14 shows the results for this chip. Figure 14a shows the measured relative attenuation of all 64 states, with no overlap. Figure 14b shows the measured insertion loss of −2.54 dB. This remarkably low insertion loss can be primarily ascribed to ILRTs. Figure 14c shows the measured RMS amplitude and phase errors, which are within 0.18 dB and 7°, respectively, exhibiting the lowest amplitude error with small phase error between in 2 to 22 GHz in the GaAs process. As expected, the phase error increases when the frequency is increased. However, the additional phase shift is greatly reduced due to the use of series inductive compensation structures. Figure 14d,e show the measured return loss, and the return loss of all 64 states is better than −17 dB, indicating excellent matching performance. Figure 14f shows the measured power capacity at the reference state with the input 1 dB compression point (IP1 dB) of 29 dBm at IF 12 GHz, indicating that the DSA has a sufficiently high-power performance to notbe a limiting factor in the linearity of the phased array system.

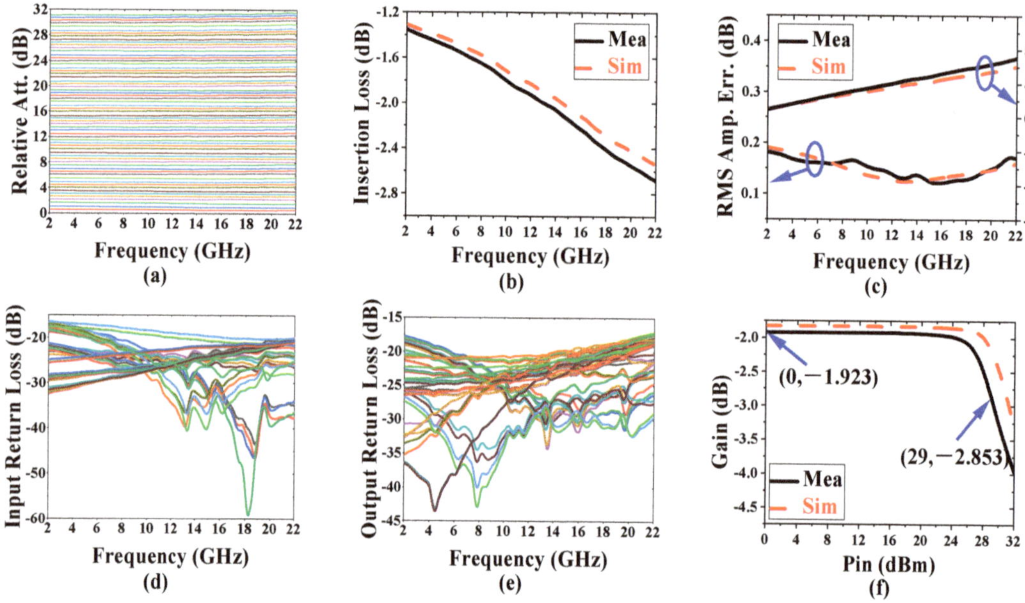

Figure 14. (**a**) Measured relative attenuation. (**b**) Simulated and measured RMS amplitude and phase errors. (**c**) Simulated and measured insertion loss. (**d**) Measured input return loss. (**e**) Measured output return loss. (**f**) Simulated and measured IP1 dB at 12 GHz.

Table 2 presents the comparison of the designed DSA with the published attenuators with state-of-the-art performance. The results show that the proposed DSA shows the best performance in terms of insertion loss, RMS amplitude error, return loss, linearity, and FOM.

Table 2. Comparison of The Prior-Art Attenuators.

Reference	[33]	[34]	[35]	[6]	This Work
Technology	GaAs	GaAs	GaAs	CMOS	GaAs
BW (GHz)	6–18	5–18	DC–23	DC–20	2–22
Range (dB)	31.75	31.5	31.5	31.5	31.5
IL (dB)	9	6.2	6.2	7.4	2.6
RMS Amp error (dB)	0.6	1.1	0.35	0.37	0.18
RMS Phase error (°)	5	2	5	4	7
RL (dB)	12	8	10	12	17
IP1dB (dBm)	N/A	24	N/A	10	29
Area (mm^2)	5.4	3.75	0.47	0.98*/0.14	0.51
FOM #	169	240	668	690	3269

* with pads and non-active space. # FOM $= \frac{\text{BW} \times \text{Range} \times \text{RL}}{\text{IL} \times \text{RMS AmpErr} \times \text{RMS PhaseErr}}$.

8. Conclusions

In this paper, insertion loss reduction techniques reduce the insertion loss of 6-bit DSA by 50%; series inductive compensation and small bit compensation structures are innovatively proposed to greatly broaden the bandwidth of the attenuator and effectively improve the return loss. To the best of our knowledge, the fabricated DSA shows the lowest insertion loss, optimal impedance matching, the highest attenuation accuracy and linearity, and the best FOM using the 0.25 µm GaAs process and is suitable for wideband phased array systems.

Author Contributions: Conceptualization, Q.L., K.W., X.W., Y.Y. and X.L.; methodology, Q.L., K.W., Y.Y. and X.L.; software, Q.L. and X.W.; validation, Q.L., K.W. and X.W.; data curation, Q.L. and X.W.; writing—original draft preparation, Q.L. and K.W.; writing—review and editing, Q.L., Y.Y. and X.L. All authors have read and agreed to the published version of the manuscript.

Funding: This research received no external funding.

Institutional Review Board Statement: Not applicable.

Informed Consent Statement: Not applicable.

Data Availability Statement: The data can be obtained from the authors on request.

Conflicts of Interest: The authors declare no conflicts of interest.

References

1. Nehring, J.; Schutz, M.; Dietz, M.; Nasr, I.; Aufinger, K.; Weigel, R.; Kissinger, D. Highly Integrated 4–32-GHz Two-Port Vector Network Analyzers for Instrumentation and Biomedical Applications. *IEEE Trans. Microw. Theory Tech.* **2017**, *65*, 229–244. [CrossRef]
2. Dogan, H.; Meyer, R.G. Intermodulation Distortion in CMOS Attenuators and Switches. *IEEE J. Solid-state Circuits* **2007**, *42*, 529–539. [CrossRef]
3. Wang, K.; Zhou, T.; Zhang, H.; Qiu, L. A 28–40-GHz Digital Step Attenuator with Low Amplitude and Phase Variations. *IEEE Microw. Wirel. Components Lett.* **2021**, *31*, 64–67. [CrossRef]
4. Cheon, C.D.; Rao, S.G.; Lim, W.; Cardoso, A.S.; Cho, M.-K.; Cressler, J.D. Design methodology for a wideband, low insertion loss, digital step attenuator in SiGe BiCMOS technology. *IEEE Trans. Circuits Syst. II Express Briefs* **2022**, *69*, 744–748. [CrossRef]
5. Zeng, J.; Yu, Z.; Li, J.; Yuan, Y. A Wideband Compact 6-bit Digital Step Attenuator. In Proceedings of the 2022 IEEE MTT-S International Microwave Workshop Series on Advanced Materials and Processes for RF and THz Applications (IMWS-AMP), Guangzhou, China, 13–15 November 2022; pp. 1–3.
6. Song, I.; Cho, M.-K.; Cressler, J.D. Design and Analysis of a Low Loss, Wideband Digital Step Attenuator with Minimized Amplitude and Phase Variations. *IEEE J. Solid-state Circuits* **2018**, *53*, 2202–2213. [CrossRef]
7. Sjogren, L.; Ingram, D.; Biedenbender, M.; Lai, R.; Allen, B.; Hubbard, K. A low phase-error 44-GHz HEMT attenuator. *IEEE Microw. Guid. Wave Lett.* **1998**, *8*, 194–195. [CrossRef]
8. Cho, M.-K.; Kim, J.-G.; Baek, D. A Broadband Digital Step Attenuator with Low Phase Error and Low Insertion Loss in 0.18-µm SOI CMOS Technology. *ETRI J.* **2016**, *35*, 638–643. [CrossRef]
9. Kim, K.; Lee, H.-S.; Min, B.-W. V-W Band CMOS Distributed Step Attenuator with low Phase Imbalance. *IEEE Microw. Wirel. Components Lett.* **2014**, *24*, 548–550. [CrossRef]
10. Min, B.-W.; Rebeiz, G.M. A 10–50-GHz CMOS distributed step attenuator with low loss and low phase imbalance. *IEEE J. Solid-state Circuits* **2007**, *42*, 2547–2554. [CrossRef]
11. Cheng, W.; Alink, M.S.O.; Annema, A.J.; Wienk, G.J.M.; Nauta, B. A Wideband IM3 Cancellation Technique for CMOS Π- and T-Attenuators. *IEEE J. Solid-state Circuits* **2013**, *48*, 358–368. [CrossRef]
12. Dogan, H.; Meyer, R.G.; Niknejad, A.M. Analysis and Design of RF CMOS Attenuators. *IEEE J. Solid-state Circuits* **2008**, *43*, 2269–2283. [CrossRef]
13. Huang, Y.-Y.; Woo, W.; Yoon, Y.; Lee, C.-H. Highly Linear RF CMOS Variable Attenuators with Adaptive Body Biasing. *IEEE J. Solid-state Circuits* **2011**, *46*, 1023–1033. [CrossRef]
14. Bae, J.; Lee, J.; Nguyen, C. A 10–67-GHz CMOS Dual-Function Switching Attenuator with Improved Flatness and Large Attenuation Range. *IEEE Trans. Microw. Theory Tech.* **2013**, *61*, 4118–4129. [CrossRef]
15. Zhang, Z.; Li, N.; Gao, H.; Li, M.; Wang, S.; Kuan, Y.-C.; Song, C.; Yu, X.; Gu, Q.J.; Xu, Z. A DC–*Ka*-Band 7-Bit Passive Attenuator with Capacitive-Compensation-Based Bandwidth Extension Technique in 55-nm CMOS. *IEEE Trans. Microw. Theory Tech.* **2021**, *69*, 3861–3874. [CrossRef]
16. He, Y.; Zhang, T.; Tang, Y.; Fan, C.; Yan, B. Wideband pHEMT Digital Attenuator with Positive Voltage Control Driver. *IEEE Microw. Wirel. Technol. Lett.* **2023**, *33*, 295–298. [CrossRef]
17. Zhang, L.; Zhao, C.; Zhang, X.; Wu, Y.; Kang, K. A CMOS K-Band 6-bit Attenuator with Low Phase Imbalance for Phased Array Applications. *IEEE Access* **2017**, *5*, 19657–19661. [CrossRef]
18. Davulcu, M.; Caliskan, C.; Kalyoncu, I.; Kaynak, M.; Gurbuz, Y. 7-Bit SiGe-BiCMOS Step Attenuator for X-Band Phased-Array RADAR Applications. *IEEE Microw. Wirel. Components Lett.* **2016**, *26*, 598–600. [CrossRef]
19. Li, N.; Zhang, Z.; Li, M.; Gao, H.; Wang, S.; Song, C.; Xu, Z. A DC–28-GHz 7-Bit High-Accuracy Digital-Step Attenuator in 55-nm CMOS. *IEEE Microw. Wirel. Components Lett.* **2021**, *32*, 157–160. [CrossRef]
20. Nguyen, L.; Nguyen, V.; Phan, K.; Vu, T.; Lee, S.; Huynh, C. Low-loss 6-bit Sub-7 GHz Digital Step Attenuator. In Proceedings of the 2019 International Symposium on Electrical and Electronics Engineering (ISEE), Ho Chi Minh City, Vietnam, 10–12 October 2019; pp. 13–18. [CrossRef]

21. Nagaveni, H.; Karthik, S. 0.5–4 GHz 7-bit GaAs MMIC Digital Attenuator with High Accuracy and Low Insertion Phase. In Proceedings of the 2022 International Conference for Advancement in Technology (ICONAT), Goa, India, 21–22 January 2022; pp. 1–4. [CrossRef]
22. Li, A.; Ding, Y.; Chen, Z.; Wang, W.; Jiang, S.; Sun, S.; Chen, Z.; Chi, B. A Fully Integrated 27.5-30.5 GHz 8-element phased-array transmit front-end module in 65 nm CMOS. In In Proceedings of the 2019 IEEE Asian Solid-State Circuits Conference (A-SSCC), Macau, China, 4–6 November 2019; pp. 153–156.
23. Suh, B.; Min, B.-W. A 20–36-GHz voltage-controlled analog distributed attenuator with a wide attenuation range and low phase imbalance. *IEEE Trans. Microw. Theory Tech.* **2021**, *69*, 2485–2493. [CrossRef]
24. Khaira, N.K.; Singh, T.; Mansour, R.R. RF MEMS Based 60 GHz Variable Attenuator. In Proceedings of the 2018 IEEE MTT-S International Microwave Workshop Series on Advanced Materials and Processes for RF and THz Applications (IMWS-AMP), Ann Arbor, MI, USA, 16–18 July 2018; pp. 1–3. [CrossRef]
25. Xiao, H.; Hu, J.; Ma, K. A DC-6 GHz 7-Bit Digital Attenuator with Low Insertion Loss. In Proceedings of the 2021 IEEE 4th International Conference on Electronics Technology (ICET), Chengdu, China, 7–10 May 2021; pp. 95–98. [CrossRef]
26. Cho, M.-K.; Han, J.-H.; Kim, J.-H.; Kim, J.-G. An X/Ku-band bi-directional true time delay T/R chipset in 0.13 μm CMOS technology. In Proceedings of the 2014 IEEE MTT-S International Microwave Symposium—IMS 2014, Tampa, FL, USA, 1–6 June 2014; pp. 1–3. [CrossRef]
27. Cho, M.-K.; Song, I.; Cressler, J.D. A True Time Delay-based SiGe Bi-directional T/R Chipset for Large-Scale Wideband Timed Array Antennas. In Proceedings of the 2018 IEEE Radio Frequency Integrated Circuits Symposium (RFIC), Philadelphia, PA, USA, 10–12 June 2018; pp. 272–275. [CrossRef]
28. Banerjee, S.; Pradhan, B. Design of S-Band Source-Controlled MMIC Voltage Variable Attenuator. In Proceedings of the 2021 IEEE Pune Section International Conference (PuneCon), Pune, India, 16–19 December 2021; pp. 1–4. [CrossRef]
29. Zhao, C.; Guo, J.; Liu, H.; Yu, Y.; Wu, Y.; Kang, K. A 33–41-GHz SiGe-BiCMOS digital step attenuator with minimized unit impedance variation. *IEEE Trans. Very Large Scale Integr. (VLSI) Syst.* **2021**, *29*, 568–579. [CrossRef]
30. Hu, Z.; Zhou, S.; He, R.; Zhang, Q. A Broadband Voltage Variable Attenuator with High-Power Tolerance and Compact Size Based on Dual-Gate GaN HEMTs. *IEEE Trans. Power Electron.* **2023**, *38*, 6108–6115. [CrossRef]
31. Park, K.; Lee, S.; Jeon, S. A new compact CMOS distributed digital attenuator. *IEEE Trans. Microw. Theory Tech.* **2020**, *68*, 4631–4640. [CrossRef]
32. Gharibdoust, K.; Mousavi, N.; Kalantari, M.; Moezzi, M.; Medi, A. A Fully Integrated 0.18- μm CMOS Transceiver Chip for X-Band Phased-Array Systems. *IEEE Trans. Microw. Theory Tech.* **2012**, *60*, 2192–2202. [CrossRef]
33. Jeong, J.-C.; Yom, I.-B.; Kim, J.-D.; Lee, W.-Y.; Lee, C.-H. A 6–18-GHz GaAs Multifunction Chip with 8-bit True Time Delay and 7-bit Amplitude Control. *IEEE Trans. Microw. Theory Tech.* **2018**, *66*, 2220–2230. [CrossRef]
34. Bangar, N.; Yadav, A.P.S.; Chaturvedi, S.; Naik, A.A. Design of Monolithic 5–18 GHz Digital Attenuator with Low Attenuation and Phase Error. In Proceedings of the 2022 IEEE Microwaves, Antennas, and Propagation Conference (MAPCON), Bangalore, India, 12–16 December 2022; pp. 796–799. [CrossRef]
35. Li, Z.; Zhang, W.; Hao, D.; Liu, X.; Liang, Y.; Liu, Y. A DC-23 GHz Broadband Digital Step Attenuator MMIC Using Capacitive-Inductive Joint Compensation. *IEEE Microw. Wirel. Technol. Lett.* **2023**, *34*, 291–294. [CrossRef]

applied sciences

MDPI

Article

Design of N-Way Wilkinson Power Dividers with New Input/Output Arrangements for Power-Halving Operations

Ceyhun Karpuz [1], Mehmet Cakir [1,*], Ali Kursad Gorur [2] and Adnan Gorur [3]

1 Department of Electrical-Electronics Engineering, Pamukkale University, Denizli 20070, Turkey; ckarpuz@pau.edu.tr
2 Department of Electrical-Electronics Engineering, Nevsehir Haci Bektas Veli University, Nevsehir 50300, Turkey; kgorur@nevsehir.edu.tr
3 Department of Electrical and Electronics Engineering, Nigde Omer Halisdemir University, Nigde 51100, Turkey; adnangorur@hotmail.com
* Correspondence: cakirmehmet@pau.edu.tr

Abstract: In this paper, new single/double-layer N-way Wilkinson power dividers (WPDs) were designed by using slow-wave structures such as narrow-slit-loaded and meandered transmission lines. For size reduction, the slit-loaded and meandered lines were used instead of the quarter-wavelength transmission lines of a conventional WPD. Based on the proposed approaches, two-, four-, and eight-way power dividers were designed, simulated, and fabricated. The fabricated 2-, 4-, and 8-way circuits were measured at the center frequencies of 2.03, 1.77, and 1.73 GHz, which are in excellent agreement with the predicted ones. The meandered transmission lines were also used to design WPD types with novel input/output port arrangements. For this purpose, two three-way WPDs were located on both sides of the same board to have different power-splitting ratios at different inputs and outputs in order to provide alternative solutions for antenna arrays. Furthermore, a five-way dual-layer WPD was introduced by locating the meandered transmission lines into two layers. The most important advantage of the proposed 3- and 5-way WPDs is that they allowed the input power at the next output port to be halved, in the order of P/2, P/4, P/8, P/16, and P/16. All the designed power-halving WPDs were simulated, fabricated, and successfully tested.

Keywords: N-way; Wilkinson power dividers (WPDs); meandered transmission line; power halving

1. Introduction

There are different design approaches for WPDs, such as tapered transmission lines [1], three-section [2], dual-band structures [3], dual-lines [4], and triple lines [5]. In addition to these, open-/short-circuited stubs and defected ground structures are used to achieve improved isolation [6–8]. Single- or multi-band WPDs operating in a wide frequency band have also been studied. Among these, semi-lumped-element power dividers with dual-band characteristics in the UHF/SHF bands [9], flexible design schemes for single− and dual−band power-dividing operations [10], dual-band unequal WPDs [11], multi-T-section characterization of high-impedance transmission lines [12], multi-band reconfigurable differential power dividers [13], multiband WPDs based on multisection LC ladder circuits [14], and single/multi-band WPDs with transversal filtering sections [15] stand out. Such WPDs may find application areas especially in multifunctional communication systems. Moreover, the placement of power dividers in antenna arrays is one of the main focuses for researchers. Therefore, size reduction is another important parameter for WPDs, since it is required for feeding the elements of the antenna array within a relatively reduced area. To date, slow-wave structures have been especially used for size reduction. For this purpose, capacitively loaded stubs [16,17] and narrow-slit-loaded and meandered transmission lines [18,19] are largely employed for WPD designs. Periodically, loaded slow-wave structures are modified for different targets, including dual-band applications [20,21], SiGe

BiCMOS processes [22], filtering power dividers having resonant LC branches [23], etc. In some structures, coupled lines are used to suppress undesirable harmonics so that the signal can be transmitted within the desired frequency region [24]. On the other hand, in order to adjust the directivity, antenna array elements may need different power ratios at different elements. Therefore, unequal power dividers may also find applications in antenna arrays to meet the asymmetric power division requirement [25–27]. There are two types of equal WPDs: N-way and cascaded models. N-way WPDs have an input port and N output ports with the impedance of $Z_0\sqrt{2}$, whereas cascaded model WPDs divide the input power by 2 at each cascaded path [28].

In this paper, two-, four-, and eight-way cascaded WPDs are designed by using slow-wave structures. For this purpose, the two-, four-, and eight-way WPDs constructed by using narrow-slit-loaded transmission lines in [28] were first taken into account. Next, they were replaced with meandered transmission lines to improve the compactness. Two-, four-, and eight-way WPDs having meandered transmission lines were also fabricated and successfully measured for demonstration. Based on the meandered transmission lines, three- and five-way WPDs with new input/output port arrangements were designed, fabricated, and successfully measured for demonstration. The proposed three- and five-way circuits allow halving the input power at the next output port.

2. Materials and Methods

2.1. Analysis of WPDs

The S-parameters of a WPD can be calculated by using even–odd mode analysis. An equivalent circuit model for a conventional 2-way WPD is shown in Figure 1a. As can be seen from the figure, quarter-wavelength transmission lines having characteristic impedances of $Z_0\sqrt{2}$ are used in both output paths. Furthermore, an isolation resistor of $2Z_0$ is located between the two output ports. Based on this model, slow-wave transmission lines can be employed instead of the conventional quarter-wavelength transmission lines. Therefore, these models are the starting point of the entire design procedure.

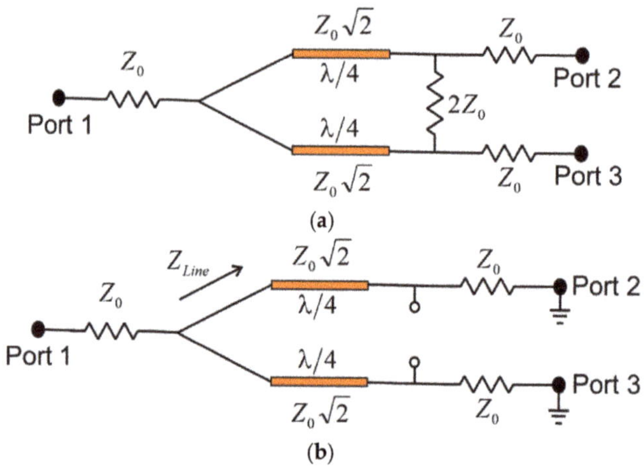

Figure 1. (**a**) Equivalent circuit model of a 2-way WPD. (**b**) Equivalent circuit model for the calculation of S_{11}.

The S parameters of the WPD given in Figure 1a can be obtained by different approaches. In order to obtain S_{11}, the circuit model should be considered as illustrated in

Figure 1b, since no current will flow through the isolation resistor. As can be seen from Figure 1b, S_{11} can be calculated from the following set of equations:

$$S_{11} = \frac{Z_{in} - Z_0}{Z_{in} + Z_0} \tag{1}$$

$$Z_{in} = \frac{Z_{Line}}{2} \tag{2}$$

$$Z_{Line} = Z_0\sqrt{2}\frac{Z_0 + jZ_0\sqrt{2}\tan\theta}{Z_o\sqrt{2} + jZ_0\tan\theta} \tag{3}$$

$$\theta = \frac{2\pi}{\lambda}\cdot\frac{\lambda_0}{4} \tag{4}$$

For equal power splitting, S_{21} and S_{31} can be obtained from S_{11}. Furthermore, the isolation between the output ports can be calculated by using even–odd mode analysis. For this purpose, the even- and odd-mode half-circuits can be obtained by locating the magnetic and electric walls, respectively, on the symmetry axis shown in Figure 2a. Figure 2b,c illustrate the even- and odd-mode half-circuit models, respectively. It should be noted that the isolation resistor can be neglected for the even-mode circuit since the symmetry axis is open-circuited. In addition, the isolation resistor is Z_0 for the odd-mode circuit due to the short-circuited symmetry axis. The isolation between the output ports, S_{32}, can be found by deriving the even-mode and odd-mode input impedances. Here, the circuit model shown in Figure 1a should be considered a 2-port network, with the input port as port 2 and the output port as port 3. Thus, under the even- and odd-mode excitations, the input impedance seen from port 2 or 3 will be the even- and odd-mode input impedances, so that the isolation can be expressed as

$$S_{32} = \frac{(Z_{even} - Z_{odd})Z_0}{(Z_{even} - Z_0)(Z_0 + Z_{odd})} \tag{5}$$

The even-mode input impedance shown in Figure 2b can be expressed as

$$Z_{even} = Z_0\sqrt{2}\frac{2Z_0 + jZ_0\sqrt{2}\tan\theta}{Z_o\sqrt{2} + jZ_0\tan\theta} \tag{6}$$

The odd-mode input impedance can be calculated as

$$Z_{odd} = \frac{Z_x Z_0}{Z_x + Z_0} \tag{7}$$

$$Z_x = jZ_0\sqrt{2}\tan\theta \tag{8}$$

Similar analyses can also be applied to 4- and 8-way WPDs. Since 2 and 3 quarter-wavelength transmission lines should be used in each output path for 4- and 8-way WPDs, respectively, they will excite 2 and 3 transmission poles within the power-dividing frequency band.

2.2. N-Way Wilkinson Power Dividers Based on Slow-Wave Structures

As is well known, 1-to-N-way WPDs consist of N branches which have characteristic impedances of $Z_0\sqrt{N}$, as shown Figure 3. Although they provide better size reduction than that of the cascaded structures, they suffer from the narrow band and poor isolation between output ports [28].

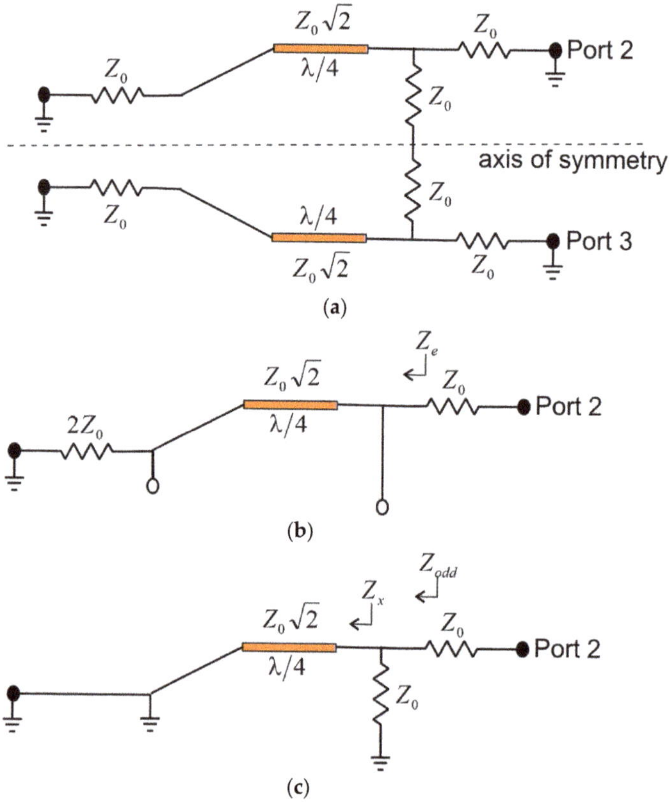

Figure 2. (**a**) Equivalent circuit model for the calculation of S_{32}. (**b**) Even-mode half-circuit model. (**c**) Odd-mode half-circuit model.

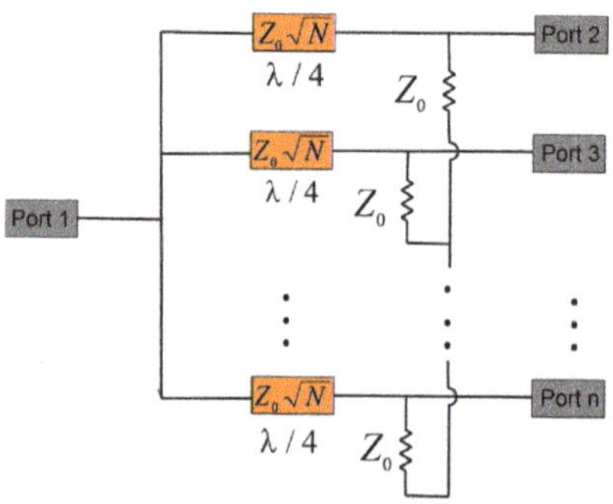

Figure 3. Conventional 1-to-N-way WPD.

Equivalent circuit models for conventional 2-, 4-, and 8-way cascaded WPDs are illustrated in Figure 4a–c, respectively. As can be seen from Figure 4, conventional 2-way WPDs comprise 2 quarter-wavelength ($\lambda/4$) transmission lines with the characteristic impedances of $Z_0\sqrt{2}$. The same transmission line length and impedance values are also valid for 4- and 8-way WPDs, as depicted in Figure 4b,c. In order to obtain high isolation between the output ports, an isolation resistor of $2Z_0$ is connected between 2 output ports.

In addition, the input/output ports must have characteristic impedances of Z_0 for good impedance matching [28].

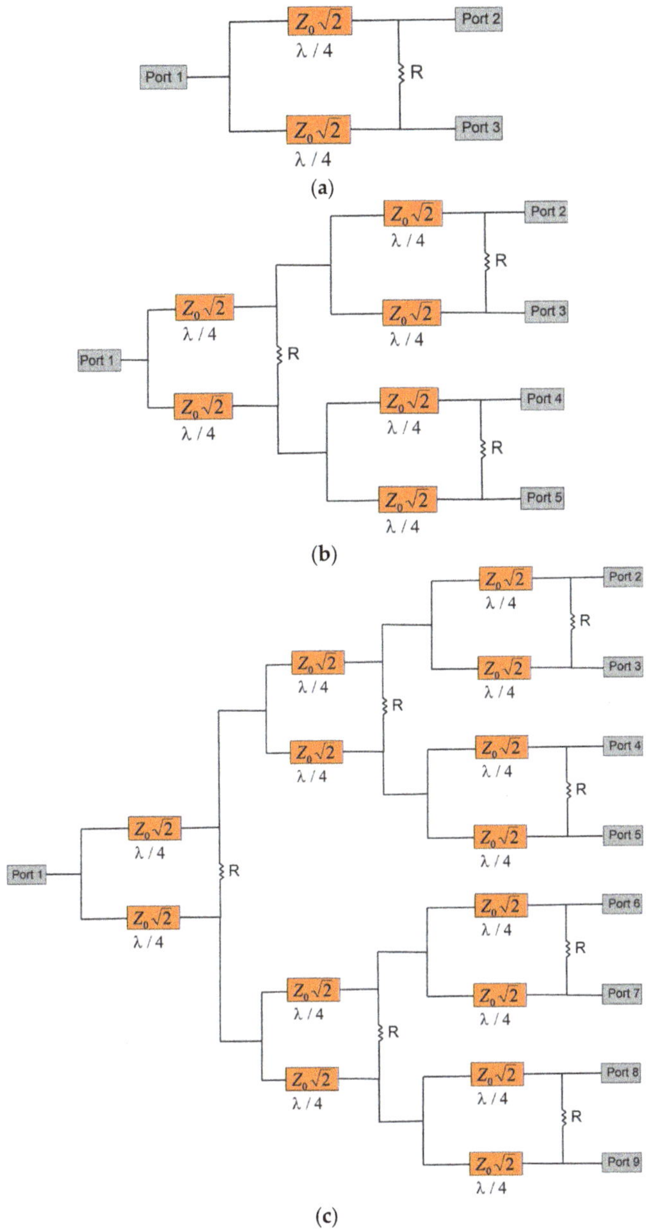

Figure 4. Conventional cascaded WPDs: (**a**) 2-way; (**b**) 4-way; (**c**) 8-way.

In order to reduce the overall circuit size, quarter-wavelength transmission lines can be capacitively or inductively loaded by different approaches, as shown in Figure 5. Due to the increase in the capacitance or inductance per unit length of the transmission line, the phase velocity on it decreases, so such a transmission line can be called a slow-wave structure. Among the slow-wave structures shown in Figure 5, slit-loaded and meandered transmission lines are widely preferred for different kinds of microwave circuits because of their easy fabrication processes. However, the fabrication of interdigital capacitor-loaded transmission lines are more difficult than the others since they may need extra via connections. The number of narrow slits and meandered sections can be determined with respect to the desired center frequency. Here, it should be taken into consideration that a transmission line with narrow slits and meandered sections should exhibit similar behavior to a quarter-wavelength transmission line having the characteristic impedance of $Z_0\sqrt{2}$. Thus, the designed WPD would have optimum performance in terms of return loss and isolation. For this purpose, all dimensions and the number of narrow slits and meandered sections were determined by using the parameter sweep in Sonnet Software. Investigations into the related parameters are described in the following sections.

(a) (b) (c)

Figure 5. Slow-wave structures. (**a**) Slit-loaded. (**b**) Meandered. (**c**) Interdigital.

2.2.1. WPDs with Narrow-Slit-Loaded Transmission Lines

A 2-way WPD was constructed by loading narrow slits as shown in Figure 6 [28]. An RT/Duroid substrate with a relative dielectric constant of 10.2 and a thickness of 1.27 mm was used in all design processes. The number of slits and the location of the isolation resistor affect the frequency response as depicted in Figure 7a,b, respectively. It is clear that the center frequency decreases as the number of narrow slits increases. Therefore, a very compact size can be achieved by using the maximum number of slits. Since 50-ohm port impedances were utilized, the isolation resistor was chosen as 100 ohm.

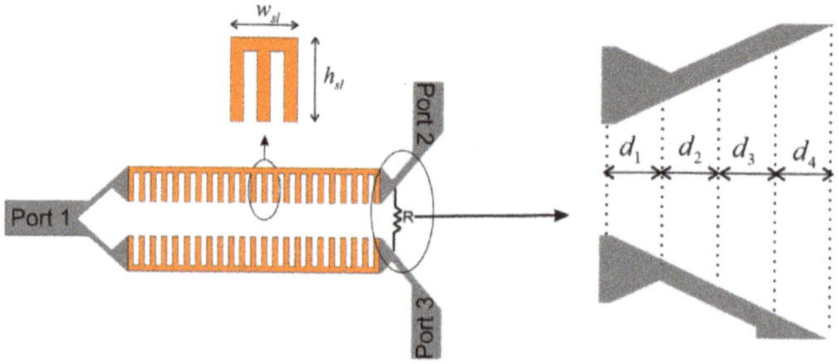

(a) (b)

Figure 6. (**a**) Conventional 2-way WPD-loading with narrow slits, and (**b**) location of the isolation resistor.

The proposed slit-loaded transmission lines in [28] were also used to design 4- and 8-way WPDs. Figure 8 illustrates the layouts of the 4- and 8-way WPDs in a cascaded model. It is clear that 2 more paths have been added instead of output ports for the 4- and 8-way power dividers.

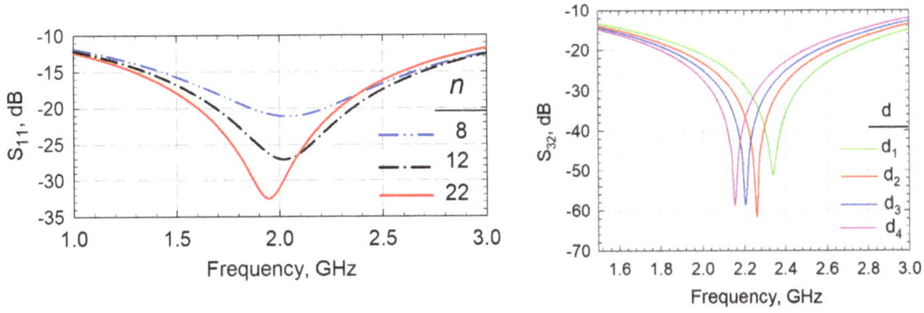

(a) (b)

Figure 7. (**a**) Effects of narrow slits on the frequency response. (**b**) Effects of locations of the isolation resistor on the frequency response.

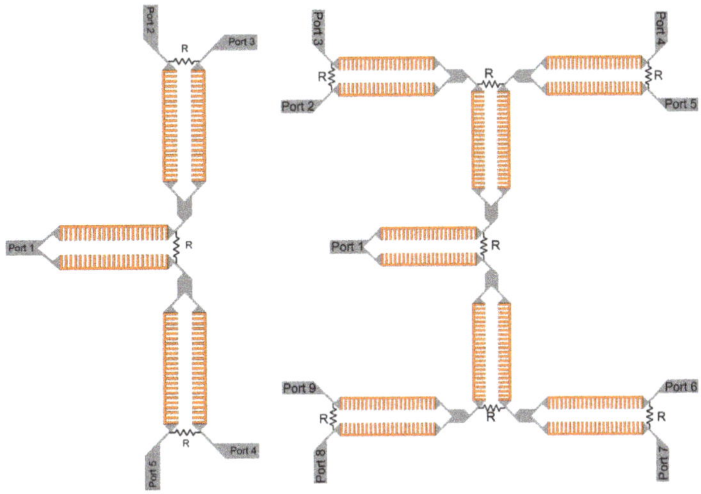

(a) (b)

Figure 8. (**a**) Four-way WPD loading with narrow slits. (**b**) Eight-way WPD loading with narrow slits.

2.2.2. WPDs with Meandered Transmission Lines

The meandered line model taken into consideration in this paper is shown in Figure 9. According to the even or odd number of meandered sections, the equivalent circuit model is depicted in Table 1 [29].

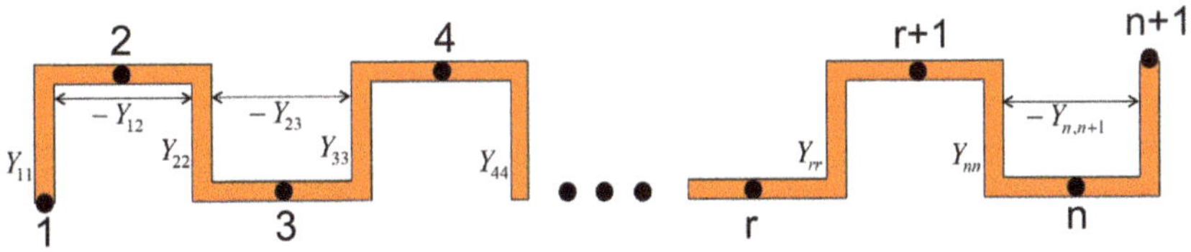

Figure 9. Meandered line model.

Table 1. Meandered line models and their equivalent circuits.

Odd line	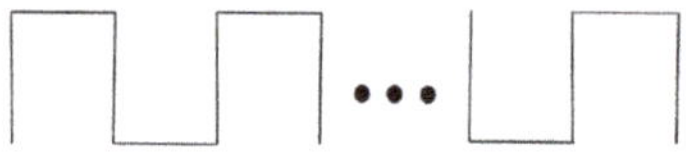	
Even line	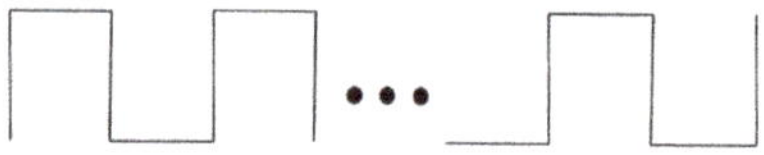	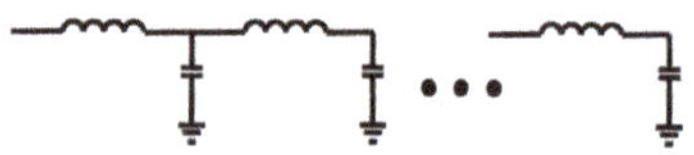

In a similar manner to the slit-loaded transmission lines, meandered sections can also be used for size reduction. Figure 10a illustrates a 2-way cascaded WPD constructed by meandered transmission lines. An RT/Duroid substrate with a relative dielectric constant of 10.2 and a thickness of 1.27 mm was used in all design processes. Effects of the number of meandered sections on the frequency response are depicted in Figure 10b. Here, the number of meandered sections is changed, starting from the middle of the transmission line. Depending on the proposed approach, a size reduction of 20.14% can be achieved.

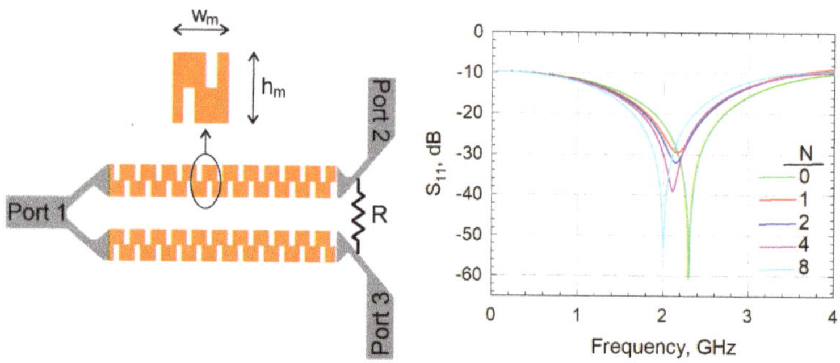

(**a**) (**b**)

Figure 10. (**a**) Conventional 2-way WPD having meandered transmission lines. (**b**) Effects of WPD meandering on the frequency response.

In addition to 2-way WPDs, more than 2 power-dividing operations can be necessary for the antenna arrays to be used in multi-way systems. For this purpose, the designed 2-way WPDs can be enhanced to 4-and 8-way circuits by means of meandered transmission lines, as shown in Figure 11a,b. As can be seen from Figure 11a, two two-way WPDs were connected in cascade for the four-way design. In this case, two reflection zeros can be observed since there are two identical transmission lines. In the eight-way WPDs, three reflection zeros appear due to the three cascaded transmission lines shown in Figure 11b. Here, the dimensions of the transmission lines, the isolation resistor and the resistor position are same as the 2-way WPDs. Accordingly, it is possible to achieve power-dividing operations in wider frequency bands by using more cascaded sections. The size reductions for the proposed 4- and 8-way power dividers are 16.62% and 9.99%, respectively. Electrical lengths of the proposed 2-, 4-, and 8-way WPDs are 0.20 $\lambda_g \times 0.078\ \lambda_g$, 0.56 $\lambda_g \times 0.25\ \lambda_g$, and 0.63 $\lambda_g \times 0.57\ \lambda_g$, respectively, where λ_g is the guided wavelength at the center frequency. Frequency responses of the designed circuits are given with the experimental studies in the following section.

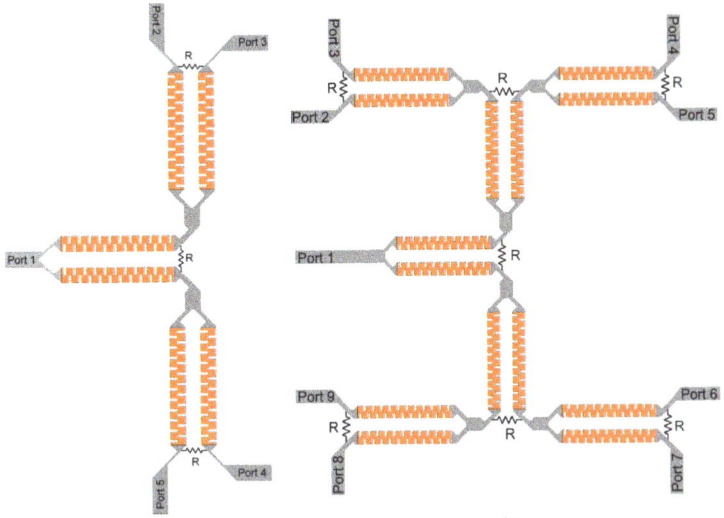

Figure 11. Conventional WPDs with meandered transmission lines: (**a**) 4-way; (**b**) 8-way.

2.2.3. Alternative Input/Output Port Arrangements for Various Power-Division Processes

In order to introduce various types of WPDs to be used for different feeding schemes of antenna arrays, new WPD configurations with different power-division ratios and input/output port arrangements were proposed by using meandered transmission lines. For this purpose, 3 different circuit topologies, including 3-way, 3-way double-layer, and 5-way double-layer WPDs were proposed. The main challenge of the proposed circuits is their power-halving property. Hence, the input power is halved at the next output ports and equal power-dividing ratios are achieved at the last two output ports.

Figure 12 illustrates the circuit for two three-way WPDs located at the same side of the board. As can be seen from Figure 12a, there are two input ports located at the upper left and bottom right. The upper output ports belong to the upper circuit, whereas the bottom output ports are for the bottom circuit. Both these WPDs perform the same operations in different paths, so the proposed arrangement actually behaves like a power-divider bank to be used for a multi-input—multi-output (MIMO) system. They are useful for different antenna arrays serving different operations. Moreover, the input power is halved at output port 2, while the power at output ports 3 and 4 is equal. The simulated frequency responses are demonstrated in Figure 12b. The magnitude and phase differences are also shown in Figure 12c,d, respectively.

The second WPD configuration was created for double-layer applications, as shown in Figure 13a. Here, a common ground plane was located between two layers, where both layers included two three-way power dividers for utilization in MIMO systems. The power divider in the upper layer is completely similar to the one in the bottom layer, so the proposed approach can be useful for bidirectional MIMO systems. The bottom and upper power dividers exhibit similar circuit performances. The S parameters of one of the two proposed WPDs are given in Figure 13b. Figure 13c,d represent the simulated magnitude and phase differences, respectively.

The last configuration was a 5-way WPD, as shown in Figure 14a. As can be seen from the figure, two layers were combined so as to have a common ground plane. The input port was located at the upper layer, and there were five output ports for power splitting. Two of the outputs are in the upper layer, whereas the remaining ports are in the bottom layer. The designed 5-way WPD halves the input power at the next port in a similar manner to the previous configuration. Hence, port 2 has 1/2, port 3 has 1/4, port 4 has 1/8, and ports 5 and 6 both have 1/16 of the input power. This phenomenon can be seen from the simulated frequency responses shown in Figure 14b. The $|S_{21}|$, $|S_{31}|$, $|S_{41}|$, $|S_{51}|$, and $|S_{61}|$ parameters were approximately 3, 6, 9, 12, and 12 dB at the center frequency, respectively. Figure 14c,d show the magnitude and phase difference of the simulated S parameters.

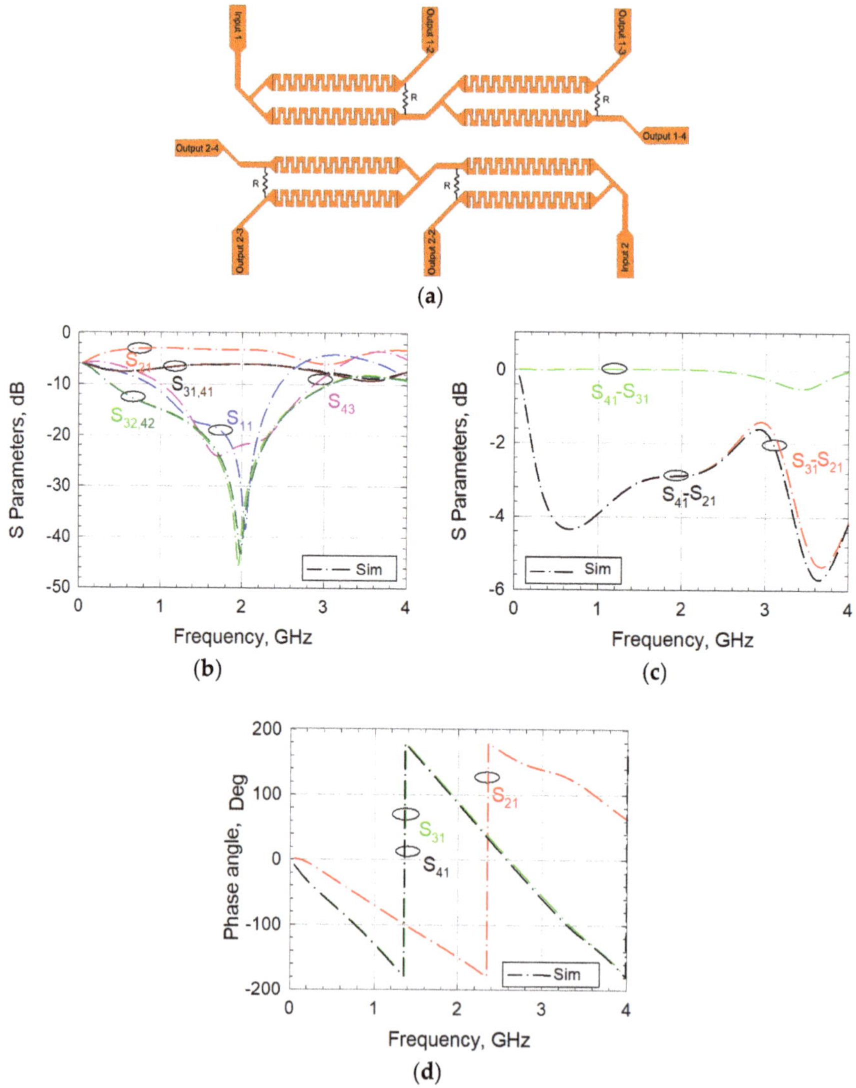

Figure 12. (**a**) Two 3-way WPDs with different inputs and outputs: (**b**) simulation results, (**c**) magnitude differences of S-parameters, and (**d**) phase responses.

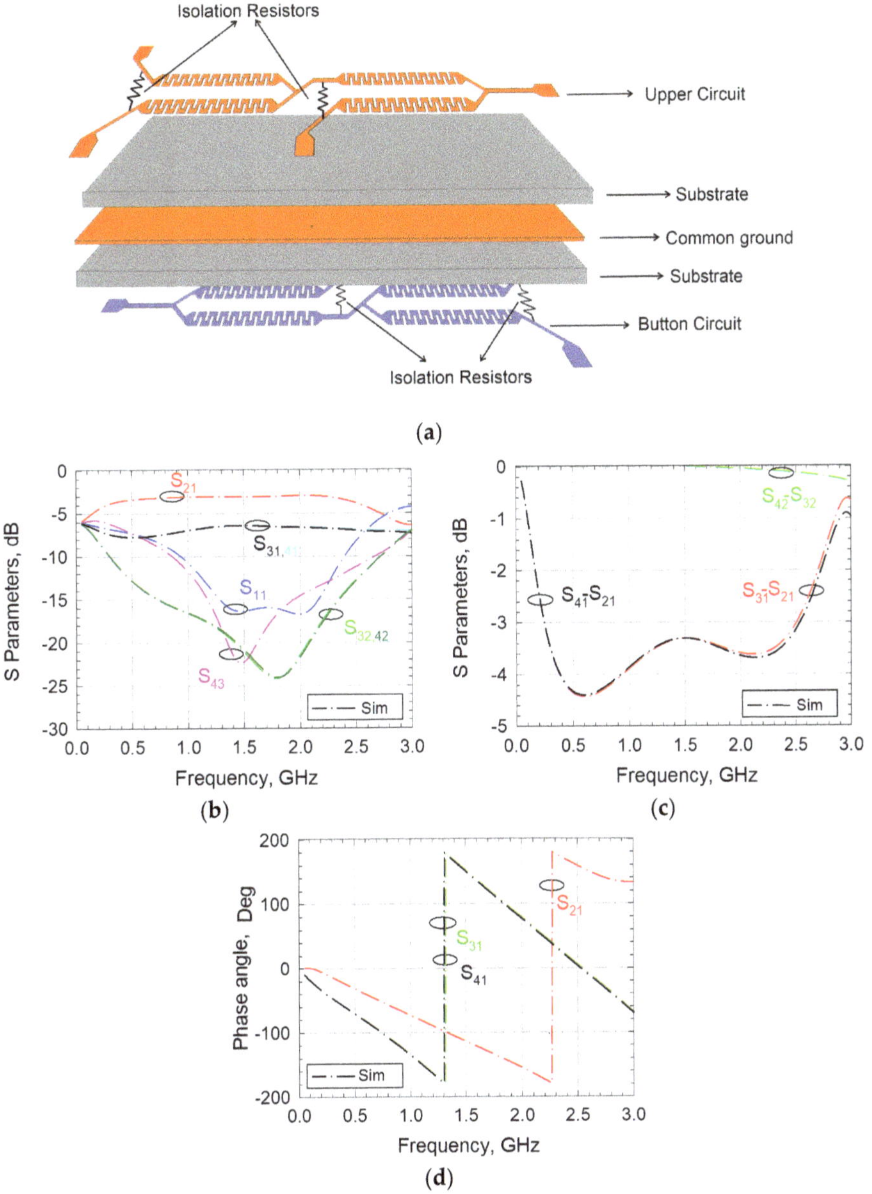

Figure 13. (**a**) Two 3-way WPDs with a common ground plane: (**b**) simulation results, (**c**) magnitude differences of S-parameters, and (**d**) phase responses.

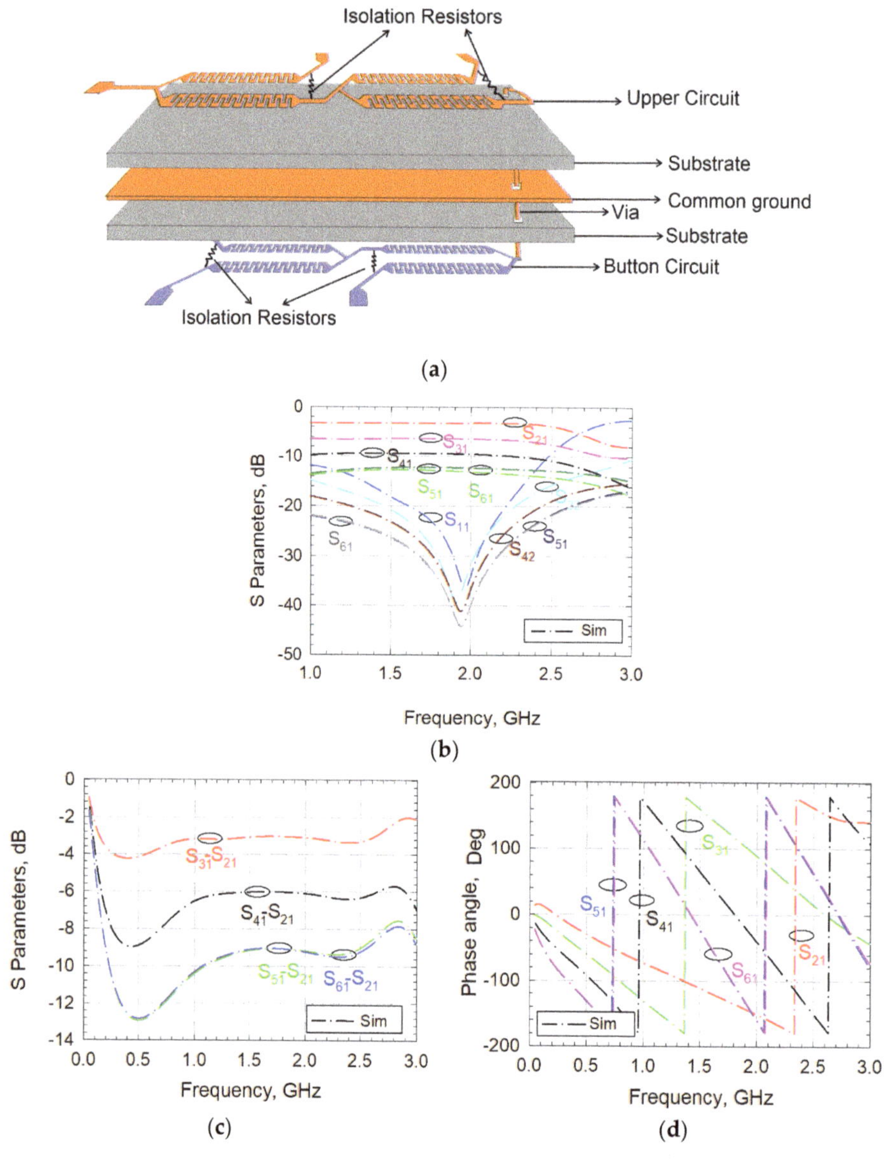

Figure 14. (**a**) Layout of a 5-way dual-layer WPD: (**b**) simulation results, (**c**) magnitude differences of S-parameters, and (**d**) phase responses.

3. Results

To demonstrate the proposed approaches, all the circuits designed with narrow-slit-loaded and meandered transmission lines were fabricated and tested. All circuits were simulated by using a full-wave electromagnetic simulator, Sonnet.

First, experimental studies of the power dividers having narrow-slit-loaded transmission lines introduced in [28] are presented. Photographs of the manufactured N-way WPDs having narrow slits in [28] are shown in Figure 15. Figure 16 illustrates a measurement view, where 50-ohm loads were used to match the idle output ports for 4- and 8-way power dividers since the measurement device has only four ports. The measured results of the power dividers having narrow-slit-loaded transmission lines were compared with the simulated results in Figure 17 [28]. It is obvious that the measured and simulated results are in very good agreement. In Figure 17a, the center frequency was 2 GHz at the 15 dB fractional bandwidth (FBW) of 36.26%. As can be seen from the figure, the return loss and isolation level were better than 15 and 20 dB within the FBW, respectively. Within the FBW, the minimum isolation level and insertion loss were approximately 29.56 dB at 2 GHz and 3.29 dB, respectively. The measured and simulated results of the 4-way WPD are illustrated in Figure 17b, with the return loss and isolation level of better than 15 and 20 dB, respectively. The center frequency was 1.8 GHz at the FBW of 94.44%, and the insertion loss was better than 3 + 0.5 dB. Moreover, the center frequency of the 8-way WPD was measured at 1.9 GHz at the FBW of 81.05%. The measured results were compared with the simulated ones in Figure 17c, where the insertion loss, return loss, and isolation level obtained were better than 9 + 0.5, 15, and 20 dB, respectively.

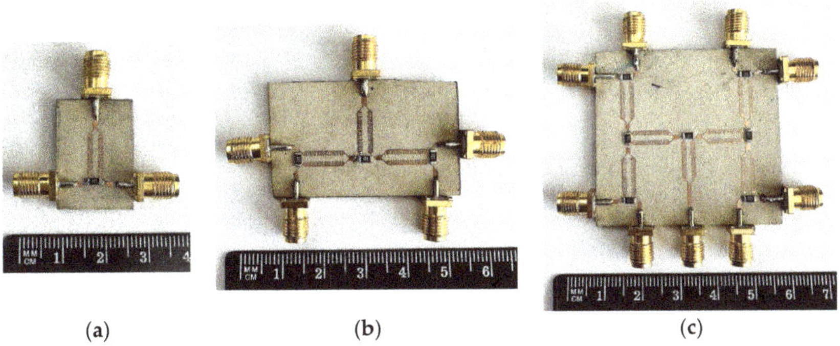

Figure 15. Photographs of N-way WPDs with narrow-slit-loaded transmission lines: (**a**) 2-way; (**b**) 4-way; and (**c**) 8-way.

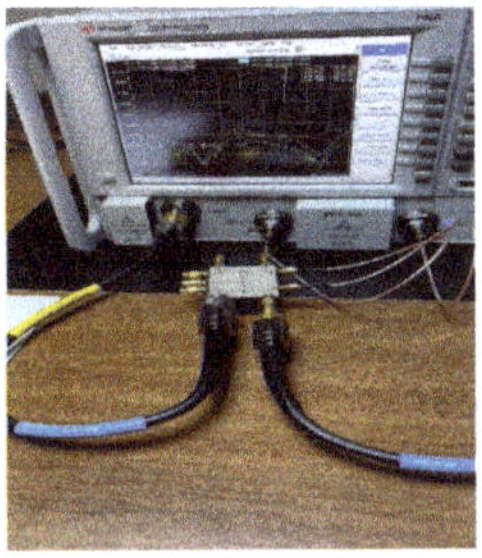

Figure 16. Measurement setup for an 8-way WPD.

Meandered-transmission-line-based two-, four-, and eight-way WPDs were also manufactured and tested. Photographs of the fabricated circuits are shown in Figure 18. The simulated and measured frequency responses for the 2-way WPD are compared in Figure 19a, where the return loss and isolation level are better than 15 and 30 dB, respectively. Frequencies of the minimum isolation and return loss levels are different because of fabrication errors. The measured center frequency was 2 GHz at the FBW of 75%. It should also be noted that there was only one reflection zero within the related frequency band. The measured magnitude and phase difference can be observed from Figure 19b,c, respectively. The center frequency of the 4-way power divider was measured at 1.77 GHz at the FBW of 73.09% as depicted in Figure 19a. At that frequency, the measured return and insertion losses were 24 and 3.1 dB, respectively. Isolation levels between the output ports were measured as better than 26 dB. In the frequency response of the four-way power divider, there are two reflection zeros, since the circuit is composed of two cascaded sections. The measured magnitude and phase differences are also shown in Figure 20b,c, respectively. The simulated and measured results of the eight-way WPD are compared in Figure 21a and were found to be in very good agreement. In this case, three reflection zeros can be observed due to the utilization of three cascaded sections. The center frequency was measured at 1.73 GHz at the 15 dB FBW of 87.86%. Isolation levels between the output ports were better than 15 dB, 26 dB, 24 dB within the FBW, where the minimum insertion losses were 3.6 dB, 6.5, and 9.6 for 2-, 4-, and 8-way WPDs, respectively. The measured magnitude differences are shown in Figure 21b, with acceptable values smaller than 1 dB, and the phase differences can be observed in Figure 21c.

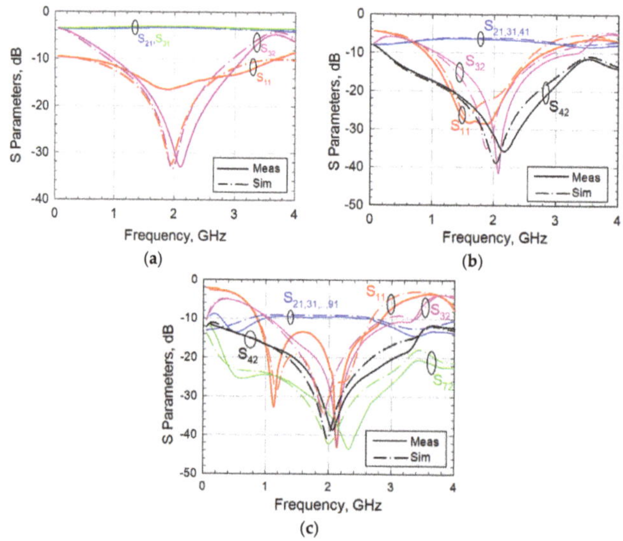

Figure 17. Comparison of the measured and simulated results of the WPDs having narrow-slit-loaded transmission lines for a (**a**) 2-way WPD; (**b**) 4-way WPD; and (**c**) 8-way WPD.

Figure 18. Meandered transmission line-based WPDs: (**a**) 2-way, (**b**) 4-way, and (**c**) 8-way.

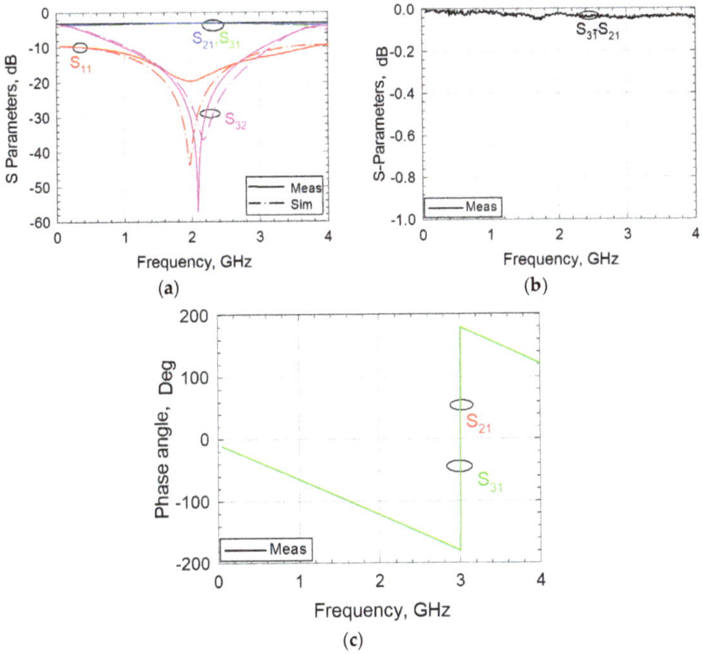

Figure 19. (**a**) Comparison of the measured and simulated results of the WPDs having meandered transmission lines for a 2-way WPD; (**b**) magnitude differences of S-parameters for a 2-way WPD; and (**c**) phase responses.

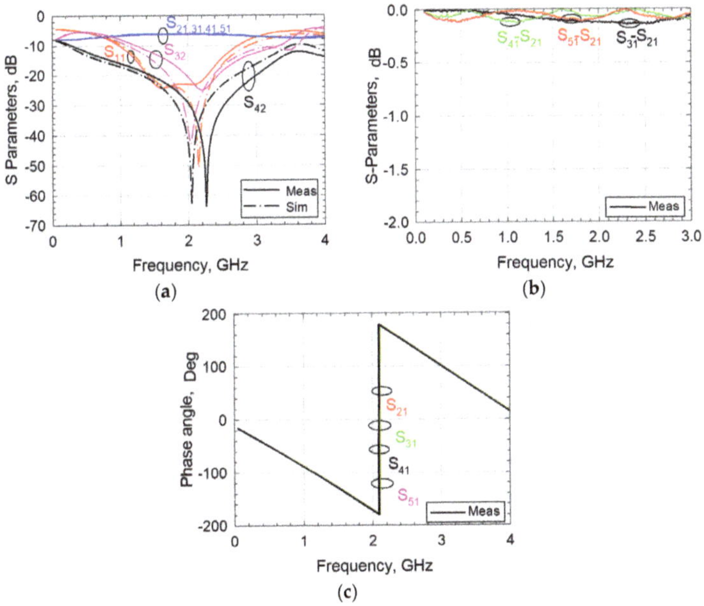

Figure 20. (**a**) Comparison of the measured and simulated results of the WPDs having meandered transmission lines for a 4-way WPD; (**b**) magnitude differences of S-parameters for a 4-way WPD; and (**c**) phase responses.

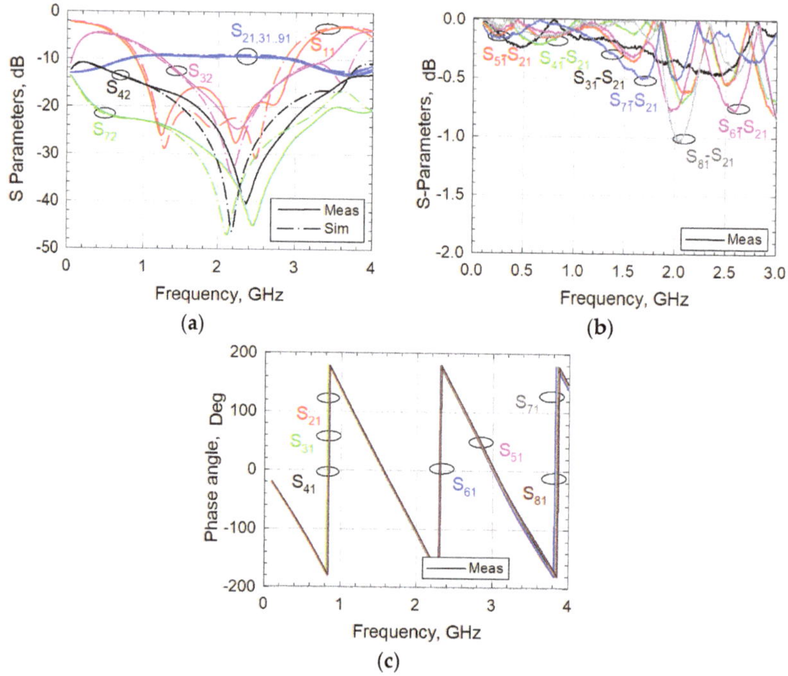

Figure 21. (**a**) Comparison of the measured and simulated results of the WPDs having meandered transmission lines for an 8-way WPD; (**b**) magnitude differences of S-parameters; and (**c**) phase responses.

Meandered-transmission-line-based WPDs with alternative input/output port configurations were manufactured and tested for demonstration. A photograph of the two fabricated three-way WPDs is shown in Figure 22a. The frequency response of the designed power divider is depicted in Figure 22b. As can be seen from the figure, the designed circuit can serve within the frequency band of 1.5–2.0 GHz. Within this frequency region, the isolation level between the output ports is better than 19.96 dB, the return loss is better than 17.71 dB, and the insertion loss is better than 3 + 0.5 dB. Figure 22c shows the magnitude differences, which are better than 4.4 dB. The phase responses corresponding to the S-parameters are also illustrated in Figure 22d.

The multilayer three-way WPD was implemented as shown in Figure 23a,b. It is obvious that the implemented circuit is different from the previous one, since two WPDs are located at both sides of the structure with a common ground plane. Comparisons of the simulated and measured results in terms of the return loss, insertion loss, and isolation level are depicted in Figure 23c. Within the frequency band of 1.5–2 GHz, the return loss was observed as better than 15 dB with an insertion loss of better than 3 + 0.5 dB at each output port. Furthermore, the isolation levels between all output ports were better than 19 dB. Figure 23d,e show the measured magnitude differences and phase responses, respectively. Unwanted differences for the measured results result from the fabrication errors which especially appeared during the combination of two layers. Two layers were combined by using copper tape to transfer the ground plane into the upper/down plane. Because of the small circuit size and the laboratory conditions, two circuit boards could not be combined by using screws.

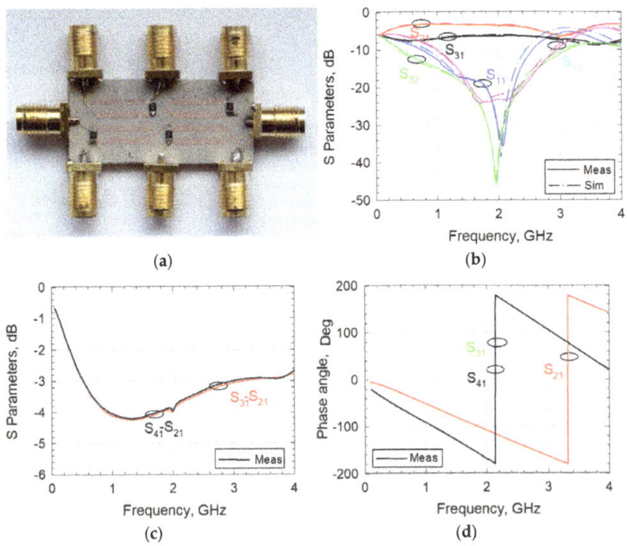

Figure 22. (**a**) Photograph of the meandered transmission lines based on two three-way WPDs. (**b**) Measured and simulated S-parameters, (**c**) magnitude differences of S-parameters, and (**d**) phase responses.

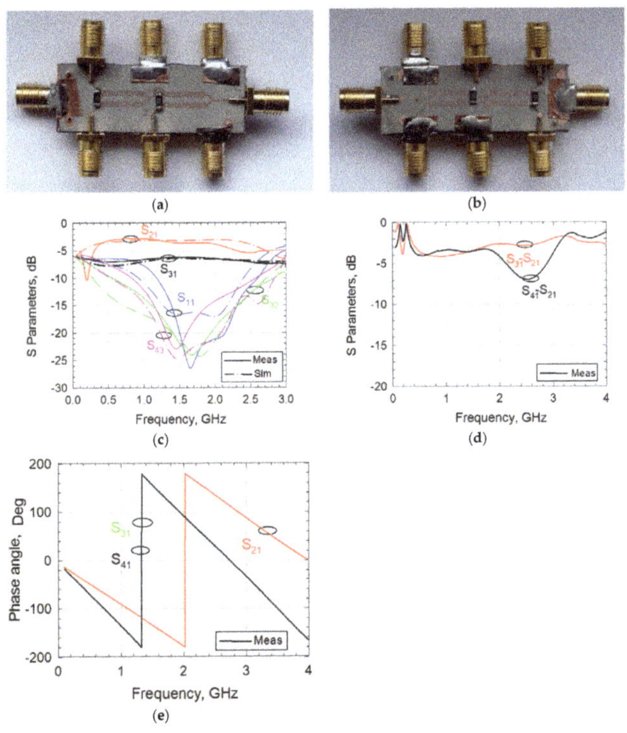

Figure 23. Meandered-transmission-line-based multilayer WPDs having a common ground plane: (**a**) front view, (**b**) back view. (**c**) Measured and simulated results. (**d**) Magnitude differences of S-parameters, and (**e**) phase responses.

A meandered-transmission-line-based multilayer five-way WPD was manufactured and successfully tested. Photographs of the front and back views of the fabricated circuits are shown in Figure 24a,b, respectively. The measured and simulated return and insertion losses of the proposed five-way power divider are compared in Figure 24c, with an acceptable agreement. The measured and simulated isolation levels between the output ports are also depicted in Figure 24d. It is clear that the proposed multilayer 5-way WPD can be used for different communication systems. Within the frequency band of 1.5–2 GHz, the return loss, insertion loss, and isolation levels are better than 16.52, 3 + 0.5, and 14.47 dB, respectively. The differences between the measured and simulated responses resulted from the fabrication errors and especially from the layer-combining process. Magnitude differences between the insertion losses and the phase responses are shown in Figure 24e,f, respectively.

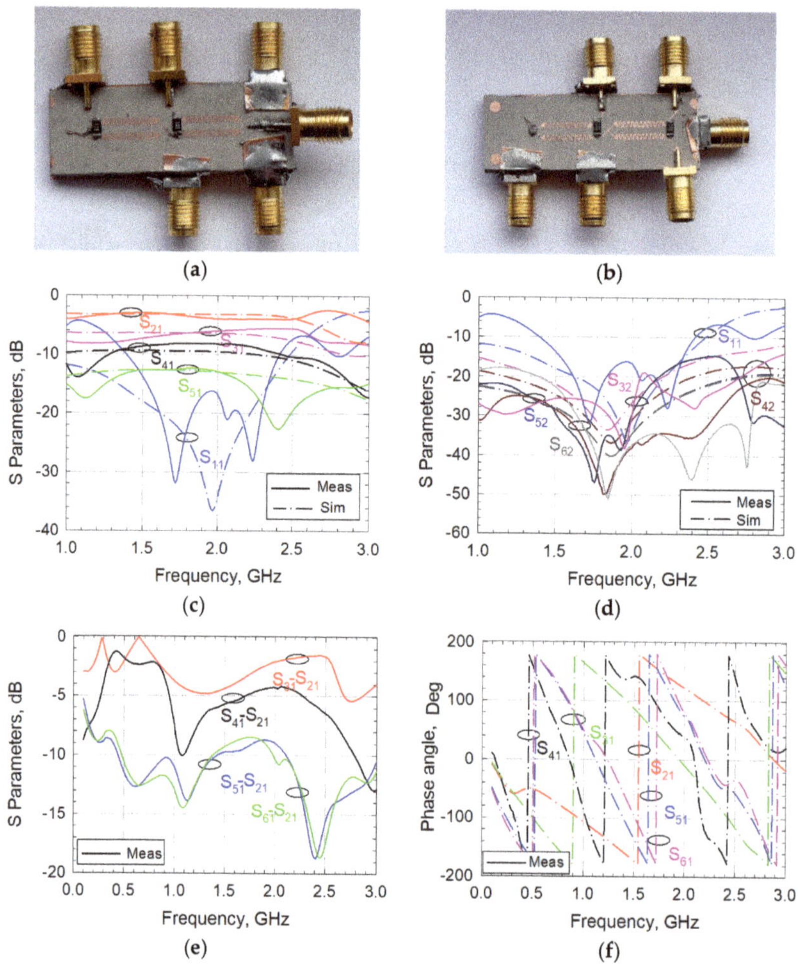

Figure 24. Meandered-transmission-line-based multilayer 5-way WPD: (**a**) front view and (**b**) back view. (**c**) Measured and simulated return and insertion losses and (**d**) isolation levels. (**e**) Magnitude differences of S-parameters and (**f**) phase responses.

Table 2 summarizes and compares the measurements of the proposed designs with the work published in the literature. The proposed configurations in this work introduce WPDs with new input/output configurations that may be used in front of antenna arrays to be used for different personal and mobile communication systems. For this purpose, WPDs based on single- and dual-layer slow-wave structures providing equal power division have been introduced for the first time in this paper. The proposed circuits in this work also have promising circuit sizes.

Table 2. Performance comparison of the proposed WPD with previous works.

Ref	f_o (GHz)	Size (λ_g^2)	FBW (%)	Structure Topology
[2]	1.32	-	103.60	Three-section TL lines & three resistors
[6]	2.1	0.6386	80	Two-stage with coupled lines
[10]	2.6	0.096	18.2	ITPD
[28]	2	0.0156	36.26	Slit-loaded transmission lines
Meandered line (this work)	2.03	0.0156	56.79	Meandered transmission lines
3-way WPDs (this work)	1.95	0.12	52.5	Microstrip transmission lines
3-way WPDs with common ground plane (this work)	1.74	0.079	52.29	Microstrip transmission lines
5-way WPDs with common ground plane (this work)	1.97	0.073	51.26	Microstrip transmission lines

4. Conclusions

In this study, various types of WPDs have been developed and experimentally investigated by using slow-wave structures. For this purpose, WPDs having narrow-slit-loaded transmission lines were investigated, and novel meandered-transmission-line-based WPDs were introduced for the first time. The proposed meandered-line-based two-, four-, and eight-way power dividers have been fabricated and measured successfully. The proposed meandered transmission lines have also been used to design new types of WPDs with new input/output port arrangements. In this context, two three-way WPDs have been located on the same plane for similar power-splitting processes. Next, this circuit has been developed by locating one of the power dividers to another layer. The proposed circuits can behave like a power-divider bank to be used for alternative antenna array applications. Moreover, a five-way WPD has been designed by using meandered transmission lines in two layers. These alternative WPD topologies allow halving of the input power at the next output port. These WPDs have been implemented and their measurements show an acceptable agreement with the simulated results.

Author Contributions: Conceptualization, M.C., A.K.G., C.K. and A.G.; methodology, M.C., A.K.G., C.K. and A.G.; software, M.C., A.K.G., C.K. and A.G.; validation, M.C., A.K.G., C.K. and A.G.; investigation, M.C., A.K.G., C.K. and A.G.; resources, M.C., A.K.G., C.K. and A.G.; writing—original draft preparation, M.C., A.K.G., C.K. and A.G.; writing—review and editing, M.C., A.K.G., C.K. and A.G.; visualization, M.C., A.K.G., C.K. and A.G.; supervision, C.K. and A.K.G.; project administration, A.G. All authors have read and agreed to the published version of the manuscript.

Funding: This work was supported by the Scientific and Technological Research Council of Turkey (TUBITAK) under Grant 120E101.

Institutional Review Board Statement: Not applicable.

Informed Consent Statement: Not applicable.

Data Availability Statement: Data and simulation files will be shared upon request from the authors.

Conflicts of Interest: The authors declare no conflict of interest.

Abbreviation

S_{mm}	Reflection coefficient at port m.
S_{mn}	Transmission coefficient from port n to port m.
WPD	Wilkinson power divider.
Meas	Measured.
Sim.	Simulated.
f_0	Center frequency.
FBW	Fractional bandwidth.
λ_g	Guided wavelength.

References

1. Razzaz, F.; Saeed, S.M.; Alkanhal, M.A.S. Compact Ultra-Wideband Wilkinson Power Dividers Using Linearly Tapered Transmission Lines. *Electronics* **2022**, *11*, 3080. [CrossRef]
2. Omi, A.I.; Zafar, Z.N.; Al-Shakhori, H.; Savage, A.N.; Islam, R.; Maktoomi, M.A.; Zakzewski, C.; Sekhar, P. A new analytical design methodology for a three-section wideband wilkinson power divider. *Electronics* **2021**, *10*, 2332. [CrossRef]
3. Gao, N.; Wu, G.; Tang, Q. Design of a novel compact dual-band wilkinson power divider with wide frequency ratio. *IEEE Microw. Wirel. Compon. Lett.* **2014**, *24*, 81–83. [CrossRef]
4. Khajeh-Khalili, F.; Honarvar, M.A.; Virdee, B.S. Miniature In-Phase Wilkinson Power Divider with Pair of Parallel Transmission-Lines for Application in Wireless Microwave Systems. *IETE J. Res.* **2018**, *67*, 227–234. [CrossRef]
5. Khajeh-Khalili, F.; Honarvar, M.A. A design of triple lines Wilkinson power divider for application in wireless communication systems. *J. Electromagn. Waves Appl.* **2016**, *30*, 2110–2124. [CrossRef]
6. Saxena, A.; Hashmi, M.; Banerjee, D.; Chaudhary, M.A. Theory and design of a flexible two-stage wideband wilkinson power divider. *Electronics* **2021**, *10*, 2168. [CrossRef]
7. Liang, W.-F.; Hong, W.; Chen, J.-X. A Q-band compact Wilkinson power divider using asymmetrical shunt-stub in 90nm CMOS technology. In Proceedings of the 2012 Asia Pacific Microwave Conference Proceedings, Kaohsiung, Taiwan, 4–7 December 2012; pp. 974–976. [CrossRef]
8. Khajeh-Khalili, F.; Honarvar, M.A.; Limiti, E. A Novel High-Isolation Resistor-Less Millimeter-Wave Power Divider Based on Metamaterial Structures for 5G Applications. *IEEE Trans. Compon. Packag. Manuf. Technol.* **2020**, *11*, 294–301. [CrossRef]
9. Kawai, T.; Nagano, K.; Enokihara, A. Dual-Band Semi-Lumped-Element Power Dividers at UHF/SHF Bands. In Proceedings of the 50th European Microwave Conference, EuMC, Utrecht, The Netherlands, 12–14 January 2021; pp. 844–847.
10. Gupta, R.; Assaad, M.; Chaudhary, M.A.; Hashmi, M. An Effective Design Scheme of Single- and Dual-Band Power Dividers for Frequency-Dependent Port Terminations. *Electronics* **2023**, *12*, 1991. [CrossRef]
11. Wu, Y.; Liu, Y.; Zhang, Y.; Gao, J.; Zhou, H. A dual band unequal wilkinson power divider without reactive components. *IEEE Trans. Microw. Theory Tech.* **2009**, *57*, 216–222. [CrossRef]
12. Al Shamaileh, K.; Dib, N.; Abushamleh, S. A Dual-Band 1:10 Wilkinson Power Divider Based on Multi-T-Section Characterization of High-Impedance Transmission Lines. *IEEE Microw. Wirel. Compon. Lett.* **2017**, *27*, 897–899. [CrossRef]
13. Lee, S.; Park, J.; Hong, S. Millimeter-Wave Multi-Band Reconfigurable Differential Power Divider for 5G Communication. *IEEE Trans. Microw. Theory Tech.* **2022**, *70*, 886–894. [CrossRef]
14. Okada, Y.; Kawai, T.; Enokihara, A. Design Method for Multiband WPDs Using Multisection LC-Ladder Circuits. *IEEE Microw. Wirel. Compon. Lett.* **2017**, *27*, 894–896. [CrossRef]
15. Gomez-Garcia, R.; Loeches-Sanchez, R.; Psychogiou, D.; Peroulis, D. Single/multi-band Wilkinson-type power dividers with embedded transversal filtering sections and application to channelized filters. *IEEE Trans. Circuits Syst. I Regul. Pap.* **2015**, *62*, 1518–1527. [CrossRef]
16. Scardelletti, M.C.; Ponchak, G.E.; Weller, T.M. Miniaturized Wilkinson power dividers utilizing capacitive loading. *IEEE Microw. Wirel. Compon. Lett.* **2002**, *12*, 6–8. [CrossRef]
17. Wu, J.; Li, E.; Guo, G.F. Microstrip power divider with capacitive stubs loading for miniaturization and harmonic suppression. In Proceedings of the ICMTCE2011-Proceedings 2011 IEEE International Conference on Microwave Technology and Computational Electromagnetics, Beijing, China, 22–25 May 2011; pp. 237–239.
18. Karpuz, C.; Gorur, A.K. Effects of narrow slits on frequency response of a microstrip square loop resonator and dual-mode filter applications. *Microw. Opt. Technol. Lett.* **2012**, *55*, 143–146. [CrossRef]
19. Santiko, A.B.; Munir, A. Development of dual band power divider using meander line technique for local oscillator system. In Proceedings of the IEEE 13th International Conference on Telecommunication Systems, Services, and Applications (TSSA), Bali, Indonesia, 3–4 October 2019; pp. 286–289.
20. Rawat, K.; Ghannouchi, F.M. A design methodology for miniaturized power dividers using periodically loaded slow wave structure with dual-band applications. *IEEE Trans. Microw. Theory Tech.* **2009**, *57*, 3380–3388. [CrossRef]
21. Park, M.J.; Lee, B. A dual-band Wilkinson power divider. *IEEE Microw. Wirel. Compon. Lett.* **2008**, *18*, 85–87. [CrossRef]
22. Kim, K.; Nguyen, C. An ultra-wideband low-loss millimeter-wave slow-wave wilkinson power divider on 0.18 μm SiGe BICMOS process. *IEEE Microw. Wirel. Compon. Lett.* **2015**, *25*, 331–333. [CrossRef]

23. Roshani, S.; Yahya, S.I.; Alameri, B.M.; Mezaal, Y.S.; Liu, L.W.Y.; Roshani, S. Filtering Power Divider Design Using Resonant LC Branches for 5G Low-Band Applications. *Sustainability* **2022**, *14*, 12291. [CrossRef]
24. Wang, Y.; Zhang, X.-Y.; Liu, F.-X.; Lee, J.-C. A Compact Bandpass Wilkinson Power Divider With Ultra-Wide Band Harmonic Suppression. *IEEE Microw. Wirel. Compon. Lett.* **2017**, *27*, 888–890. [CrossRef]
25. Sallam, T.; Attiya, A.M. Neural network inverse model for multi-band unequal Wilkinson power divider. *COMPEL-Int. J. Comput. Math. Electr. Electron. Eng.* **2022**, *41*, 1604–1617. [CrossRef]
26. Cheng, J.-D.; Xia, B.; Xiong, C.; Wu, L.-S.; Mao, J.-F. An Unequal Wilkinson Power Divider Based on Integrated Passive Device Technology and Parametric Model. *IEEE Microw. Wirel. Compon. Lett.* **2022**, *32*, 281–284. [CrossRef]
27. Peng, C.; Ye, Z.; Wu, J.; Chen, C.; Wang, Z. Design of a wide-dynamic rf-dc rectifier circuit based on an unequal wilkinson power divider. *Electronics* **2021**, *10*, 2815. [CrossRef]
28. Karpuz, C.; Gorur, A.K.; Cakir, M.; Cengiz, A.; Gorur, A. Compact Wilkinson Power Dividers Based on Narrow Slits Loaded Transmission Lines. In Proceedings of the 2023 IEEE Radio and Wireless Symposium (RWS), Las Vegas, NV, USA, 22–25 January 2023; pp. 174–177. [CrossRef]
29. Sato, R. A Design Method for Meander-Line Networks Using Equivalent Circuit Transformations. *IEEE Trans. Microw. Theory Tech.* **1971**, *19*, 431–442. [CrossRef]

applied sciences

MDPI

Article

Compliance Assessment of the Spatial Averaging Method for Magnetic Field Leakage from a Wireless Power Transfer System in Electric Vehicles

Masanori Okada [1], Keishi Miwa [1], Sachiko Kodera [1] and Akimasa Hirata [1,2,*]

[1] Department of Electrical and Mechanical Engineering, Nagoya Institute of Technology, Nagoya 466-8555, Japan; m.okada.596@nitech.jp (M.O.); kodera@nitech.ac.jp (S.K.)
[2] Center of Biomedical Physics and Information Technology, Nagoya Institute of Technology, Nagoya 466-8555, Japan
* Correspondence: ahirata@nitech.ac.jp

Abstract: Wireless power transfer (WPT) via magnetic resonance offers efficient electrical power transfer, making it an increasingly attractive option for charging electric vehicles (EVs) without conventional plugs. However, EV charging requires a transfer power in order of kW or higher, resulting in a higher-leaked magnetic field than conventional wireless systems. The leaked magnetic field is nonuniform, and the assessment in terms of the limit prescribed in the guideline is highly conservative because it assumes that a person standing in free space is exposed to a uniform field. In such cases, an assessment should be performed using the limits of the internal electric field, as it is more relevant to the adverse health effects, whereas its evaluation is time-consuming. To mitigate this over-conservativeness, international product standards introduce a spatial averaging method for nonuniform exposure assessment. In this study, we investigate assessment methods, especially for measurement points of nonuniform magnetic field strength leaked from the WPT system. Various spatial averaging methods are correlated with the internal electric field derived from electromagnetic field analysis using an anatomically based human body model. Our computational results confirm a good correlation between the spatially averaged magnetic and internal electric fields. Additionally, these methods provide an appropriate compliance assessment with the exposure guidelines. This study advances our understanding of the suitability of spatial averaging methods for nonuniform exposure and contributes to the smooth assessment in WPT systems.

Keywords: compliance assessment; electric vehicles; human protection from electromagnetic field; wireless power transfer

1. Introduction

Wireless power transfer (WPT) technology relies on the principle of magnetic resonant coupling between two coils, enabling the efficient transmission of electrical power over distances ranging from several tens of centimeters to a few meters [1–5]. This technology is increasingly gaining interest as a streamlined method for charging electric vehicles (EVs) without conventional plugs [6]. However, EV charging requires a significant amount of power, reaching several kilowatts, leading to the leakage of strong electromagnetic fields during charging [7]. Concerns about the potential adverse effects for WPT continue to exist, which require further investigation [8].

The International Commission on Non-Ionizing Radiation Protection (ICNIRP) [9–11] and IEEE International Committee on Electromagnetic Safety Technical Committee 95 [12,13] have published international guidelines and standards (hereafter referred to as "guidelines") limiting exposures to electromagnetic fields for human protection. These guidelines were designed to protect humans from electrostimulation up to 5–10 MHz and heating above 100 kHz. In these guidelines, the operational threshold for adverse health effects is identified [9–12]. A reduction (safety) factor [11], basic restriction (BR) [9,10], or dosimetric reference limit [12] (hereafter referred to as BR), which is the electrical quantity related to the health effect, is then derived. For practical assessment, permissible external field strength (hereafter referred to as reference level (RL)) [10] or exposure RL is derived under a worst-case exposure scenario. In the IEEE C95.6-2002 standard [13], the RL, permissible external field strength is derived based on an analytical formula using homogeneous ellipsoidal induction models, which differ between the limb and trunk [13].

International standards for product conformity [14], such as International Electrotechnical Commission (IEC) 61980-1 [15], SAE J2954 [16], and ISO 19363 [17], have been established for WPT in EVs, all of which consider the international exposure guidelines. Computational assessments using human body models have been widely used, including international standardization, because in vivo measurements cannot be performed for the assessment of an induced electric field (e.g., IEC 61980 and SAE J2954). Numerous studies have pointed out the limitation of RL in highly nonuniform environments, such as those emitted from WPT [18–21]. These results consistently showed that, although the spatial peak magnetic fields at the human body parts exceeded RL, the internal electric fields in the human body were one or two orders of magnitude smaller than the limit of induced physical quantities (see Section 2).

The IEC Technical Committee 106 is an international standardization body that provides compliance assessment methods corresponding to international exposure guidelines/standards. The spatially averaged method defined in IEC 62110 was discussed and extended to assess nonuniform fields [22]. The spatial averaging for the magnetic field strength measured (computed) at some points confirms compliance with RL. The assessment of the spatial averaging method is also included in IEC TR 62905 and IEC PAS 63184 [23,24], wherein various measurement points inside and outside the vehicle are considered.

Several studies have evaluated the location of the magnetic field measurement point in realistic vehicle models [25–32]. As summarized in Section 2, recent studies conducted measurements at designated test points near the vehicle and within the driver's seat. However, the relationship between these measurement points and the internal electric field, which should be assessed computationally, has not been clarified. Appropriate measurement locations that can effectively capture their coupling quantities have not been considered in the current product standards. In addition, the exposure scenarios are different for different studies, and thus, straightforward comparison was difficult.

At the IEC JWG 63184 meeting, the methods for assessing EV WPT are discussed. For practical assessment, the correlation between the spatially averaged magnetic field strength and permissible exposure levels defined by ICNIRP and IEEE is required for justification. However, in [18], this correlation was only discussed between the spatially averaged magnetic field value and internal electric fields assuming a standing posture. Additionally, there has been no study discussed measurement point for compliance assessment to relate the induced electric field and magnetic field in the cabin for realistic postures such as those of a driver situated inside the vehicle. Scientific data considering the scenarios discussed in the standardization meeting would provide insight for reliable measurement procedure.

This study aims to clarify the measurement (evaluation) points for nonuniform magnetic field strength leaked from the WPT system for a realistic EV cabin environment. To provide the rationale for the averaging points, we considered different spatial averaging methods to correlate with the internal electric field, which is the metric in the human exposure guidelines.

2. Related Studies

2.1. Evaluation of Leaked Magnetic Field

Laakso validated the leaked magnetic field from WPT coil for a simplified vehicle model [33]. Che et al. evaluated the conformity at several measurement points inside and outside the vehicle according to the electromagnetic field test procedure in SAE J2954 [25].

Mohamed et al. conducted testing on a WPT system installed in a heavy-duty electric shuttle [26]. El-Shahat evaluated the external magnetic field from different WPT coils and derive a distance compliant with the ICNIRP guidelines [27]. The magnetic field inside and outside the vehicle cabin has been measured [28]. The magnetic field leaked from the wireless power transfer system is analyzed for heavy duty vehicles [29].

2.2. Computational Dosimetric Study for EV WPT Systems

Numerous compliance assessments have been conducted for exposure resulting from the WPT system. For example, Laakso et al. evaluated the compliance of three human body models beside a vehicle at different human–vehicle angles [18]. Shimamoto et al. considered various postures, such as lying on the ground with the right arm extended toward the coils, to consider worst-case scenarios [19]. Wen et al. predicted the magnetic field leakage generated from parallel WPT systems operating with a phase difference, including assessing the internal electric field and the specific absorption rate in a human body near these systems [34]. De Santis et al. investigated the influence of vehicle materials and human body posture on compliance assessment in compact EVs [21,35,36]. Liorni et al. developed and validated a numerical model of inductive power transfer systems in a heavy vehicle and compared the exposure of passengers in public vehicles with light vehicles [37]. Lan et al. investigated the effect of ferrite materials, such as loudspeakers and electromotors, on the magnetic field distribution inside the vehicle [38]. Wang et al. conducted uncertainty analysis on dosimetric measures, i.e., quantified the impact of different parameters on power absorption [39]. In most studies, the number of scenarios and conditions considered were less than 25 cases, except for [18] (144 cases) and [34] (120 cases). To discuss the measurement methods in the product safety standardization, enough number of cases should be considered to ensure the safety based on the human exposure guidelines [9–11].

2.3. Exposure Assessment Method for Nonuniform Magnetic Field

Several studies have investigated methodologies for a realistic and comprehensive assessment of nonuniform magnetic fields resulting from the WPT system. Wake et al. derived coupling factors for EVs and home appliances with WPT to assess compliance with BR equivalently in terms of the measured external field strengths. They also compared these factors across multiple institutions [40]. Chakarothai et al. proposed an experimental approach for determining the internal electric field without prior knowledge of the WPT system's design using measured magnetic near-field data [41]. Miwa et al. assessed adherence to international standards using a coupling factor, focusing on the exposure of human subjects positioned in both the driver's seat and the rear passenger compartment of a vehicle [20]. Park et al. examined WPT using the minimum accessible distances in compliance with RL in instances of proximity exposure, considering various exposure patterns [42]. Ahn et al. proposed an improved coupling factor calculation method for removing computational artifacts to achieve uniform assessment results in various human body models [43]. These studies assessed compliance procedures under these specific conditions, making it challenging to make overarching generalizations.

2.4. Comparison of Exposure Assessment in Human Exposure Inside and Outside the Vehicle

As listed in above subsections, several studies have been conducted to assess the relationship between the leaked magnetic field and induced electric field. The frequency for WPT has been now standardized at 85 kHz [16]. The limit of induced electric field (BR) and external magnetic field (RL) in the international guidelines are (11.5 V/m, 21 A/m) and (17.8 V/m, 163 A/m) for ICNIRP and IEEE, respectively. Thus, their relationship and over-conservativeness are assessed in different manners. Table 1a,b lists the key computational results in previous studies for scenarios outside and inside the vehicles. In these studies, the number of vehicle models was one. For proper comparison, the coupling factor [44] was derived.

Table 1. Comparison of the related research (**a**) outside and (**b**) inside the vehicle.

(a)

Year	**2015** [19]	**2017** [40]	**2018** [41]	**2022** [43]
Considered exposure scenario	Standing near to a vehicle model.	Standing 650 mm away from coils.	Standing near the coil, human-coil distance is 235 mm *.	Standing near to a vehicle mimic steel plate *.
Operating frequency [kHz]	85	85	125	85
Transferred power [kW]	7	7	7.7	7.7
Exposed magnetic field strength (max) [A/m]	56.7 excluding the region under the vehicle	N/A	28	N/A
Maximum electric field [V/m]	0.4 (ankle)	N/A	0.42 (leg)	N/A
Coupling factor [44] for the ICNIRP guidelines [V/A]	0.013 (99.9th percentile)	0.038–0.054 ** for different coil positions of a vehicle model	0.027 (averaged over 2 × 2 × 2 mm^3 cube and 99.9th percentile)	0.018 (99.9th percentile)

(b)

Year	**2018** [36]	**2019** [20]	**2020** [38]
Considered exposure scenario	driving passenger.	driving passenger and other passenger.	driving passenger.
Operating frequency [kHz]	85	85	85
Transferred power [kW]	7.7	3.7	3.7
Exposed magnetic field strength (max) [A/m]	347.4	16.3 (driver's buttocks) 23.6 (passenger's feet)	46.5
Maximum electric field [V/m]	19.9 (feet)	0.53 (driver's buttocks) 0.35 (passenger's feet)	0.61 (buttocks)
Coupling factor [44] for the ICNIRP guidelines [V/A]	0.105 (averaged over 2 × 2 × 2 mm^3 cube)	0.060 (driver) 0.027 (passenger) (99.9th percentile)	0.024 (averaged over 2 × 2 × 2 mm^3 cube and 99.9th percentile)

* Ferrite tiles and a metal plate as a vehicle body are considered. ** No clear definition for the method processing the electric field induced in the body.

3. Models and Methods

Our computational approach to compute the external magnetic field and induced electric field (Section 3.3), as well as human body models (Section 3.1), is the same as in our previous studies [19,20]. Our approach is briefly summarized below.

3.1. Human Body Models

In this study, we employed an anatomical human body model TARO developed at the National Institute of Information and Communications Technology (NICT) [45]. The height and mass of the model is 173.2 cm and 65 kg, respectively. The number of tissues considered in TARO is 51. The voxel resolution of the human model was set at 2.0 mm. Two distinct postures were considered for the models, including the original standing posture and a seated posture simulating a driver's position, as illustrated in Figure 1. The seated model was created utilizing software provided by NICT [46]. The posture of the driving seat has been generated considering [47].

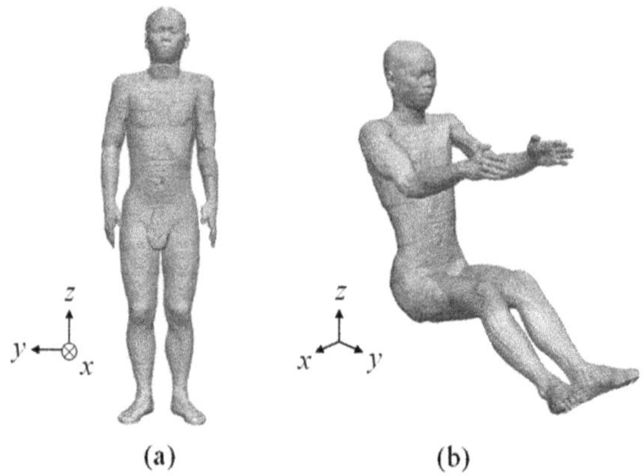

Figure 1. Japanese adult male model TARO: (**a**) standing and (**b**) sitting postures.

3.2. Exposure Scenarios

SAE J2954 [16] recommends three standard power levels: WPT1 at 3.7 kW, WPT2 at 7.7 kW, and WPT3 at 11 kW. In this study, the WPT system integrated into the vehicle adhered to the specifications outlined in SAE J2954 at 3.7 kW (WPT1). Note that the computational results are scalable to other transfer power. Figure 2 shows a detailed schematic of the WPT coils. The transmitting coil was rectangular, with a length of 580 mm and a width of 420 mm following to SAE J2954 [16]. It consisted of 15 turns, with a width of 3 mm. The receiving coil was a 320 mm square. It equipped with 20 turns and had a width of 2 mm. The distance between the top of the transmitting coil and the bottom of the receiving coil was 150 mm, which is in accordance with WPT1. The two coils were assumed to be composed of a perfect conductor for conservative assessment. The thicknesses of shield plate and ferrite core are 2 mm and 3 mm, respectively. The transmitter and receiver current sources I_1 and I_2, were set to 16 and 17 A, respectively, to achieve a transmitting power of 3.7 kW. The phase difference of the currents between the transmitting and receiving coils was 90°. The frequency of the WPT system was 85 kHz. The following relationship holds for current source I_1 and I_2 [48].

$$I_2 \approx -j\sqrt{\frac{L_1}{L_2}}I_1 \tag{1}$$

where L_1 and L_2 represent the inductance of the transmitter and receiver coils, respectively. Note that the assumption of coil material as perfect conductor would result in a slightly larger leaked magnetic field of 1% or less, under the condition that the transmitting power is fixed at 3.7 kW [49].

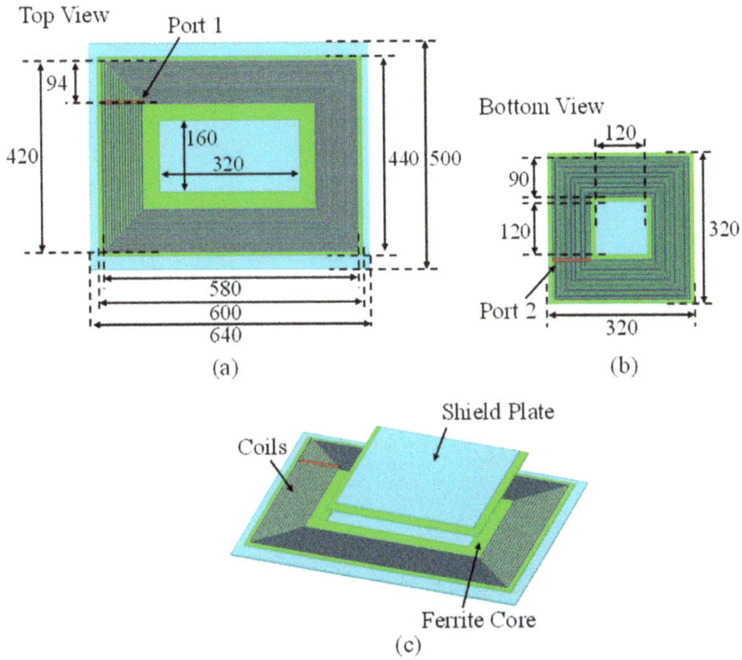

Figure 2. Schematic of WPT coils (unit: mm): (**a**) transmitting coils, (**b**) receiving coils, and (**c**) positions.

Figure 3 shows a vehicle model and two realistic exposure scenarios for the assessment. The simplified vehicle's geometry was developed using the Toyota Motor Corporation's Prius model and was positioned in free space. Note that the effect of the ground and surrounding objects marginally influences the magnetic field where the quasi-magnetostatic approximation is valid. The receiving coil was positioned beneath the center of the vehicle body. Misalignment between the receiving and transmitting coils were chosen as 75 mm and 100 mm in the *x* and *y* directions, respectively, following SAE J2954 [16]. We considered two vehicle body materials, including iron and carbon fiber-reinforced plastic (CFRP). The relative permittivity, permeability, and conductivity values for iron and CFRP were 1, 4000, and 10.3×10^6 S/m and 1, 1, and 0.25×10^6 S/m, respectively. The thicknesses of iron and CFRP were 0.5 and 2.0 mm, respectively. In this study, two scenarios discussed in IEC JWG 63184 were considered. In the first scenario, a human body model stands outside the vehicle body where the highest exposure to the magnetic field from the WPT system was computed. Meanwhile, the human body in the second scenario was located in the driver's seat, where the distance from the floor to the driver's buttocks was 200 mm. Note that the position of the human body in the vehicle cabin has been standardized for such purposes [23,24,50].

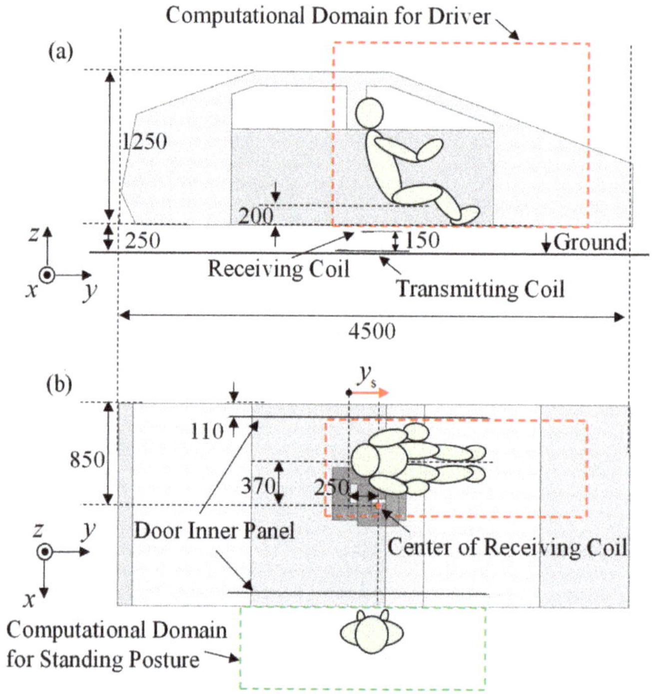

Figure 3. Cross-section of the vehicle model (unit: mm): (**a**) side view and (**b**) top view. y_s represents the distance in the front–back direction starting from the seat position of the vehicle.

3.3. Computational Methods

In a frequency range lower than the MHz range, it is feasible to utilize a quasi-static approximation to calculate the internal electric field within biological tissue [51–53]. In the regime in which the quasi-static approximation is valid, the conduction current in the body is dominant rather than the displacement current. This assumption causes the external electromagnetic fields to be decoupled into electric and magnetic fields. Additionally, the external magnetic field can be assumed to be unperturbed by the presence of the human body.

The magnetic field leaked from the WPT system was computed using a commercial electromagnetic simulator (HFSS, Ver. 2023 (R1), ANSYS, Canonsburg, PA, USA) without considering the human body model. The magnetic field was extracted from HFSS in several rectangular volumes with a grid resolution of 50 mm × 50 mm × 50 mm. These data were used to construct a magnetic vector potential with a grid resolution of 2 mm × 2 mm × 2 mm.

The next step involves computing the internal electric field by substituting the vector potential distribution into an electromagnetic solver developed at the Nagoya Institute of Technology [54]. The solver used to compute the internal electric field is based on the scalar-potential finite-difference (SPFD) method [55].

The SPFD method discretizes the human model using cubical voxels and generates simultaneous linear equations for all contacts, with the electric scalar potential as an unknown variable. The internal electric field is then computed through matrix calculation. To accelerate computation, the geometric multigrid method is employed as a preconditioner [54]. In the SPFD method, tissue conductivities in the human body model are taken from a technical report in [56], which is consistent with [57,58]. In addition, the leaked magnetic field from WPT coil has been validated for the simplified vehicle model with same computational approach [33]. Our computational approach here has been validated by intercomparison [40].

3.4. Compliance Assessment Procedure

Several product assessment standards and technical reports define spatial averaging methods for exposure to nonuniform magnetic fields. The spatial averaging method involves averaging the magnetic field strength measured (computed) at specific spatial points. For instance, IEC TR 62905 [23] specified the measurement of magnetic fields on a plane located 20 cm from three points at different heights (0.5, 1, and 1.5 m) for humans standing by the side of the vehicle. This reference is derived from IEC 62110 [22]. IEC PAS 63184 [24] extended this by including the closest accessible point to the three points. IEC PAS 63184 also outlined the measurement of the magnetic field at the center position of the floor flat surface and the points at the center position of the headrest, backrest, and seat for sitting on the driver's seat.

We outlined three definitions for each posture, as shown in Table 2 and Figure 4. In Definition 1 for outside the vehicle, the measurement point was defined at the heights of 0.5, 1.0, and 1.5 m from the ground based on IEC TR 62905. In Definition 1 for inside the vehicle, the measurement heights are 0.5, 1.0, and 1.2 m from the floor, corresponding to the cushion, chest, and head of the seat, respectively. In Definition 2, the closest accessible point was incorporated into Definition 1 based on IEC PAS 63184, considering the foot. In Definition 3, the measurement point was defined at the heights of the closest accessible point and two points, each 25 cm higher from it, aiming to be more stringent than Definitions 1 and 2. The measurement points of 0.35, 0.6, 1.0, and 1.2 were defined along a 15-degree angle of the seat to align with the central axis of the human body, considering the human body model sits in the driver's seat in a realistic posture.

Table 2. Definition of the heights of magnetic field strength measurement used for averaging.

Definitions	Heights from The Ground [m]	Heights from The Floor [m]
1: TR 62905	0.5, 1.0, 1.5	0.5, 1.0, 1.2
2: Maximum point + PAS 63184	0.06, 0.5, 1.0, 1.5	0.1, 0.5, 1.0, 1.2
3: Every 25 cm from maximum point	0.06, 0.31, 0.56	0.1, 0.35, 0.6

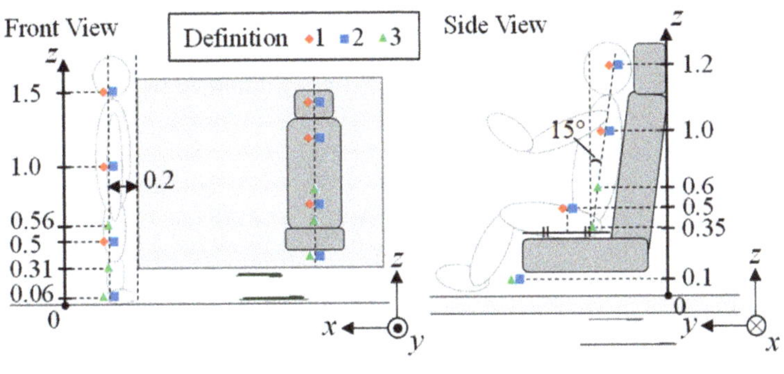

Figure 4. Measurement points outlined in each definition for outside and inside the vehicle body (unit: m).

Using these definitions, we calculated the spatial average of the magnetic fields (H_{ave}) and compared them with the limit prescribed in the exposure guidelines [7,9]. In this study, we compared the ratio of H_{ave} to RL (H_{RL}) and that of the maximum internal electric field (E_{max}) to BR (E_{BR}). The spatial averaged method was evaluated using the following equation:

$$a_c = \left(\frac{E_{max}}{E_{BR}}\right) / \left(\frac{H_{ave}}{H_{RL}}\right) \quad (2)$$

where a_c represents an index of the conformity ratio of external magnetic and internal electric fields to the guideline. If a_c is less than 1, it indicates that the assessment based on the spatial average of the magnetic field in the corresponding definition is more stringent than that based on the maximum electric field.

The magnetic field strength was computed in the domain averaged over 100 cm^2, assuming the 100 cm^2 loop antenna prescribed in the IEC standard [44,59]. This averaging area was considered as a postprocessing step for the computed field distribution.

In the evaluation of the internal electric field in anatomical models, the skin-to-skin contact is not negligible [12], which is inherent to the discretization of a real human model into a finite-resolution model. This computational weakness results in a higher internal electric field at the corresponding part. Therefore, the electric field at skin-to-skin contact was excluded, as mentioned in IEEE C95.1 standard [12]. Different postprocessing method has been proposed to reduce numerical artifacts. This study employed the peak voxel value for 99.9th percentile values for body parts in the anatomical human body as a conservative approach. We express the external magnetic and internal electric fields in root-mean-square values for comparison with international standards.

4. Results

In this section, the internal electric field computation has been evaluated for exposure scenarios in which the human stands outside and sits inside the vehicle (Sections 4.1 and 4.2). Then, we have evaluated the averaging method of magnetic field for the scenario outside (Section 4.3) and inside the vehicle (Sections 4.4 and 4.5). For the exposure scenario inside the vehicle, the correlation between the averaged magnetic field and internal electric field is evaluated to provide the scientific rationale for the measurement point in the standardization.

4.1. Distribution of Magnetic Field Outside and Inside the Vehicle

Figures 5 and 6 show the magnetic field strength distributions outside and inside the vehicle, as viewed from the side of the vehicle, for different vehicle body materials. The computational domain is illustrated in Figure 3. As shown in Figure 5, there was no significant difference in the distribution between iron and CFRP outside the vehicle. In contrast, the magnetic field inside the vehicle was scarcely distributed for a body made of iron due to the shielding effect shown in Figure 6. In the case of the CFRP body, the magnetic field generated by the WPT system leaks directly into the cabin beneath the floor.

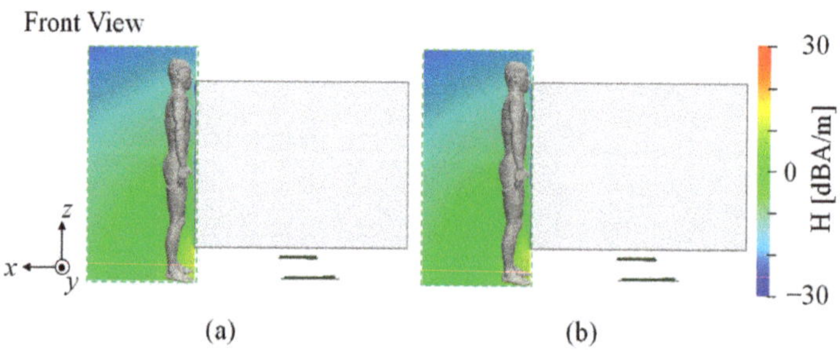

Figure 5. Distributions of the magnetic field outside the vehicle with different materials: (**a**) iron and (**b**) CFRP.

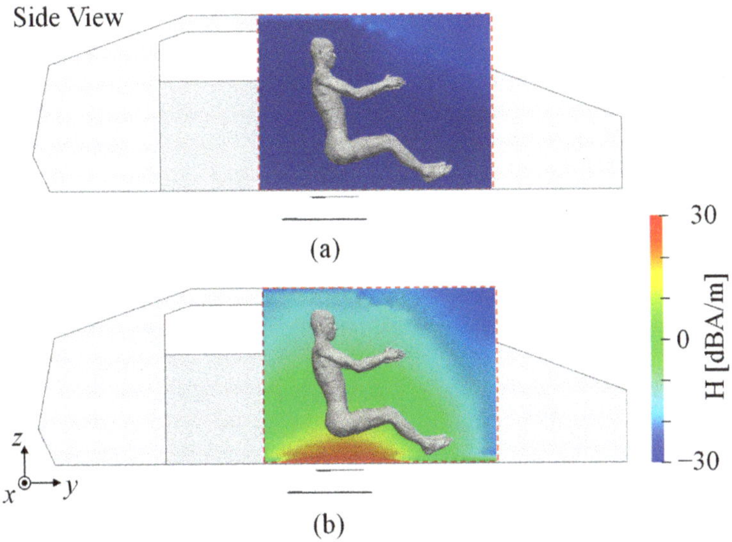

Figure 6. Distributions of the magnetic field inside the vehicle with different materials: (**a**) iron and (**b**) CFRP.

4.2. Internal Electric Field in the Human Body

Figures 7 and 8 show the distributions of the internal electric field in the human body models corresponding to outside and inside the vehicle for different vehicle body materials, respectively. As shown in Figure 7, the maximum internal electric field was observed in the lower limb near the WPT system for both materials when humans were standing outside the vehicle. The 99.9th percentile values of the electric field were 0.05 V/m. Figure 8 shows that the electric field was primarily induced in the torso and head due to the magnetic field leakage from the window, albeit with a 99.9th percentile value of 1.6×10^{-3} V/m. In contrast, a high internal electric field was distributed in the buttocks for the CFRP body, with the 99.9th percentile value of 0.20 V/m.

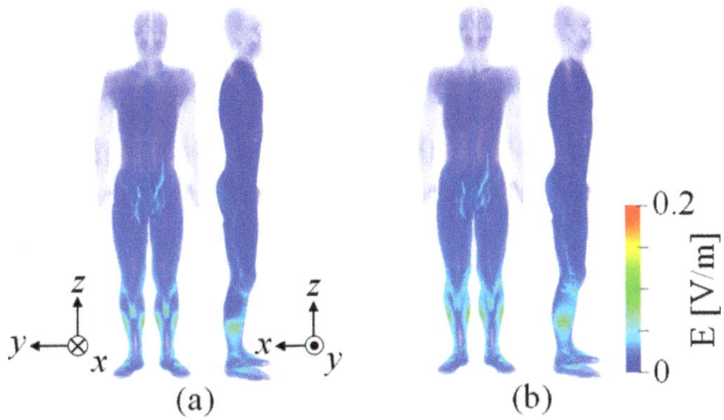

Figure 7. Internal electric field distributions in standing human body model outside the vehicle with different materials: (**a**) iron and (**b**) CFRP.

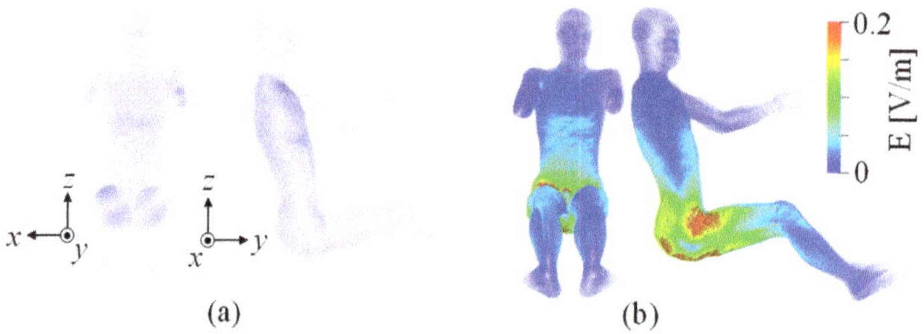

Figure 8. Internal electric field distributions in human body model inside the vehicle sitting on the driver's seat for different materials: (**a**) iron and (**b**) CFRP.

4.3. Assessment with Spatial Averaged Magnetic Field Outside the Vehicle

The relationship between the external magnetic and internal electric fields was evaluated outside the vehicle [18]. The averaged magnetic field strength (H_{ave}) was calculated based on the measurement points and averaging scenario defined in Table 2. Table 3 presents the magnetic field strength outside the vehicle at the measurement points and H_{ave} for vehicle body with different materials. As shown in Table 3, the magnetic field strengths for different materials are consistent. Figure 9 shows the conformity assessment in each definition against the ICNIRP guidelines and IEEE standard in terms of a_c outside the vehicle. As shown in the figure, the stringer assessment methods were in the order of Definitions 3, 2, and 1 because a high magnetic field was distributed near the ground. All definitions exhibited a conservativeness with more than a margin of a factor of four with respect to the ICNIRP guidelines, whereas only Definition 1 was an unconservative assessment for the IEEE standard. The magnetic field limits for the ICNIRP guidelines and IEEE standard are 21 and 163 A/m for public exposure.

Table 3. Heights of magnetic field strength measurement points and average magnetic field strength averaged for each definition with outside the vehicle body made of (**a**) iron and (**b**) CFRP.

(a)

Setting	Magnetic Field Strength [A/m]				
	Measurement Points				H_{ave}
Definition 1	0.5 m 0.78	1.0 m 0.28	1.5 m 0.13	-	0.40
Definition 2	0.06 m 1.54	0.5 m 0.78	1.0 m 0.28	1.5 m 0.13	0.68
Definition 3	0.06 m 1.54	0.31 m 1.23	0.56 m 0.67	-	1.15

(b)

Setting	Magnetic Field Strength [A/m]				
	MeasurementPoints				H_{ave}
Definition 1	0.5 m 0.82	1.0 m 0.27	1.5 m 0.13	-	0.41
Definition 2	0.06 m 1.55	0.5 m 0.82	1.0 m 0.27	1.5 m 0.13	0.69
Definition 3	0.06 m 1.55	0.31 m 1.27	0.56 m 0.72	-	1.18

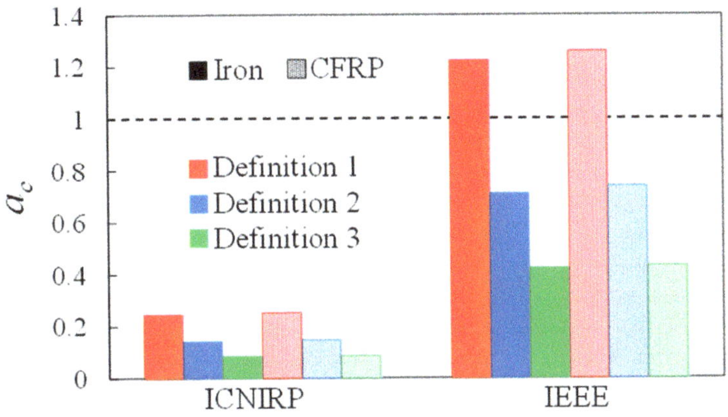

Figure 9. Assessment of three definitions for the ICNIRP guidelines and IEEE standard in terms of a_c outside the vehicle with different materials.

4.4. Correlation between E_{max} and H_{ave} Inside the Vehicle Made of CFRP

The average of three or more measurement points can provide a reliable estimate of the internal electric field when the human body model stands by the side of the vehicle [18]. However, we did not consider a scenario in which a person sat in the driver's seat [18]. We evaluated the relationship between the external magnetic and internal electric fields for different vehicle seat positions. The seat position was varied by $y_s = \pm 100$ mm (in 20-mm increments) in the y-direction based on the adjustment range of a typical seat. This section focuses on the CFRP body since the magnetic field was almost shielded for an iron body.

Figure 10a shows the relationship between the average magnetic field strength within the volume containing the human body model and the internal electric field. The average magnetic field was determined by calculating the arithmetic mean of the absolute magnetic field values over the whole body. The correlation has been evaluated in terms of coefficient of determination (R^2) [60,61] assuming the linear relationship. If they are correlated linearly, the induced electric field can be equivalently evaluated in terms of average magnetic field without detailed computation. There was a strong correlation between the whole-body H_{ave} and the internal electric field. Figure 10b shows the relationship between the magnetic field strength at a single measurement point defined in Table 2 and the internal electric field. As shown in Figure 10b, good correlations were observed at specific points, such as 0.1, 0.5, and 1.2 m. However, there was no consistent trend for the other measurement points. Figure 10c shows the relationship between the average magnetic and internal electric fields at three or four points outlined for each definition. As shown in Figure 10c, the average magnetic field reasonably correlates with the internal electric field across all definitions. In Definitions 2 and 3, E_{max} reached its peak because the exposure area from the buttocks to the feet approaches a certain level when the seat is moved backward.

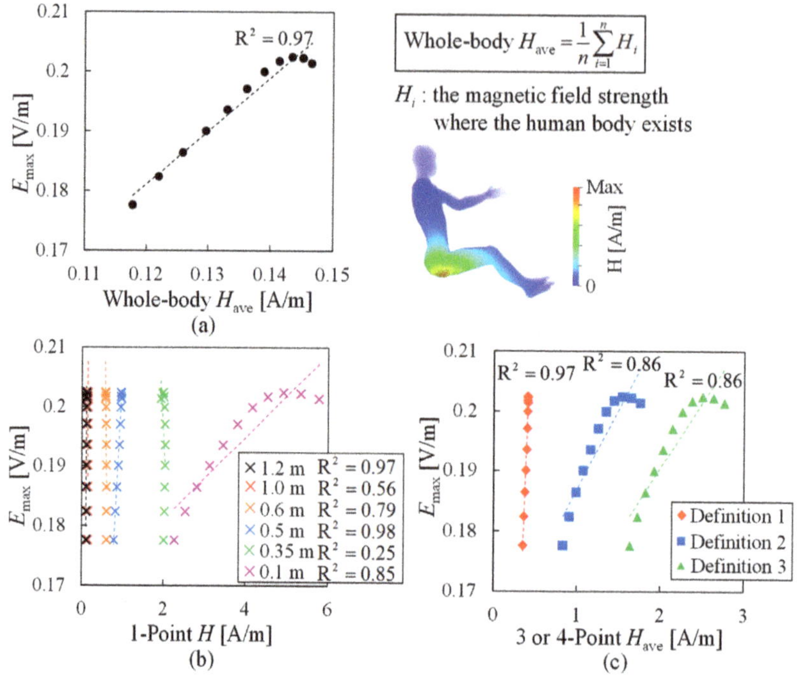

Figure 10. Correlation between the maximum internal electric field and magnetic field strength for different vehicle seat positions: (**a**) whole-body H_{ave}, (**b**) 1-Point H, and (**c**) 3 or 4-Point H_{ave}. R^2 represents the coefficient of determination. (**a**) was evaluated in the volume where the human body model exists in realistic posture (the number of voxels was 8,193,397).

4.5. Assessment with Spatial Averaged Magnetic Field Inside the Vehicle

Table 4 presents the magnetic field strength inside the vehicle at the measurement points and H_{ave} for different materials ($y_s = 0$). As shown in Table 4, the magnetic field strength within the iron body was two orders of magnitude lower than that within the CFRP body. Figure 11 shows the conformity assessment in each definition for the ICNIRP guidelines and IEEE standard in terms of a_c inside the vehicle ($y_s = 0$). As shown in the figure, all definitions in the case of the iron body exhibited a conservativeness with more than a margin of a factor of four with respect to the ICNIRP guidelines, whereas they were either comparable or unconservative with respect to the IEEE standard. In contrast, for the CFRP body, Definition 1 provided a level of protection equivalent to the ICNIRP guidelines. Definitions 2 and 3 displayed a conservativeness with more than a margin of a factor of 3.4 with respect to the ICNIRP guidelines. In the IEEE standard, Definitions 1 and 2 were unconservative, whereas Definition 3 exhibited a conservativeness with more than a margin of a factor of 1.2.

Table 4. Heights of magnetic field strength measurement points and average magnetic field strength averaged for each definition with inside the vehicle body made of (**a**) iron and (**b**) CFRP.

(**a**)

Setting	Magnetic Field Strength [A/m]				
	Measurement Points				H_{ave}
Definition 1	0.5 m 0.015	1.0 m 0.017	1.2 m 0.013	-	0.015
Definition 2	0.1 m 0.009	0.5 m 0.015	1.0 m 0.017	1.2 m 0.013	0.013
Definition 3	0.1 m 0.009	0.35 m 0.011	0.6 m 0.016	-	0.012

(**b**)

Setting	Magnetic Field Strength [A/m]				
	Measurement Points				H_{ave}
Definition 1	0.5 m 0.94	1.0 m 0.15	1.2 m 0.11	-	0.40
Definition 2	0.1 m 3.81	0.5 m 0.94	1.0 m 0.15	1.2 m 0.11	1.25
Definition 3	0.1 m 3.81	0.35 m 2.03	0.6 m 0.60	-	2.14

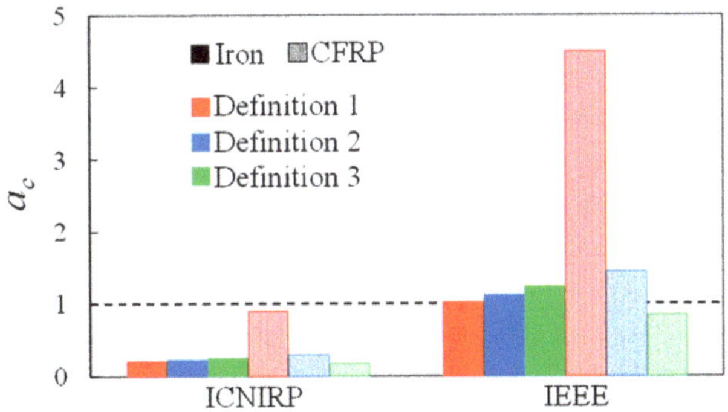

Figure 11. Assessment of three definitions for the ICNIRP guidelines and IEEE standard in terms of a_c inside the vehicle with different materials.

Figure 12 shows the assessment of each definition for compliance with the ICNIRP guidelines and IEEE standard using a_c when the vehicle seat position was changed in the front–back direction (y_s = ±100 mm). As shown in Figure 12, the variation in a_c in the case of the iron body remained relatively insignificant due to the weak nonuniformity of the magnetic field distribution. In contrast, for the CFRP body, a_c varied from approximately −0.1% to 3%, −27% to 36%, and −21% to 18% in Definitions 1, 2, and 3, respectively, regardless of the target guidelines. The variability in Definitions 2 and 3 is influenced by the observation that the variation in H_{ave} considering the measurement point at the lower limb is larger than that of E_{max} appearing at the buttocks. However, by adding a reduction factor smaller than two, such an effect of variations is sufficiently avoided.

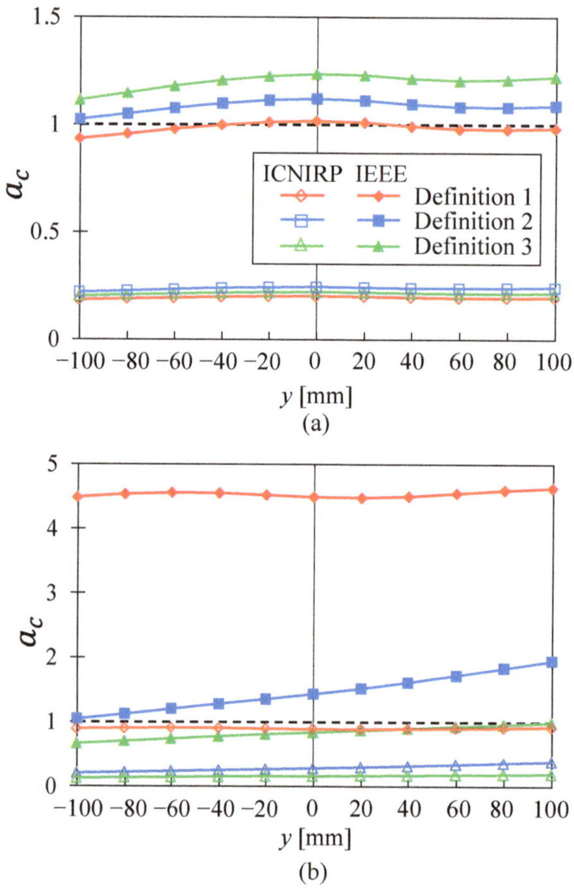

Figure 12. Variation of the compliance assessment for each definition when the vehicle seat position was changed in the front–back direction with the vehicle body made of (**a**) iron and (**b**) CFRP.

5. Discussion

This study evaluated compliance assessment using the spatial averaging method as a scheme to evaluate human exposure to nonuniform fields at low frequencies. The evaluation index for this method was the spatially averaged magnetic field strength at three or four heights from the ground or floor and the internal electric field in the anatomical human model placed near or inside the vehicle.

As shown in Figures 5 and 6, the distributions of the magnetic field were not significantly different between iron and CFRP outside the vehicle. However, the magnetic field was scarcely distributed for a body made of iron placed inside the vehicle. This is because the conductive chassis shields the magnetic field. In contrast, the magnetic field generated by the WPT system for a body made of CFRP leaks directly into the cabin beneath the floor. This resulted in a marginal difference in the internal electric field in the human body outside the vehicle because of the body material, but a significant difference inside the vehicle.

We calculated the magnetic field averaged over the heights defined in this study and compared it with the ICNIRP guidelines and IEEE standard in terms of a_c. When a_c is less than one, the BR would still be satisfied even with an increased coil output, provided that H_{ave} of the definition satisfies the RL. Laakso et al. reported a ratio of the internal electric field to the three-point average magnetic field strength of 0.20 V/A with three human body models at several angles outside the vehicle [18], which is close to 0.13 V/A obtained for a similar averaging definition in this study. These values are close but somewhat different from other studies (see Table 1), which is attributable to different postures and target areas discussed in previous studies. Note that the discussion here follows the discussion for the product safety standard (See Section 3.4). As shown in Figure 9, Definition 1 provided a conservative assessment of the ICNIRP guidelines with each body material. However, Definition 1 did not meet the requirements of the IEEE standard, whereas Definitions 2 and 3 were satisfied. This is because RL prescribed in the IEEE standard are relatively tolerant of a permissible external magnetic field. Using Definition 1 and following the ICNIRP guidelines, the maximum permissible transferred power for outside the vehicle made of iron and CFRP was 196 and 191 kW, respectively. Moreover, it was 884 and 871 kW using Definition 2 and the IEEE standard, respectively.

The whole-body H_{ave}, H at specific points, and three- or four-point H_{ave} provided good correlation for the scenario inside the vehicle, whereas no consistent trend was discernible for the other one-point measurement points (Figure 10). This can be attributed to the difficulty of capturing the electric field at a single point in a highly nonuniform environment with a realistic human body shape, such as the driver's posture. Additionally, measuring the magnetic field over the volume of the human body is practically difficult. Therefore, the spatially averaged method is a good index to reasonably simplify the compliance assessment procedure inside the vehicle. As shown in Figure 11, all definitions for the iron body provided a sufficiently conservative assessment of the ICNIRP guidelines, whereas they were either comparable or unconservative with respect to the IEEE standard. In contrast, for the CFRP body, Definition 1 provided an equivalent level of protection in the ICNIRP guidelines. However, Definitions 1 and 2 did not meet the requirements of the IEEE standard, whereas Definition 3 met this requirement. When the vehicle seat position was changed in the front–back direction, the absolute changes in a_c with respect to the ICNIRP guidelines were quite small, and the impacts on the conformity assessment were marginal (Figure 12). However, some reduction factors may need to be applied to Definition 1 for the iron body and Definition 3 for the CFRP body as they may not meet the IEEE standard. Even considering additional reduction factor, its magnitude is two, which is much smaller than the one order of magnitude or more, which is inherent to apply the RL in the exposure guidelines to nonuniform field exposure such as electric vehicle (Section 2). Although it is necessary to introduce an additional reduction factor for the IEEE standard, Definition 1 for the ICNIRP guidelines is satisfied. Using Definition 2, following the ICNIRP guidelines and IEEE standard, the maximum permissible transferred power for inside the vehicle made of CFRP was 62 and 241 kW, respectively.

6. Conclusions

The current product safety standards for WPT have not well considered the induced electric field in the human body for compliance assessment, which is more essential metric for human protection. Instead, the external magnetic field is used for practical purpose. To fill this gap, different scenarios were considered to correlate the induced electric field and external magnetic field in realistic scenarios. As the exposure scenarios can be limited around the vehicle cabin, different practical definitions were considered for assessment. From our finding, compliance assessment with exposure guidelines can be performed using appropriate definitions and averaging of magnetic field under realistic conditions discussed in the product safety standards. Spatial averaging of the magnetic field rather than single-point measurement is appropriate when considering the correlation with the internal electric field. In general, conservativeness is confirmed for most definitions considered here. If it is not satisfied, some adjustments should be introduced, such as an additional reduction factor of two to comply with the IEEE standard. Overconservativeness (by the one order of magnitude or more) may be avoided unlike the measurement of peak magnetic field for nonuniform field exposures.

Author Contributions: Conceptualization, A.H., S.K. and K.M.; methodology, K.M.; software, M.O. and K.M.; validation and formal analysis, M.O. and K.M.; investigation, K.M., M.O., S.K. and A.H.; writing—original draft preparation, M.O., A.H. and K.M.; writing—review and editing, K.M., M.O., S.K. and A.H.; visualization, M.O. and K.M.; supervision, A.H. All authors have read and agreed to the published version of the manuscript.

Funding: This research received no external funding.

Institutional Review Board Statement: Not applicable.

Informed Consent Statement: Not applicable.

Data Availability Statement: Raw data in this study are available based on reasonable request. The data are not publicly available due to the restricted use of anatomically based models.

Conflicts of Interest: The authors declare no conflicts of interest.

References

1. Karalis, A.; Joannopoulos, J.D.; Soljačić, M. Efficient Wireless Non-Radiative Mid-Range Energy Transfer. *Ann. Phys.* **2008**, *323*, 34–48. [CrossRef]
2. Vishnuram, P.; Panchanathan, S.; Rajamanickam, N.; Krishnasamy, V.; Bajaj, M.; Piecha, M.; Blazek, V.; Prokop, L. Review of Wireless Charging System: Magnetic Materials, Coil Configurations, Challenges, and Future Perspectives. *Energies* **2023**, *16*, 4020. [CrossRef]
3. Mohamed, A.A.S.; Shaier, A.A.; Metwally, H.; Selem, S.I. An Overview of Dynamic Inductive Charging for Electric Vehicles. *Energies* **2022**, *15*, 5613. [CrossRef]
4. Mohanarangam, K.; Palagani, Y.; Choi, J.R. Evaluation of Specific Absorption Rate in Three-Layered Tissue Model at 13.56 MHz and 40.68 MHz for Inductively Powered Biomedical Implants. *Appl. Sci.* **2019**, *9*, 1125. [CrossRef]
5. Zhou, Y.; Liu, C.; Huang, Y. Wireless Power Transfer for Implanted Medical Application: A Review. *Energies* **2020**, *13*, 2837. [CrossRef]
6. Chen, X.L.; Umenei, A.E.; Baarman, D.W.; Chavannes, N.; De Santis, V.; Mosig, J.R.; Kuster, N. Human Exposure to Close-Range Resonant Wireless Power Transfer Systems as a Function of Design Parameters. *IEEE Trans. Electromagn. Compat.* **2014**, *56*, 1027–1034. [CrossRef]
7. Yang, Y.; El Baghdadi, M.; Lan, Y.; Benomar, Y.; Van Mierlo, J.; Hegazy, O. Design Methodology, Modeling, and Comparative Study of Wireless Power Transfer Systems for Electric Vehicles. *Energies* **2018**, *11*, 1716. [CrossRef]
8. Brecher, A.; Arthur, M.D. *Review and Evaluation of Wireless Power Transfer (WPT) for Electric Transit Applications*; U.S. Department of Transportation: Washington, DC, USA, 2014.
9. ICNIRP Guidelines for Limiting Exposure to Time-Varying Electric and Magnetic Fields (1 Hz to 100 kHz). *Health Phys.* **2010**, *99*, 818–836. [CrossRef]
10. ICNIRP Guidelines for Limiting Exposure to Electromagnetic Fields (100 kHz to 300 GHz). *Health Phys.* **2020**, *118*, 483–524. [CrossRef]
11. ICNIRP Principles for Non-Ionizing Radiation Protection. *Health Phys.* **2020**, *118*, 477–482. [CrossRef]
12. *IEEE Std C95.1-2019*; IEEE Standard for Safety Levels with Respect to Human Exposure to Electric, Magnetic, and Electromagnetic Fields, 0 Hz to 300 GHz. IEEE: Piscataway Township, NJ, USA, 2019.
13. *IEEE C95.1-2005*; IEEE Standard for Safety Levels with Respect to Human Exposure to Electromagnetic Fields, 0–3 kHz. Institute of Electrical and Electronics Engineers: Piscataway Township, NJ, USA, 2002; ISBN 0738133892.
14. Grazian, F.; Shi, W.; Dong, J.; van Duijsen, P.; Soeiro, T.B.; Bauer, P. Survey on Standards and Regulations for Wireless Charging of Electric Vehicles. In Proceedings of the 2019 AEIT International Conference of Electrical and Electronic Technologies for Automotive (AEIT AUTOMOTIVE), Turin, Italy, 2–4 July 2019; pp. 1–5.
15. *IEC 61980-1*; Electric Vehicle Wireless Power Transfer (WPT) Systems—Part 1: General Requirements. International Electrotechnical Commission: Geneva, Switzerland, 2020.
16. *SAE J2954*; Wireless Power Transfer for Light-Duty Plug-In/Electric Vehicles and Alignment Methodology. SAE International: Warrendale, PA, USA, 2016.
17. *ISO 19363*; Electrically Propelled Road Vehicles—Magnetic Field Wireless Power Transfer—Safety and Interoperability Requirements. International Organization for Standardization: Beijing, China, 2020.
18. Laakso, I.; Hirata, A. Evaluation of the Induced Electric Field and Compliance Procedure for a Wireless Power Transfer System in an Electrical Vehicle. *Phys. Med. Biol.* **2013**, *58*, 7583–7593. [CrossRef]
19. Shimamoto, T.; Laakso, I.; Hirata, A. In-Situ Electric Field in Human Body Model in Different Postures for Wireless Power Transfer System in an Electrical Vehicle. *Phys. Med. Biol.* **2015**, *60*, 163–173. [CrossRef]
20. Miwa, K.; Takenaka, T.; Hirata, A. Electromagnetic Dosimetry and Compliance for Wireless Power Transfer Systems in Vehicles. *IEEE Trans. Electromagn. Compat.* **2019**, *61*, 2024–2030. [CrossRef]

21. De Santis, V.; Giaccone, L.; Freschi, F. Influence of Posture and Coil Position on the Safety of a WPT System While Recharging a Compact EV. *Energies* **2021**, *14*, 7248. [CrossRef]
22. *IEC 62110*; Electric and Magnetic Field Levels Generated by AC Power Systems—Measurement Procedures with Regard to Public Exposure. IEC: Geneva, Switzerland, 2009.
23. *IEC TR 62905*; Exposure Assessment Methods for Wireless Power Transfer Systems. IEC: Geneva, Switzerland, 2018.
24. *IEC PAS 63184*; Assessment Methods of the Human Exposure to Electric and Magnetic Fields from Wireless Power Transfer Systems—Models, Instrumentation, Measurement and Numerical Methods and Procedures (Frequency Range of 1 kHz to 30 MHz). IEC: Geneva, Switzerland, 2021.
25. Che, K.; Yu, J.; Yang, P.; Wei, M.; Liu, S.; Li, D. Limits of Electromagnetic Environment for Electric Vehicle Wireless Power Transfer. In Proceedings of the Asia-Pacific Power and Energy Engineering Conference, APPEEC, Xi'an, China, 17–19 April 2021; IEEE Computer Society: Washington, DC, USA, 2021.
26. Mohamed, A.S.A.; Meintz, A.; Schrafel, P.; Calabro, A. In-Vehicle Assessment of Human Exposure to EMFs from 25-kW WPT System Based on Near-Field Analysis. In Proceedings of the 2018 IEEE Vehicle Power and Propulsion Conference (VPPC), Chicago, IL, USA, 27–30 August 2018; pp. 1–6.
27. El-Shahat, A.; Danjuma, J.; Abdelaziz, A.Y.; Abdel Aleem, S.H.E. Human Exposure Influence Analysis for Wireless Electric Vehicle Battery Charging. *Clean Technol.* **2022**, *4*, 785–805. [CrossRef]
28. Liu, S.; Li, D.; Chen, C.; Jia, W.; Che, K.; Yu, J. Electromagnetic Field Safety Analysis of a 7.7 kW Wireless Power Transfer System for Electric Vehicles. *Prog. Electromagn. Res. M* **2023**, *117*, 1–12. [CrossRef]
29. David, A.; Tiemann, M.; Clemens, M.; Schmuelling, B. Magnetic Field Analysis of 50 kW Wireless Power Transfer System for Heavy Duty Vehicles. *IEEE Trans. Magn.* **2023**, *60*, 3306809. [CrossRef]
30. Park, S.W. Influence of Fields and SAR Evaluation for 13.56 MHz EV Resonance-Based Wireless Power Charging Systems. *Microw. Opt. Technol. Lett.* **2017**, *59*, 937–941. [CrossRef]
31. Elnail, K.E.I.; Huang, X.; Xiao, C.; Tan, L.; Haozhe, X. Core Structure and Electromagnetic Field Evaluation in WPT Systems for Charging Electric Vehicles. *Energies* **2018**, *7*, 1734. [CrossRef]
32. Hirata, A.; Diao, Y.; Onishi, T.; Sasaki, K.; Ahn, S.; Colombi, D.; De Santis, V.; Laakso, I.; Giaccone, L.; Wout, J.; et al. Assessment of Human Exposure to Electromagnetic Fields: Review and Future Directions. *IEEE Trans. Electromagn. Compat.* **2021**, *63*, 1619–1630. [CrossRef]
33. Laakso, I.; Hirata, A.; Fujiwara, O. Computational Dosimetry for Wireless Charging of an Electrical Vehicle. *Int. Symp. Electromagn. Compat.* **2014**, 202–205.
34. Wen, F.; Huang, X.L. Human Exposure to Electromagnetic Fields from Parallel Wireless Power Transfer Systems. *Int. J. Environ. Res. Public Health* **2017**, *14*, 157. [CrossRef]
35. De Santis, V.; Giaccone, L.; Freschi, F. Chassis Influence on the Exposure Assessment of a Compact EV during WPT Recharging Operations. *Magnetochemistry* **2021**, *7*, 25. [CrossRef]
36. De Santis, V.; Campi, T.; Cruciani, S.; Laakso, I.; Feliziani, M. Assessment of the Induced Electric Fields in a Carbon-Fiber Electrical Vehicle Equipped with a Wireless Power Transfer System. *Energies* **2018**, *11*, 684. [CrossRef]
37. Liorni, I.; Bottauscio, O.; Guilizzoni, R.; Ankarson, P.; Bruna, J.; Fallahi, A.; Harmon, S.; Zucca, M. Assessment of Exposure to Electric Vehicle Inductive Power Transfer Systems: Experimental Measurements and Numerical Dosimetry. *Sustainability* **2020**, *12*, 4573. [CrossRef]
38. Lan, J.; Hirata, A. Effect of Loudspeakers on the in Situ Electric Field in a Driver Body Model Exposed to an Electric Vehicle Wireless Power Transfer System. *Energies* **2020**, *13*, 3635. [CrossRef]
39. Wang, T.; Yu, Q.; Li, B.; Lv, G.; Wu, Y.; Guan, S. Uncertainty Quantification of Human Electromagnetic Exposure From Electric Vehicle Wireless Power Transfer System. *IEEE Trans. Intell. Transp. Syst.* **2023**, *24*, 8886–8896. [CrossRef]
40. Wake, K.; Laakso, I.; Hirata, A.; Chakarothai, J.; Onishi, T.; Watanabe, S.; De Santis, V.; Feliziani, M.; Taki, M. Derivation of Coupling Factors for Different Wireless Power Transfer Systems: Inter-and Intralaboratory Comparison. *IEEE Trans. Electromagn. Compat.* **2017**, *59*, 677–685. [CrossRef]
41. Chakarothai, J.; Wake, K.; Arima, T.; Watanabe, S.; Uno, T. Exposure Evaluation of an Actual Wireless Power Transfer System for an Electric Vehicle with Near-Field Measurement. *IEEE Trans. Microw. Theory Tech.* **2018**, *66*, 1543–1552. [CrossRef]
42. Park, S. Evaluation of Electromagnetic Exposure during 85 kHz Wireless Power Transfer for Electric Vehicles. *IEEE Trans. Magn.* **2018**, *54*. [CrossRef]
43. Ahn, J.; Hong, S.E.; Kim, H.; Song, K.; Choi, H.D.; Ahn, S. Improved Calculation Method of Coupling Factors for Low-Frequency Wireless Power Transfer Systems. *Int. J. Environ. Res. Public Health* **2022**, *19*, 44. [CrossRef]
44. *IEC 62233*; Measurement Methods for Electromagnetic Fields of Household Appliances and Similar Apparatus with Regard to Human Exposure. IEC: Geneva, Switzerland, 2005.
45. Nagaoka, T.; Watanabe, S.; Sakurai, K.; Kunieda, E.; Watanabe, S.; Taki, M.; Yamanaka, Y. Development of Realistic High-Resolution Whole-Body Voxel Models of Japanese Adult Males and Females of Average Height and Weight, and Application of Models to Radio-Frequency Electromagnetic-Field Dosimetry. *Phys. Med. Biol.* **2004**, *49*, 1. [CrossRef]
46. Nagaoka, T.; Watanabe, S. Voxel-Based Variable Posture Models of Human Anatomy. *Proc. IEEE* **2009**, *97*, 2015–2025. [CrossRef]
47. Reed, M.P.; Manary, M.A.; Flannagan, C.A.C.; Schneider, L.W. Effects of Vehicle Interior Geometry and Anthropometric Variables on Automobile Driving Posture. *Hum. Factors* **2000**, *42*, 541–552. [CrossRef]

48. Hori, Y.; Yokoi, Y. *Wireless Power Transfer and Infrastructure Construction for Electric Vehicle (Popular Edition)*; CMC Publishing Co., Ltd.: Tokyo, Japan, 2011. (In Japanese)
49. Kim, H.; Song, C.; Kim, D.-H.; Jung, D.H.; Kim, I.-M.; Kim, Y.-I.; Kim, J.; Ahn, S.; Kim, J. Coil Design and Measurements of Automotive Magnetic Resonant Wireless Charging System for High-Efficiency and Low Magnetic Field Leakage. *IEEE Trans. Microw. Theory Tech.* **2016**, *64*, 383–400. [CrossRef]
50. *IEC 62764-1*; Measurement Procedures of Magnetic Field Levels Generated by Electronic and Electrical Equipment in the Automotive Environment with Respect to Human Exposure—Part 1: Low-Frequency Magnetic Fields. IEC: Geneva, Switzerland, 2022.
51. Hirata, A.; Ito, F.; Laakso, I. Confirmation of Quasi-Static Approximation in SAR Evaluation for a Wireless Power Transfer System. *Phys. Med. Biol.* **2013**, *58*, N241. [CrossRef]
52. Barchanski, A.; De Gersem, H.; Gjonaj, E.; Weiland, T. Impact of the Displacement Current on Low-Frequency Electromagnetic Fields Computed Using High-Resolution Anatomy Models. *Phys. Med. Biol.* **2005**, *50*, N243. [CrossRef]
53. Van Oosterhout, K.; Paulides, M.; Pflug, H.; Beumer, S.; Mestrom, R. An Approximate Electromagnetic Model for Optimizing Wireless Charging of Biomedical Implants. *IEEE Trans. Biomed. Eng.* **2022**, *69*, 1954–1963. [CrossRef]
54. Laakso, I.; Hirata, A. Fast Multigrid-Based Computation of the Induced Electric Field for Transcranial Magnetic Stimulation. *Phys. Med. Biol.* **2012**, *57*, 7753–7765. [CrossRef]
55. Dawson, T.W.; Stuchly, M.A. Analytic Validation of a Three-Dimensional Scalar-Potential Finite-Difference Code for Low-Frequency Magnetic Induction. *Appl. Comput. Electromagn. Soc. J.* **1996**, *11*, 72–81.
56. Gabriel, S.; Lau, R.W.; Gabriel, C. The Dielectric Properties of Biological Tissues: III. Parametric Models for the Dielectric Spectrum of Tissues. *Phys. Med. Biol.* **1996**, *41*, 2271. [CrossRef] [PubMed]
57. Sasaki, K.; Porter, E.; Rashed, E.A.; Farrugia, L.; Schmid, G. Measurement and Image-Based Estimation of Dielectric Properties of Biological Tissues—Past, Present, and Future. *Phys. Med. Biol.* **2022**, *67*, 14TR01. [CrossRef] [PubMed]
58. Wake, K.; Sasaki, K.; Watanabe, S. Conductivities of Epidermis, Dermis, and Subcutaneous Tissue at Intermediate Frequencies. *Phys. Med. Biol.* **2016**, *61*, 4376. [CrossRef]
59. Yamazaki, K. Assessment Methods for Electric and Magnetic Fields in Low and Intermediate Frequencies Related to Human Exposures and the Status of Their Standardization. *IEEJ Trans. Fundam. Mater.* **2019**, *139*, 649–656. [CrossRef]
60. Sachs, L. *Applied Statistics: A Handbook of Techniques*; Springer Science & Business Media: Berlin, Germany, 2012.
61. Chicco, D.; Warrens, M.J.; Jurman, G. The Coefficient of Determination R-Squared Is More Informative than SMAPE, MAE, MAPE, MSE and RMSE in Regression Analysis Evaluation. *PeerJ Comput. Sci.* **2021**, *7*, e623. [CrossRef] [PubMed]

applied sciences

MDPI

Article

Discrete Wavelet Transform—Based Metal Material Analysis Model by Constant Phase Angle Pulse Eddy Current Method

Yong Xie [1,2], Yating Yu [1,3,*] and Liangting Li [2]

1 School of Mechanical and Electrical Engineering, University of Electronic Science and Technology, Chengdu 611731, China
2 Shanghai Lanbao Sensing Technology Co., Ltd., Shanghai 201404, China
3 Key Laboratory of Traction Power, Southwest Jiaotong University, Chengdu 610031, China
* Correspondence: yuyating-uestc@hotmail.com

Abstract: Traditional eddy current technology identifies metal information with information of single frequency of limited frequency spectrum. To solve existing problems, this paper proposes a discrete wavelet transform-based metal material analysis model by using a constant phase angle pulse eddy current (CPA-PEC) sensor which collects and depicts metal feature information from multiple dimensions; then, the quantification calculation model of metal material by CPA-PEC feature is presented; finally, an experimental platform is built to collect the CPA-PEC features of various metal samples and verify recognition accuracy of the proposed metal material analysis model. In the investigation, 1000 eddy current signals from four standard metals (Cu, Fe, Al, St) and three types of metallic irons (Fe-K162, Fe-K163, Fe-K240) are measured and the features are identified by discrete wavelet transform. The feature correlation and significance are determined by regression analysis. Finally, the calculation model of feature evaluation index is present. The experimental analysis indicates that the stability of the quantitative evaluation index of eddy current features reaches 97.1%, the comprehensive accuracy error is less than 0.32% and the average measurement speed is about 50 ms for 1000 random sampling tests on standard metals.

Keywords: constant phase angle pulse eddy current (CPA-PEC) sensor; discrete wavelet transform; eddy current feature recognition; feature evaluation model

1. Introduction

Materials are an important support for industrial development; metallic materials are particularly important. The consistency of metallic materials seriously affects the development of high-end industries. In order to test the consistency of metallic materials, titrimetric [1] and spectroscopic methods [2] are usually presented to detect the elemental composition of samples and thus determine the material properties. However, these methods are complex in operation process, destructive, costly in instrument and inconvenient for applications in the material production site, which greatly constrains the rapid identification of metallic materials.

The eddy current technique is one of the nondestructive testing methods that is widely applied in metal material inspection, such as material integrity testing and material characterization [3–8]. The single-frequency eddy current technique is suitable for classifying metals because different metals are made of materials with different electrical conductivity and magnetic permeability; thus, their absorptions of eddy currents are different. Consequently, eddy current detection for metals can replace the expensive spectral scanning method, since it reduces the detection time and improves the detection efficiency [9]. However, compared with the pulsed eddy current technique, single-frequency eddy current signals cannot express frequency domain information. The pulsed eddy current technique has a wide frequency spectrum, and the time domain signal can be filtered for multiple estimation of the electromagnetic parameters of the material from different frequency characteristics during the measurement. Yu et al. put forward the method of quantitative calculation of metal defects by using eddy current nondestructive testing to establish a finite element model, semi-analytical method, etc. [10–13]; Chen et al. developed a model of the time domain induced voltage and the electromagnetic parameters of the material based on the pulsed signal. They demonstrated the effectiveness of the multi-frequency method in detecting the electromagnetic parameters of the material [14,15]. Wang et al. investigated the phase log-based of conductivity measurements. The results showed that the relationship between the logarithm of the phase and the variation of the conductivity was linear [16]. Lu et al. proposed a model to reconstruct the magnetic permeability, conductivity, and thickness of the sample. The reconstruction error was 5% [17].

When using the eddy current technique to detect the electromagnetic properties of metallic materials, lift-off is one of the important factors affecting the accuracy of the measurement [18–20]. As such, eliminating the effect of lift-off is of great importance for metallic material identification. By varying the relative spatial positions of the receiving and transmitting coils, Jin et al. achieved accurate measurements of the sample conductivity within a few millimeters, with a maximum error of 2.68% [21]. Ivan Rep et al. proposed a stochastic inversion method with multi-frequency excitation to simultaneously determine the magnitude of lift-off as well as the conductivity and permeability of the sample; the inversion of these data resulted in a significant reduction in the a priori uncertainty of all model parameters [22]. Lu et al. found that the dual-frequency signal was linearly related to lift-off. By applying different frequencies through the calculation, it was possible to reconstruct the variation of the lift-off of the signal and to reconstruct the electromagnetic properties of the material with an error of 1.4% within 12 mm [23].

In addition, the signal processing method also determines the accuracy of the eddy current method to identify material properties. Yang et al. presented the wavelet transform method to process the eddy current signal of ferromagnetic metal. The weak defective magnetic field signal inside the ferromagnetic metal plate can be identified because of the interference of magnetic domains [24]. Bolos et al. introduced a new wavelet tool, windowed scale index, that is suitable for studying the aperiodic degree of signal, such as the volatility in crude oil and gold prices [25]. Lepicka et al. applied windowed scale index into tribological behavior analysis of titanium nitride-coated stainless steel, successfully proving that the windowed scale index has a great ability to distinguish different tribological states [26]. Silipigni et al. adopted the pulse compression algorithm to compress other forms of signals into a signal similar to the impulse response. They then processed the corresponding signal of the pulse using the time domain algorithm. Compared to the frequency domain signal, the time domain signal is much smoother and less noisy, facilitating signal analysis [27]. Tamhanne et al. analyzed the degree of corrosion of a 20 mm diameter steel bar by the principal component analysis, allowing the lift-off of 55 mm for corrosion detection [28]. Empirical modal decomposition and a non-smooth signal processing method can efficiently remove the noise from the signal. Liu et al. proposed an improved differential empirical modal decomposition method that can effectively suppress signal aliasing and distortion and significantly improve signal accuracy [29]. Likewise, neural network-based classification has better recognition of signals with low signal-to-noise ratios [30–33].

However, for the metal material consistency issue, metal alloy classification with similar components is still a problem to be solved. More attention should be paid on metal alloy identification with similar components using multi-frequency or magnetic spectrum information.

This paper proposes a technical method of metal material analysis model based on the discrete wavelet transform of eddy current method. The eddy current characteristic information of the metals is collected and described in multiple dimensions, the quantitative calculation model of metal material eddy current characteristic is proposed, and a standard metal material eddy current feature sample library is established to provide a new method for accurate description of metal material features.

2. Principle and Methods

The material characteristics technology with eddy current method is based on the constant phase angle pulse eddy current (CPA-PEC) sensor. The metal material properties sample is formed by extracting the frequency variation and attenuation characteristics of the eddy current field using a discrete wavelet transform model.

The CPA-PEC sensor is an eddy current sensor with a dual coil probe structure, as shown in Figure 1, in which the excitation coil is fed with a periodic ramp signal with a constant change in current to generate the magnetic field. The receiving coil is placed outside the excitation coil to measure the change in the magnetic field. According to the law of electromagnetic induction, it is known that the voltage across the receiving coil is proportional to the rate of change of the magnetic flux at the coil location. The magnetic flux at the receiving coil is related to the material, thickness, lift-off and shape of the detected metal. Different frequencies of the magnetic field in the metals to be measured attenuate the energy differently, so that the corresponding time domain signal in the receiving coil changes. It can then be extracted from the detection target material, thickness, lift-off, shape and other information through data processing. The working principle is described as follows.

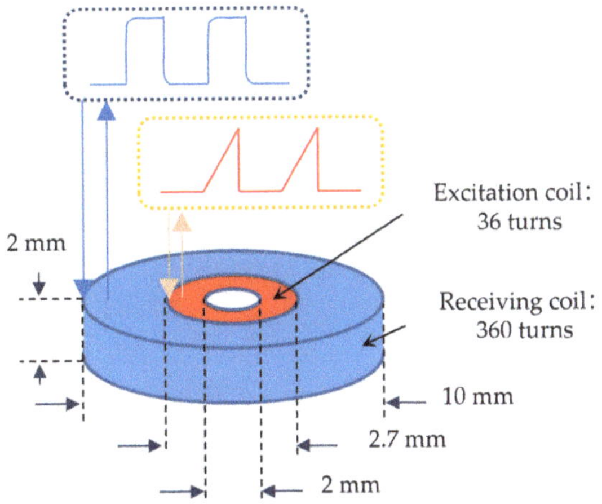

Figure 1. The structure of CPA-PEC sensor.

The model is simplified for better analysis: the coils are considered as a clingy excitation-receiving system without volume. The excitation source used in this paper is a ramp signal as shown in Figure 1, i.e., $I_{SE} = k_1 t$, where k_1 is constant. Then, the magnetic field at its axial position x from the center is

$$B = \frac{N_1 \mu_0 I_{SE} R^2}{2(R^2 + x^2)^{3/2}} \tag{1}$$

where N_1 is the turns of the excitation coil, μ_0 is the vacuum magnetic permeability and R is the coil radius; I_{SE} is the current amplitude of the excitation source.

Then, the voltage across the induction coil U_s should be

$$U_s = \frac{N_2 d\varnothing}{dt} = \frac{N_1 N_2 S \mu_0 R^2 k_2}{2(R^2 + x^2)^{3/2}} \tag{2}$$

where Φ is the magnetic flux flows through the coil, S is the induction area of the coil and N_2 is the number of turns of the receiving coil. k_2 is a scale factor of the central magnetic field to the detected magnetic field. From Equation (2), it can be seen that the voltage in the induction coil should be constant when the excitation field is a ramp signal. That is, in free space, the voltage across the induction coil is a constant value.

Assuming an infinite plane current magnetic field source with a magnetic field H_{0z}, the solution for the magnetic field in the conductor located at x is

$$H_z = H_{0z} e^{-\frac{1+j}{\sqrt{2}}\sqrt{\omega\mu\sigma}x} \tag{3}$$

where H_z is the z-directional component of the magnetic field in the metal, x is the location of H_z in the conductor, ω is the angular frequency, μ is the magnetic permeability and σ is the electrical conductivity.

It is clear that the current inside the conductor is related to the magnetic field frequency ω and the conductor's electromagnetic properties μ, σ. The intensity of eddy currents in the conductor increases with frequency within a certain range, but the eddy currents become more and more concentrated in the center of the magnetic field with increasing frequency. Since the magnitude of the magnetic field is proportional to the product of the induced current and its loop area, the magnetic field decreases with increasing frequency within a certain range. It means that there is an optimal magnetic field frequency in detection. The optimal frequency point is dependent of the conductor properties μ, σ in Equation (1).

The wavelet decomposition of the ramp signal shows that the ramp signal corresponds to different spectra at different times in the time domain. The frequency spectrum of the induced voltage of the coil is different at different moments. By combining Equations (1)–(3), the optimal magnetic field frequency for detecting different metals is different, meaning that the time domain signal intensity of the induced voltage at both ends of the coils should be different at different time points when detecting different metals. By analyzing the time domain characteristics of the induction signal of the coil, the material type of the metal can be identified.

The constant parameters used in this paper are shown in Table 1.

Table 1. The constant parameters used in this paper.

Constant Parameter	Value
k_1 (A/us)	~0.09
μ_0 (N/A^2)	$4\pi \times 10^{-7}$
N_1	36
N_2	360
e	~2.718

3. Experiment and Results

3.1. Experimental Setup

In this section, a system with CPA-PEC is built and shown in Figure 2. The eddy current metal material analysis experimental platform mainly consists of the CPA-PEC probe, the signal generation module, the signal amplification module, the signal receiving module, the 2D scanning platform, the computer and the analysis software of the upper computer.

The signal generation module is used to control the excitation coil to send out the ramp excitation signal. A microcontroller-controlled gate circuit is used for generating a periodic square wave. Then, the CPA-PEC voltage signal is generated by charging and discharging to a capacitor. After that, the voltage on the capacitor is converted to a current signal which is connected to the excitation coil. The receiving coil is used to collect the characteristic signal of metal induced eddy current field when the probe is close to the samples. The signal is fed into a signal amplification and filtering system and then converted into a digital signal by a 10-bit AD.

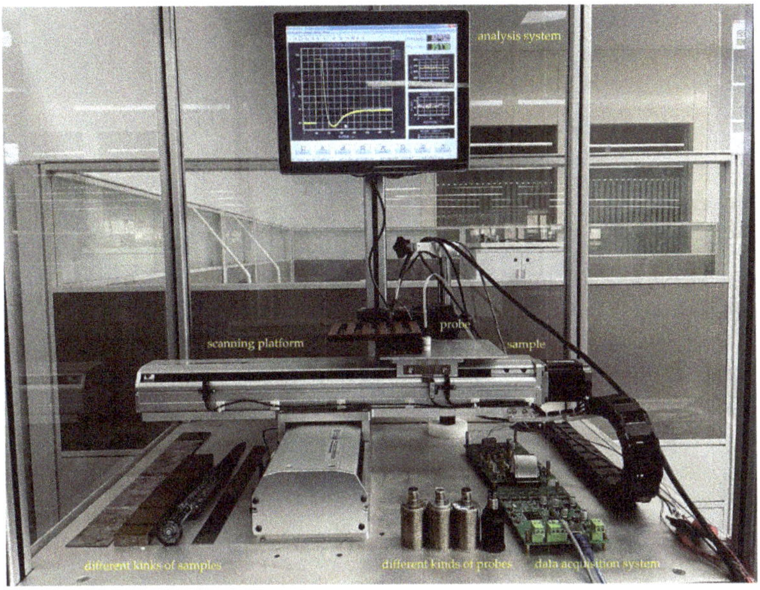

Figure 2. Photograph of the experimental setup.

A discrete wavelet transform model is used to transform the signal. The eddy current characteristic index is calculated by the discrete signal characteristic analysis model, which can quickly identify the metal to be measured. The experimental data are based on 100 replicate measurement sampling with the CPA-PEC probe. The detection results of metal samples Cu, Fe, Al, St, Fe-K162, Fe-K163 and Fe-K240 were analyzed by feature extraction of the data. The proportion of the elements in metallic alloys Fe-K162, Fe-K163 and Fe-K240 are listed in Table 2.

Table 2. Proportion of the elements in metallic alloys.

	Fe-K162	**Fe-K163**	**Fe-K240**
Fe	93.97%	93.994%	94.326%
C	3.21%	3.30%	3.81%
Si	1.28%	1.25%	1.23%
Mn	1.05%	1.16%	0.319%
P	0.075%	0.065%	0.044%
S	0.045%	0.055%	0.026%
Cr	0.24%	0.1%	0.037%
Ni	0.09%	0.036%	0.009%
Mo	0.001%	0.002%	0.026%
V	0.012%	0.011%	0.017%
Cu	0.014%	0.012%	0.142%
Ti	0.013%	0.015%	0.014%

The amplitude of the ramp signal is 825 mV, the width of the signal is 20 us and the repeat frequency is 1 kHz.

3.2. Discrete Wavelet Transform—Based Metal Material Analysis Model

The identification method and steps of the metal material analysis model are shown in Figure 3.

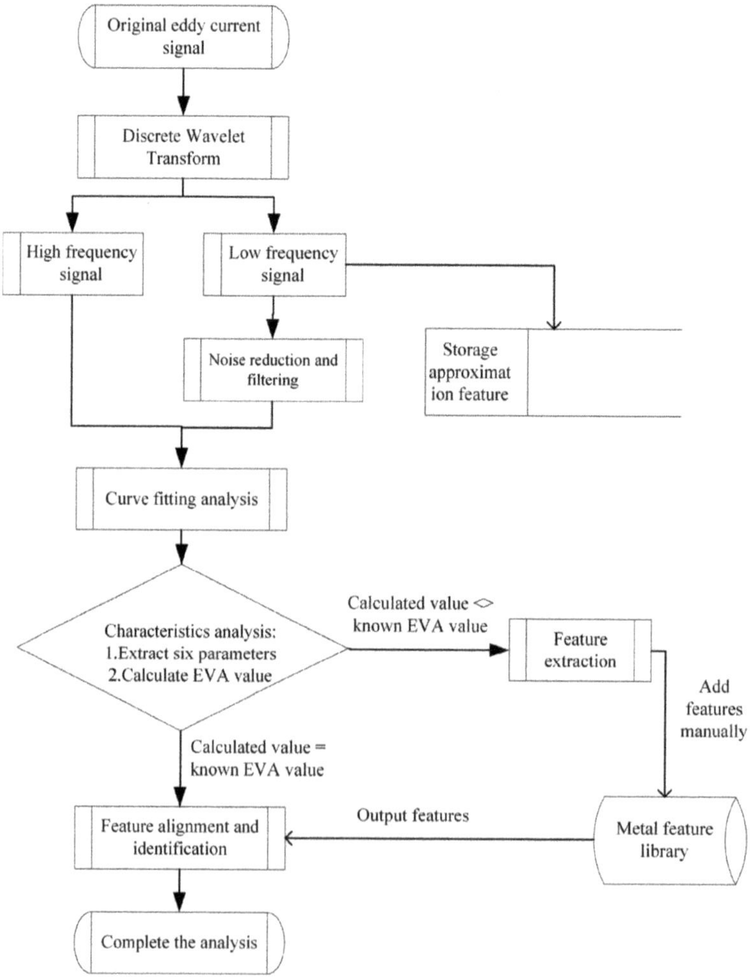

Figure 3. Identification method and steps of the eddy current method metal material analysis model.

In this paper, the characteristics of the signal in the time domain and frequency domain are analyzed, different eddy current characteristics are extracted from time and frequency domain, and the calculation model of the characteristic evaluation index is constructed to calculate the eddy current characteristic index. The consistency between the measured metal and the known metal materials is judged by comparing the metal eddy current characteristic database.

Wavelet transform is an algorithm suitable for non-smooth signal analysis. It can decompose a given time-domain signal into several sets of wavelets carrying different frequency information. That means that the wavelet transform analysis method can be used to control the frequency components in the signal very well. It should be noticed that, when the wavelet reconstruction is performed, the reconstruction coefficients should be optimized for the specific length of the data. If too much low-frequency information is retained in the reconstructed signal, the jitter from signal acquisition cannot be removed. If the reconstructed signal retains too much high-frequency information, the effect of high frequency noise on the signal cannot be removed. However, if too little high-frequency or low-frequency information is retained, the signal will be severely distorted, affecting the feature extraction of the signal.

The dbN wavelet has a better regularity: that is, the smooth error introduced by this wavelet is not easily detected as a sparse basis, making the signal reconstruction process relatively smooth. The characteristic of the dbN wavelet is that the order of vanishing moment increases with the increase of order (sequence N). The higher the vanishing moment, the better the smoothness, the stronger the localization ability of frequency domain, and the better the division effect of the frequency band [34]. In this paper, the wavelet base adopted in the experiment is db4; the approximate coefficient in level six is used for signal reconstruction with better signal-to-noise ratio (SNR). All the parameters used in the following analysis are based on the approximate coefficient at level six in db4. A comparison chart of the raw signal measured in air and the signal after wavelet transformed is show in Figure 4. It can be seen that high frequency noise has been filtered; this will affect classification accuracy.

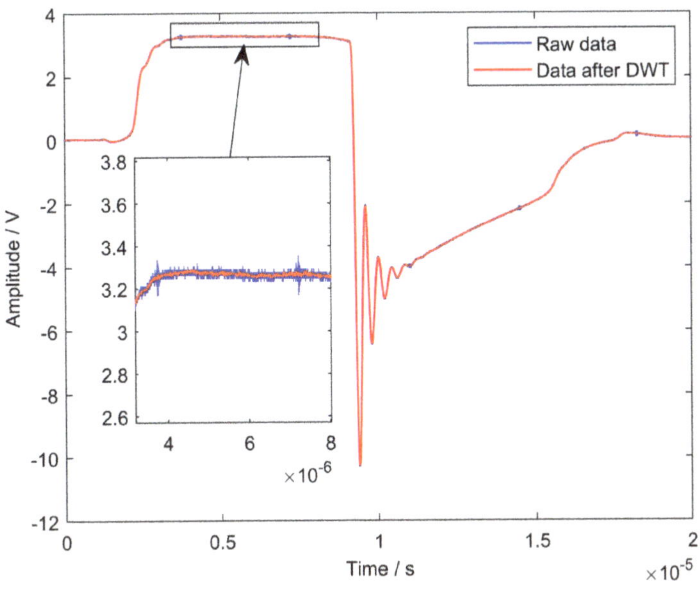

Figure 4. Comparison of the signals before and after wavelet transformed.

In this paper, the signals are evaluated in terms of peak frequency, peak time, zero crossing time, time of decay, main peak area and amplitude of attenuation, respectively. In selecting the characteristic parameters, 1000 independent measurements of the metal under test were performed in a random scan to obtain eddy current characteristic samples for the four standard metals (Cu, Fe, Al and St) and three standard metals of Fe (Fe-K162, Fe-K163 and Fe-K240) studied in this paper.

Six extracted parameters of Fe samples are shown in Figure 5. It can be seen that Fe-K162, Fe-K163 and Fe-K240 can be well classified with parameters of zero crossing time, time of decay and main peak area. However, Fe-K162, Fe-K163 and Fe-K240 have a light overlap in the feature of zero crossing time and time of decay. It is seen that, for each extracted feature, the tendency of the feature values varies with the material is different. This may be affected by the content of the elements in metallic alloys. From Table 1, it is seen that the iron element proportions of Fe-K162, Fe-K163 and Fe-K240 are 93.97%, 93.994% and 94.326%, respectively. The average values of the time of delay feature corresponding with iron element proportions are 318, 324 and 335, respectively; these have the same monotonicity compared to iron element proportions. The average values of the main peak area feature corresponding with iron element proportions are 803, 770 and 730, respectively; these have an opposite monotonicity compared to iron element proportions. It seems that the features of time of decay and main peak area are highly related with the iron element proportions and have high sensitivity to small content changes in ferroalloys.

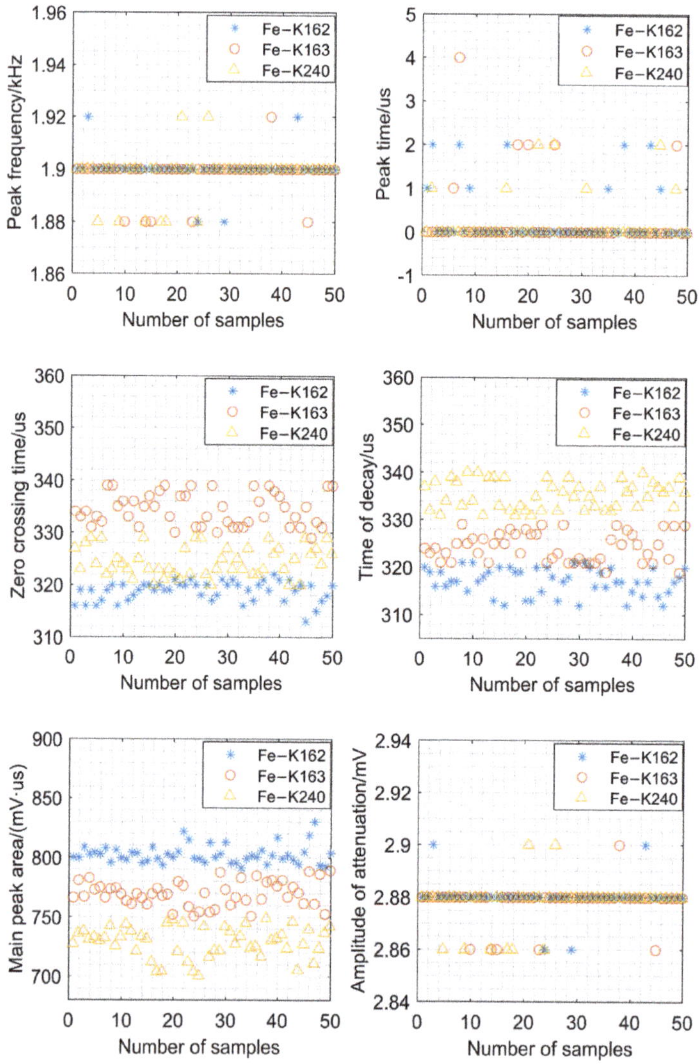

Figure 5. Signal characteristic distribution of Fe-K162, Fe-K163 and Fe-K240.

Similar experiments have been applied to Cu, Fe, Al and St, as shown in Figure 6. It can be seen that the features of peak frequency and peak time are meaningless for metal recognition combined with the results of Figure 4. Different from Figure 4, the feature of amplitude of attenuation has contribution in Cu, Fe, Al and St identification. It should be noticed that Cu and Al are easily distinguished by the zero crossing time feature; Fe and St are also easily distinguished by the zero crossing time feature. However, the results in the time of decay feature are the opposite; the feature value is far less distinguishable than the feature of zero crossing time. However, the feature of main peak area is perfect for metal classification: there is no need to make a joint judgment with other features. That is because the four metals differ greatly in electrical and magnetic permeability.

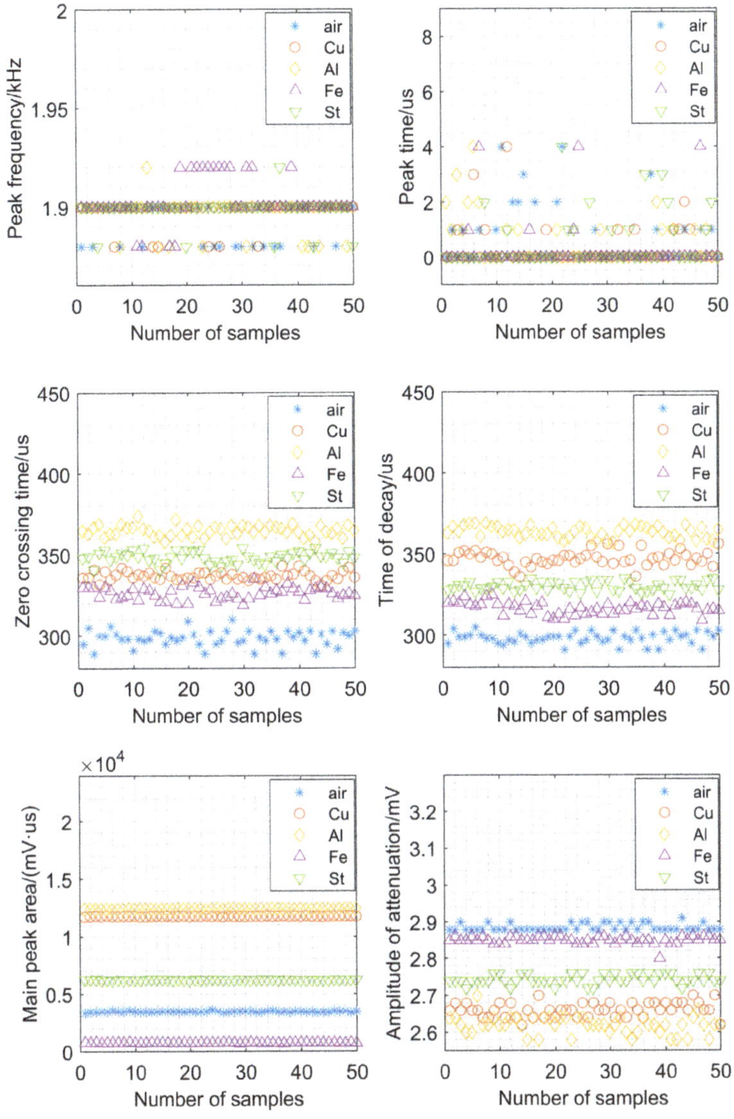

Figure 6. Signal characteristic distribution of air, Cu, Fe, Al and St.

A fast logistic regression analysis was performed using 50 independent data samples to calculate feature correlations and significance to verify the accuracy of eddy current feature parameter selection. A binary regression analysis of eddy current signal feature correlation was established to validate the data samples using 50 groups of Fe-K162 gray cast iron metal samples against 50 groups of non-Fe-K162 metal samples, and the results of the regression analysis are shown in Figure 7.

Table of classification [a,b]

		Actual measurement	Predict: Fe–K162 0	Predict: Fe–K162 1	Correct percentage
Step 0	Fe–K162	0	0	50	0.0
		1	0	50	100.0
	Overall percentage				50.0

a. Constants are included in the model.
b. The cut-off value is 0.500

Constant in an equation

		B	Standard error	Degree of freedom	Significance	Exp (B)
Step 1	Constant	0.000	0.200	1	1.000	1.000

Variables in an equation [a]

			Score	Degree of freedom	Significance
Step 2	Variable	Peak time (us)	0.159	1	0.690
		Zero crossing time (us)	88.101	1	0.000
		Main peak area (mV.us)	47.304	1	0.000
		Amplitude of attenuation (mV)	96.037	1	0.000
		Time of decay (us)	89.974	1	0.000
		Peak (mV)	1.955	1	0.162
		Peak frequency (us)	11.246	1	0.001

a. The residual chi-square was not calculated due to redundancy.

Figure 7. IBM Statistical Product Service Solutions Software statistics binary logistic regression analysis results.

The classification map and the Hosmer–Lemeshaw goodness of fit were performed according to the six features mentioned previously. The number of iterations was set to 20; the predicted value probability, group membership, influence factor and leverage values were turned on; and the residuals of the criteria are analyzed. Then, the Statistical Product Service Solutions (SPSS) correlation and significance of metal signal characteristics can be analyzed.

This paper adopts the binary logistic regression model of the IBM SPSS software and takes the characteristic data of Fe-K162 eddy current signal as an example to identify the correlation and significance of the characteristic parameters of the above samples.

According to the significance analysis setting, if the significant p value of variable X shows significance, it means that X has an influential relationship on Y. If it is not significant, the variable should be excluded. From the results of the grouped experimental data, it can be seen that there are four explanatory variables in the model with significance analysis p-values less than 0.001 for Fe-K162 recognition rate, indicating that all X factors have a significant influence relationship on Y.

The impact of the correlation of X–Y was compared based on the regression analysis scores. From the regression analysis scores, the four explanatory variables in the model, such as zero crossing time, main peak area, amplitude of attenuation and time of decay, had scores of 88.101, 47.304, 96.037 and 89.974, respectively, indicating that all variables showed significant positive influence relationships on the Fe-K162 recognition rate.

In summary, combining the feature screening results and correlation verification conclusions, it is determined that the main feature parameters such as main peak area, time of decay and zero crossing time are selected. The amplitude of attenuation is selected as an auxiliary feature parameter to establish a functional mapping relationship with the metal eddy current signal. Multiple groups of feature parameters can be combined for simultaneous discrimination, which can improve the metal identification accuracy and expand the identification range.

In order to analyze the effectiveness of the described method more intuitively, an evaluation index is proposed and its stability and accuracy are analyzed based on experiments. The evaluation index is defined as

$$EVA_{metal} = \sum_{k=1}^{j} (\frac{1}{m} \sum_{n=1}^{m} (c_k \times PCA(DWT(S_n, db4), character_k))) \tag{4}$$

where $character_k$ is the extracted parameter defined from peak frequency, peak time, zero crossing time, time of decay, main peak area and amplitude of attenuation; S_n is the signal amplitude in time domain; m is the number of times the experiment was repeated; and c_k is a balanced value for each extracted parameter, which is positively correlated with the score in Figure 7.

The model and calculation equation are tested by 1000 independent sampling, from which 100 random samples are used in the accuracy verification. The results are shown in Table 3. It is clear that the eight chosen kinds of metal are all well categorized. The EVA value for each metal is extremely stable, except the first value for air condition. It can be seen that the EVA values of Fe-K162, Fe-K163, Fe-K240 and Fe are very close, which means that iron-based metal materials are more difficult to classify. When using the EVA value in air as the standard, it can be seen that the EVA values of metals with magnetism are bigger than the EVA value of air. The EVA values of metals without magnetism are smaller than the EVA value of air.

Table 3. The quantization index of the model was verified repeatedly with 100 random samples per group.

	Repeat 1	Repeat 2	Repeat 3	Repeat 4	Repeat 5
K162	2966	2966	2965	2965	2966
K163	2956	2956	2955	2956	2955
K240	2946	2946	2945	2945	2946
Fe	2951	2951	2951	2951	2951
Air	3080	3090	3090	3090	3090
Cu	3210	3210	3211	3211	3210
Al	3221	3221	3221	3225	3221
St	3125	3125	3125	3126	3125

The sensor, model and calculation equations were tested by 1000 independent samples, among which 100 random samples were selected to verify the identification rate, stability and accuracy. The results are shown in Table 4. It can be seen that the identification rates for all the metals are 100%. The stability of the sensor for all metals classification are greater than 97.10%. The metal material with the least stability during measurement is Fe, and the metal material with the greatest stability during measurement is Fe-K240. The confidence interval has a mean value of $\pm$ 0.33‰. According to the statistics without anomalies, the relative error of the EVA value is less than 0.32%. The minimum relative error is only 0.03%.

Table 4. The identification rate of the sensor, the stability of the calculation model and the accuracy of the evaluation index.

	Identification Rate *	Stability *	Relative Error *
K162	100%	98.80%	0.03%
K163	100%	98.90%	0.03%
K240	100%	99.12%	0.03%
Fe	100%	97.10%	0.31%
Air	100%	97.45%	0.32%
Cu	100%	98.11%	0.03%
Al	100%	98.10%	0.12%
St	100%	97.45%	0.03%

* Identification rate = (identified samples/total samples) × 100%. * Stability is the samples that were within the confidence interval. * Relative error = $(|EVA_{metal} - mean(EVA_{metal})|/max(EVA_{metal})) \times 100\%$.

Based on the experimental data validation results, when the excitation parameters are fixed, four kinds of standard metal materials and three kinds of gray cast iron materials of the same type are measured, respectively, and the evaluation indexes of metal eddy current characteristics are calculated. The results can be distinguished one by one. Through 1000 repeated experiments, the recognition rate of the sensor reaches 100%, the stability of the calculation model is more than 97.1%, the comprehensive accuracy error is less than 0.32%, and the average measurement speed is about 50 ms.

4. Conclusions

To deal with the issue of metal alloys classification with similar components, a discrete wavelet transform-based metal material analysis model by CPA-PEC method is proposed in this paper, and performances are verified by a series of experiments.

The CPA-PEC sensor is excited by a constant phase angle pulse eddy current, and the eddy current signal is analyzed by the discrete wavelet transform. Compared with conventional sine wave excitation, constant phase angle pulse eddy current has a wider spectrum, which may have benefit for spectrum measurement of different metals. Therefore, the proposed method has fast discrimination capability for various metal materials and has good discrimination capability for comprehensive evaluation and quantitative analysis of the subtle differences in metal composition through the discrete signal feature analysis model.

Experimental investigations show that the CPA-PEC sensor with metal eddy current feature library and metal material feature identification algorithm can effectively improve the capability of eddy current sensor to identify metal material consistency in application environments such as qualitative determination of special metal material components and dynamic monitoring of metal consistency. The proposed evaluation index can accurately discriminate metals. The identification rates are 100% for all the metals, the stability is greater than 97.10% for all metals, and the relative errors are less than 0.32% for all metals.

Compared with other metal identification techniques, such as spectral analysis, the CPA-PEC technique can quickly determine the metal material characteristic index by a simple detection device and achieve higher accuracy and a wider range of metal identification by continuously expanding the eddy current characteristic library. This has higher practicality and a wide application value in the application scenario of rapid nondestructive testing of metal material composition. However, an obvious limitation exists: the CPA-PEC technique is more suitable for qualitative analysis of alloy components, such as measurement of the change in the content of a component in the production of an alloy. In the future, the influence of different components on the electromagnetic properties of the alloy should be fully analyzed, more data should be expanded to the metal eddy current feature library by collecting the experimental signals from more types of the samples, and the performance of the proposed metal material feature identification algorithm should be verified by the different metal categories.

Author Contributions: Conceptualization, Y.Y.; Formal analysis, Y.X.; Investigation, L.L.; Methodology, Y.X. and Y.Y.; Software, L.L.; Validation, Y.X. and L.L.; Writing—original draft, Y.X. and L.L.; Writing—review & editing, Y.Y. All authors have read and agreed to the published version of the manuscript.

Funding: This work was supported by the National Nature Science Foundation of China [grant numbers 52275522 and 61960206010], the Sichuan Science and Technology Program [grant number 2022NSFSC032], and the Open Fund of Key Laboratory of Traction Power [grant number TPL2206].

Institutional Review Board Statement: Not applicable.

Informed Consent Statement: Not applicable.

Data Availability Statement: No new data were created or analyzed in this study. Data sharing is not applicable to this article.

Conflicts of Interest: The authors declare no conflict of interest.

References

1. Ďuriš, S.; Ďurišová, Z.; Wimmer, G.; Dovica, M. Stability Check of Certified Reference Materials of Carbon and Sulphur Content in Steel Used for Analysis of Low-Alloyed Steels. In Proceedings of the 2019 12th International Conference on Measurement, Smolenice, Slovakia, 27–29 May 2019; pp. 331–334. [CrossRef]
2. Liu, L.; Huan, H.; Zhang, M.; Shao, X.; Zhao, B.; Cui, X.; Zhu, L. Photoacoustic Spectrometric Evaluation of Soil Heavy Metal Contaminants. *IEEE Photonics J.* **2019**, *11*, 1–7. [CrossRef]
3. Ye, C.; Laureti, S.; Malekmohammadi, H.; Wang, Y.; Ricci, M. Swept-Frequency eddy current excitation for TMR array sensor and Pulse-Compression: Feasibility study and quantitative comparison of time and frequency domains processing. *Measurement* **2022**, *187*, 110249. [CrossRef]
4. Wang, Y.; Niu, Y.; Wei, Y.; Ye, C. Multi-frequency imaging with non-linear calibration of magnetoresistance sensors for surface and buried defects inspection. *NDT E Int.* **2022**, *132*, 102706. [CrossRef]
5. Avila, J.R.S.; How, K.Y.; Lu, M.; Yin, W. A Novel Dual Modality Sensor With Sensitivities to Permittivity, Conductivity, and Permeability. *IEEE Sens. J.* **2018**, *18*, 356–362. [CrossRef]
6. Yuan, F.; Yu, Y.; Wang, W.; Xue, K.; Tian, G. Pulsed eddy current array design and electromagnetic imaging for defects detection in metallic materials. *Nondestruct. Test. Eval.* **2021**, *37*, 81–99. [CrossRef]
7. Yuan, F.; Yu, Y.; Wang, W.; Tian, G. A Novel Probe of DC Electromagnetic NDT Based on Drag Effect: Design and Application in Crack Characterization of High-Speed Moving Ferromagnetic Material. *IEEE Trans. Instrum. Meas.* **2021**, *70*, 1. [CrossRef]
8. Yuan, F.; Yu, Y.; Li, L.; Tian, G. Investigation of DC Electromagnetic-Based Motion Induced Eddy Current on NDT for Crack Detection. *IEEE Sens. J.* **2021**, *21*, 1. [CrossRef]
9. Wang, D.; Zhang, Z.; Yin, W.; Chen, H.; Ma, H.; Zhou, G.; Zhang, Y. Linear Characteristics of the Differences in Phase Tangents of Triple-Coil Electromagnetic Sensors and Their Application in Nonmagnetic Metal Classification. *Sensors* **2022**, *22*, 7511. [CrossRef]
10. Yu, Y.; Li, X.; Simm, A.; Tian, G. Theoretical model-based quantitative optimisation of numerical modelling for eddy current NDT. *Nondestruct. Test. Eval.* **2011**, *26*, 129–140. [CrossRef]
11. Yu, Y.; Li, Y.; Qin, H.; Cheng, X. Microwave measurement and imaging for multiple corrosion cracks in planar metals. *Mater. Des.* **2020**, *196*, 109151. [CrossRef]
12. Yu, Y.; Gao, K.; Theodoulidis, T.; Yuan, F. Analytical solution for magnetic field of cylindrical defect in eddy current nondestructive testing. *Phys. Scr.* **2020**, *95*, 015501. [CrossRef]
13. Yu, Y.; Gao, K.; Liu, B.; Li, L. Semi-analytical method for characterization slit defects in conducting metal by Eddy current nondestructive technique. *Sens. Actuators Phys.* **2020**, *301*, 111739. [CrossRef]
14. Chen, X.; Liu, X. Pulsed Eddy Current-Based Method for Electromagnetic Parameters of Ferromagnetic Materials. *IEEE Sens. J.* **2021**, *21*, 6376–6383. [CrossRef]
15. XChen, X.; Lei, Y. Electrical conductivity measurement of ferromagnetic metallic materials using pulsed eddy current method. *NDT E Int.* **2015**, *75*, 33–38. [CrossRef]
16. Wang, C.; Fan, M.; Cao, B.; Ye, B.; Li, W. Novel Noncontact Eddy Current Measurement of Electrical Conductivity. *IEEE Sens. J.* **2018**, *18*, 9352–9359. [CrossRef]
17. Lu, M.; Xie, Y.; Zhu, W.; Peyton, A.J.; Yin, W. Determination of the Magnetic Permeability, Electrical Conductivity, and Thickness of Ferrite Metallic Plates Using a Multifrequency Electromagnetic Sensing System. *IEEE Trans. Ind. Inform.* **2019**, *15*, 4111–4119. [CrossRef]
18. Sreevatsan, S.; George, B. Simultaneous Detection of Defect and Lift-off Using a Modified Pulsed Eddy Current Probe. *IEEE Sens. J.* **2020**, *20*, 2156–2163. [CrossRef]
19. Meng, X.; Lu, M.; Yin, W.; Bennecer, A.; Kirk, K.J. Inversion of Lift-Off Distance and Thickness for Nonmagnetic Metal Using Eddy Current Testing. *IEEE Trans. Instrum. Meas.* **2021**, *70*, 1–8. [CrossRef]

20. Long, Y.; Huang, S.; Peng, L.; Wang, S.; Zhao, W. A New Dual Magnetic Sensor Probe for Lift-off Compensation in Magnetic Flux Leakage Detection. In Proceedings of the 2020 IEEE International Instrumentation and Measurement Technology Conference (I2MTC), Dubrovnik, Croatia, 25–28 May 2020; pp. 1–6. [CrossRef]
21. Jin, Z.; Meng, Y.; Yu, R.; Huang, R.; Lu, M.; Xu, H.; Meng, X.; Zhao, Q.; Zhang, Z.; Peyton, A.; et al. Methods of Controlling Lift-Off in Conductivity Invariance Phenomenon for Eddy Current Testing. *IEEE Access* **2020**, *8*, 122413–122421. [CrossRef]
22. Rep, I.; Spikic, D.; Vasic, D. Eddy Current Inversion of Lift-off, Conductivity and Permeability Relaxation. In Proceedings of the 2021 44th International Convention on Information, Communication and Electronic Technology (MIPRO), Opatija, Croatia, 27 September–1 October 2021; pp. 136–140. [CrossRef]
23. Lu, M.; Meng, X.; Huang, R.; Chen, L.; Peyton, A.; Yin, W. Measuring Lift-Off Distance and Electromagnetic Property of Metal Using Dual-Frequency Linearity Feature. *IEEE Trans. Instrum. Meas.* **2021**, *70*, 1–9. [CrossRef]
24. Wang, Y.; Nie, Y.; Qi, P.; Zhang, N.; Ye, C. Inspection of Defect Under Thick Insulation Based on Magnetic Imaging With TMR Array Sensors. *IEEE Trans. Magn.* **2022**, *58*, 1–10. [CrossRef]
25. Bolós, V.J.; Benítez, R.; Ferrer, R. A New Wavelet Tool to Quantify Non-Periodicity of Non-Stationary Economic Time Series. *Mathematics* **2020**, *8*, 844. [CrossRef]
26. Łępicka, M.; Górski, G.; Grądzka-Dahlke, M.; Litak, G.; Ambrożkiewicz, B. Analysis of tribological behaviour of titanium nitride-coated stainless steel with the use of wavelet-based methods. *Arch. Appl. Mech.* **2021**, *91*, 4475–4483. [CrossRef]
27. Silipigni, G.; Burrascano, P.; Hutchins, D.A.; Laureti, S.; Petrucci, R.; Senni, L.; Torre, L.; Ricci, M. Optimization of the pulse-compression technique applied to the infrared thermography nondestructive evaluation. *NDT E Int.* **2017**, *87*, 100–110. [CrossRef]
28. Tamhane, D.; Patil, J.; Banerjee, S.; Tallur, S. Feature Engineering of Time-Domain Signals Based on Principal Component Analysis for Rebar Corrosion Assessment Using Pulse Eddy Current. *IEEE Sens. J.* **2021**, *21*, 22086–22093. [CrossRef]
29. Liu, X.; Shi, G.; Liu, W. An Improved Empirical Mode Decomposition Method for Vibration Signal. *Wirel. Commun. Mob. Comput.* **2021**, *2021*, 5525270. [CrossRef]
30. Wang, S.; Xiao, S.; Huang, Z.; Xu, Z.; Chen, W. Improved One-Dimensional Convolutional Neural Networks for Human Motion Recognition. In Proceedings of the 2020 IEEE International Conference on Bioinformatics and Biomedicine (BIBM), Seoul, Korea, 16–19 December 2020; pp. 2544–2547. [CrossRef]
31. Huang, W.; Oh, S.-K.; Pedrycz, W. Hybrid Fuzzy Wavelet Neural Networks Architecture Based on Polynomial Neural Networks and Fuzzy Set/Relation Inference-Based Wavelet Neurons. *IEEE Trans. Neural Netw. Learn. Syst.* **2018**, *29*, 3452–3462. [CrossRef]
32. Subbotin, A.N.; Zhukova, N.A.; Anaam, F. Application of Recurrent Neural Networks with Controlled Elements for Accuracy Enhancement in Recognition of Sound Events in a Fog Computing Environment. In Proceedings of the 2022 III International Conference on Neural Networks and Neurotechnologies (NeuroNT), Saint Petersburg, Russian Federation, 16 June 2022; pp. 47–50. [CrossRef]
33. Yu, Y.; Cheng, X.; Wang, L.; Wang, C. Convolutional Neural Network-Based Quantitative Evaluation for Corrosion Cracks in Oil/Gas Pipeline by Millimeter-Wave Imaging. *IEEE Trans. Instrum. Meas.* **2022**, *71*, 1–9. [CrossRef]
34. Shah, M.; Vakharia, V.; Chaudhari, R.; Vora, J.; Pimenov, D.Y.; Giasin, K. Tool wear prediction in face milling of stainless steel using singular generative adversarial network and LSTM deep learning models. *Int. J. Adv. Manuf. Technol.* **2022**, *121*, 723–736. [CrossRef]

MDPI AG
Grosspeteranlage 5
4052 Basel
Switzerland
Tel.: +41 61 683 77 34

Applied Sciences Editorial Office
E-mail: applsci@mdpi.com
www.mdpi.com/journal/applsci

www.ingramcontent.com/pod-product-compliance
Lightning Source LLC
Chambersburg PA
CBHW061452210926
51850CB00008B/1692